Glucose Syrups

Technology and Applications

Glucose Syrups

Technology and Applications

Peter Hull, MIFST

A John Wiley & Sons, Ltd., Publication

This edition first published 2010
© 2010 Peter Hull

Blackwell Publishing was acquired by John Wiley & Sons in February 2007. Blackwell's publishing programme has been merged with Wiley's global Scientific, Technical, and Medical business to form Wiley-Blackwell.

Registered office
John Wiley & Sons Ltd, The Atrium, Southern Gate, Chichester, West Sussex, PO19 8SQ, United Kingdom

Editorial offices
9600 Garsington Road, Oxford, OX4 2DQ, United Kingdom
2121 State Avenue, Ames, Iowa 50014-8300, USA

For details of our global editorial offices, for customer services and for information about how to apply for permission to reuse the copyright material in this book please see our website at www.wiley.com/wiley-blackwell.

The right of the author to be identified as the author of this work has been asserted in accordance with the Copyright, Designs and Patents Act 1988.

All rights reserved. No part of this publication may be reproduced, stored in a retrieval system, or transmitted, in any form or by any means, electronic, mechanical, photocopying, recording or otherwise, except as permitted by the UK Copyright, Designs and Patents Act 1988, without the prior permission of the publisher.

Wiley also publishes its books in a variety of electronic formats. Some content that appears in print may not be available in electronic books.

Designations used by companies to distinguish their products are often claimed as trademarks. All brand names and product names used in this book are trade names, service marks, trademarks or registered trademarks of their respective owners. The publisher is not associated with any product or vendor mentioned in this book. This publication is designed to provide accurate and authoritative information in regard to the subject matter covered. It is sold on the understanding that the publisher is not engaged in rendering professional services. If professional advice or other expert assistance is required, the services of a competent professional should be sought.

Library of Congress Cataloging-in-Publication Data

Hull, Peter, 1934–
  Glucose syrups : technology and applications / Peter Hull.
     p. cm.
  Includes bibliographical references and index.
  ISBN 978-1-4051-7556-2 (hardback : alk. paper) 1. Corn syrup.   2. Glucose.   I. Title.
  TP405.H85 2010
  664′.133–dc22
                                                                                    2009016430

A catalogue record for this book is available from the British Library.

Set in 10/13pt Times New Roman by Aptara® Inc., New Delhi, India
Printed and bound in Singapore by Fabulous Printers Pte Ltd

1   2010

This book is dedicated to all those who have been or are involved in the glucose industry and its esteemed customers, without whom there would be no glucose industry, and to Ivan, Alexei, Anatoliy and Dmitry for persuading me to write this book.

Contents

Preface   xv
A note on nomenclature   xvii
Acknowledgements   xix

Chapter 1   **History of glucose syrups**   1
    1.1   Historical developments   1
    1.2   Analytical developments   4
    1.3   Process developments   6

Chapter 2   **Fructose containing syrups**   9
    2.1   Introduction   9
    2.2   Commercial development   11
    2.3   Europe and the HFGS (isoglucose) production quota   13
    2.4   Inulin   17
    2.5   Conclusion   17

Chapter 3   **Glucose syrup manufacture**   19
    3.1   Introduction   19
    3.2   Reducing sugars   20
    3.3   Starch   21
    3.4   Enzymes   23
        3.4.1 α-amylase   24
        3.4.2 β-amylase   24
        3.4.3   Glucoamylase   24
        3.4.4   Pullulanase   25
        3.4.5   Isomerase   25
        3.4.6   Lesser enzymes   25
    3.5   The process   25
    3.6   Acid hydrolysis   27
    3.7   Acid enzyme hydrolysis   33
    3.8   Paste Enzyme Enzyme hydrolysis (PEE)   34
    3.9   Crystalline dextrose production   36
    3.10   Total sugar production   38
    3.11   Enzyme enzyme hydrolysis (E/E)   39
    3.12   Isomerisation   40
    3.13   Syrups for particular applications   43
    3.14   Summary of typical sugar spectra produced by different processes   43

Chapter 4 Explanation of glucose syrup specifications — 45
- 4.1 Introduction — 45
- 4.2 What specification details mean — 45
- 4.3 Dry products — 53
- 4.4 Syrup problems and their possible causes — 54
- 4.5 Bulk tank installation — 57
- 4.6 Bulk tank design — 58

Chapter 5 Application properties of glucose syrups — 61
- 5.1 Introduction — 61
- 5.2 Summary of properties — 63
- 5.3 Bodying agent — 64
- 5.4 Browning reaction — 64
- 5.5 Cohesiveness — 65
- 5.6 Fermentability — 65
- 5.7 Flavour enhancement — 65
- 5.8 Flavour transfer medium — 66
- 5.9 Foam stabilisers — 66
- 5.10 Freezing point depression — 66
- 5.11 Humectancy — 67
- 5.12 Hygroscopicity — 68
- 5.13 Nutritive solids — 68
- 5.14 Osmotic pressure — 68
- 5.15 Prevention of sucrose crystallisation — 70
- 5.16 Prevention of coarse ice crystal formation — 70
- 5.17 Sheen producer — 71
- 5.18 Sweetness — 71
- 5.19 Viscosity — 72
- 5.20 Summary of properties — 73
- 5.21 Differences between glucose syrups and sucrose — 74

Chapter 6 Syrup applications: an overview — 77
- 6.1 Introduction — 77
- 6.2 42 DE Glucose Syrup — 77
- 6.3 28 and 35 DE Glucose Syrup — 79
- 6.4 Glucose syrup solids — 80
- 6.5 Maltose and high maltose syrups — 80
- 6.6 63 DE Glucose Syrup — 82
- 6.7 95 DE Glucose Syrup — 84
- 6.8 Dextrose monohydrate — 88
- 6.9 HFGS and fructose syrups — 91
- 6.10 Maltodextrins — 95

Chapter 7	Trehalose		101
	7.1	Introduction	101
	7.2	Production	101
	7.3	Properties	103
	7.4	Applications	104
		7.4.1 Confectionery	105
		7.4.2 Dairy	105
		7.4.3 Jams and fruit fillings	106
		7.4.4 Cosmetic and personal hygiene products	106
		7.4.5 Pharmaceuticals	106
		7.4.6 Medical applications	106
Chapter 8	Sugar alcohols: an overview		107
	8.1	Introduction	107
	8.2	Production	108
	8.3	Overview of polyol properties	111
	8.4	Applications overview	113
		8.4.1 Sorbitol	113
		8.4.2 Maltitol	115
		8.4.3 Mannitol	116
		8.4.4 Erythritol	117
Chapter 9	Glucose syrups in baking and biscuit products		119
	9.1	Introduction	119
	9.2	Fermented goods	120
	9.3	Non-fermented goods	121
	9.4	Biscuits	123
	9.5	Biscuit fillings	124
	9.6	Wafer fillings	124
	9.7	Bakery sundries	125
		9.7.1 Fondant	125
		9.7.2 Hundred and thousands	126
		9.7.3 Icings	126
		9.7.4 Marshmallows	127
		9.7.5 Marzipan	128
		9.7.6 Fruit flavoured pieces	128
		9.7.7 Piping jelly	129
		9.7.8 Bakery glaze	129
	9.8	Reduced calorie products	130
	9.9	Breakfast cereals	131
Chapter 10	Glucose syrups in brewing		133
	10.1	Introduction	133
	10.2	Brewing process	134
	10.3	Historical use of glucose syrups	135

10.4	The role of glucose syrups	137
10.5	Low alcohol and low calorie beer	138
10.6	De-ionised glucose syrups	139
10.7	High gravity brewing	140
	10.7.1 High gravity brewing calculations	140
10.8	Brewer's extract – cost calculations	141
	10.8.1 Typical extract values (hot water)	143
	10.8.2 Brewing syrup addition calculations	145
10.9	Chip sugar	146
	10.9.1 How to make chip sugar	147

Chapter 11 Glucose syrups in confectionery 149

11.1	Introduction	149
11.2	What can glucose syrups offer the confectioner?	149
	11.2.1 Control of sucrose crystallisation and graining	150
	11.2.2 Reduce moisture pickup	151
	11.2.3 Reduce cold flow	151
	11.2.4 Improve processing	152
	11.2.5 Modify the sweetness	155
	11.2.6 Modifying texture	155
11.3	Which glucose syrup to use?	155
11.4	Typical glucose syrup inclusion rates	158
11.5	Some basic confectionery recipes	161
	11.5.1 High boilings	161
	11.5.2 Pulled sugar confectionery	161
	11.5.3 Fondant	162
	11.5.4 Toffee and caramel	163
	11.5.5 Fudge	164
	11.5.6 Gums and jellies	165
	11.5.7 Chews	167
	11.5.8 Marshmallows	168
	11.5.9 Turkish delight	168
	11.5.10 Muesli bars	170
	11.5.11 Confectionery centres	171
11.6	Calorie reduced products	171

Chapter 12 Glucose syrups in fermentations: an overview 175

12.1	Introduction	175
12.2	Choice of substrate	176
12.3	Basic fermentation process	177
12.4	Products of fermentation	178
	12.4.1 Pharmaceutical	178
	12.4.2 Enzymes	179
	12.4.3 Food grade products	180
	12.4.4 Industrial products	183

Chapter 13		**Glucose syrups in ice creams and similar products**	**185**
	13.1	Introduction	185
	13.2	Ingredients and process	185
		13.2.1 Fats	185
		13.2.2 Milk solids	187
		13.2.3 Sugars	187
		13.2.4 Emulsifiers and stabilisers	187
		13.2.5 Solids	187
		13.2.6 Pasteurisation	188
		13.2.7 Homogenisation	188
		13.2.8 Cooling, ageing and freezing	188
	13.3	Glucose syrups – freezing point and relative sweetness values	188
		13.3.1 How to reformulate using glucose syrups	190
		13.3.2 How sweeteners can be re-balanced	
	13.4	Quick process checks	194
		13.4.1 Viscosity	194
		13.4.2 Overrun	194
		13.4.3 Solids	194
		13.4.4 Fats	194
	13.5	Soft serve ice creams	194
	13.6	Other types of frozen dessert	195
	13.7	Yogurts	195
	13.8	Sorbet	196
	13.9	Mousse	196
	13.10	Ice lollies	197
	13.11	Fruit lollies	197
	13.12	Ripple syrups	197
	13.13	Topping or dessert syrup	198
		13.13.1 A simple economy topping syrup	199
		13.13.2 Fruit-flavoured topping syrup	199
		13.13.3 All syrup fruit-flavoured topping syrup	200
		13.13.4 Chocolate topping	200
		13.13.5 All syrup chocolate topping	200
		13.13.6 Caramel topping	200
		13.13.7 All syrup caramel topping syrup	201
	13.14	Reduced calorie products	201
Chapter 14		**Glucose syrups in jams**	**203**
	14.1	Introduction	203
	14.2	Effects of boiling	203
	14.3	Use of glucose syrups	205
	14.4	Domestic jam	208
	14.5	Jelly jams	209
	14.6	Honey type spread	209

14.7	Chocolate spread	210
14.8	Peanut spread	211
14.9	Industrial jams	211
	14.9.1 Bake-stable jams	212
	14.9.2 Biscuit jams	212
	14.9.3 Spreadable jams	212
	14.9.4 Jam fillings	212
	14.9.5 Flan jellies	212
	14.9.6 Fruit and pie fillings	214
	14.9.7 Tablet jellies	214
	14.9.8 Mincemeat	215
	14.9.9 Fruit curds	216
14.10	Diabetic and reduced calorie products	217
14.11	How to calculate a recipe?	217

Chapter 15 Glucose syrups in tomato products and other types of dressings and sauces — 221

15.1	Introduction	221
15.2	Which glucose syrup to use?	221
15.3	Tomato products	222
15.4	Other dressings	224
15.5	Other sauces, marinades and pickles	225
15.6	Reduced calorie products	226

Chapter 16 Glucose syrups in soft drinks — 227

16.1	Introduction	227
16.2	Ingredients	228
16.3	Effect of process inversion	228
16.4	Use of glucose syrups	231
16.5	Quality considerations	233
16.6	Laboratory evaluation of glucose syrups in soft drinks	233
	16.6.1 Water	234
	16.6.2 Sweeteners	234
	16.6.3 Acidulants	235
16.7	Soft drink recipes	236
	16.7.1 Carbonated drinks, for example lemonade	237
	16.7.2 Dilutable drinks, for example orange squash	237
16.8	Powdered drinks	238
16.9	Reduced calorie drinks	238

Chapter 17 Glucose syrups in health and sports drinks — 239

17.1	Introduction	239
17.2	The energy source	239
17.3	Classification of health drinks	240
17.4	Osmotic pressure of health drinks	241

17.5	Sucrose in sports or health drinks	243
17.6	Formulating a sports drink	244
17.7	Energy values	246
17.8	Oral rehydration	247
17.9	Geriatric drinks and liquid foods	248
17.10	Slimming foods	249

Chapter 18 Carbohydrate metabolism and caloric values — 251

- 18.1 Introduction — 251
- 18.2 Human digestive system — 252
- 18.3 Carbohydrate absorption — 253
- 18.4 Summary of carbohydrate metabolism — 254
  - 18.4.1 Sugars — 254
  - 18.4.2 Starch — 255
  - 18.4.3 Fibre (that is roughage) and non-starch polysaccharides (that is cellulose) — 255
  - 18.4.4 Polyols — 256
- 18.5 Carbohydrate metabolic problems — 257
  - 18.5.1 Diabetes mellitus — 257
  - 18.5.2 Fructose and diabetes — 257
  - 18.5.3 Lactose intolerance — 258
  - 18.5.4 Fructose intolerance — 258
  - 18.5.5 Galactosaemia — 258
  - 18.5.6 Gaucher's disease — 259
  - 18.5.7 Coeliac disease (also known as gluten intolerance) — 259
  - 18.5.8 Phenylketonuria and aspartame — 259
- 18.6 Caloric values — 260

Chapter 19 Caramel – the colouring — 261

- 19.1 Introduction — 261
- 19.2 Process — 261
- 19.3 Properties — 262
- 19.4 Applications — 262

Glossary — 265

Appendix A Simple analytical information — 297

- A.1 Introduction — 297
- A.2 The ingredient declaration panel — 297
- A.3 Does it contain glucose syrup? — 299
- A.4 What HPLC sugar analysis can tell? — 301

Appendix B Simple calculations — 305

- B.1 Introduction — 305
- B.2 Adjusting syrup solids — 305

	B.3	Altering the sugar spectra of a glucose syrup blend	308
	B.4	How to calculate equivalent sweetness values?	311
	B.5	Relationship between density, volume and weight of glucose syrups	312
	B.6	How much syrup is required to obtain a given weight of syrup solids?	312
	B.7	Brix, RI and RI Solids, % Solids and Baumé	313
	B.8	Recipe costings	314
	B.9	Colligative properties	315
	B.9.1	How to calculate boiling point elevation?	315
	B.9.2	How to calculate freezing point depression?	317
	B.9.3	How to calculate osmotic pressure?	318

Appendix C Sugars data — 321

	C.1	Approximate % sugar spectra of different glucose syrups	321
	C.2	Theoretical molecular weights	322
	C.3	Sweetness values	323
	C.4	Approximate sugar spectra of domestic sweeteners	324
	C.5	Typical particle size for different grades of sucrose	326
	C.6	Melting points	327
	C.7	Glass transition temperatures – T_g values	327
	C.8	Solubility – grams per 100 ml	328
	C.8.1	In water	328
	C.8.2	In 80% alcohol at 20°C	328

Appendix D Tables — 329

	D.1	Temperature conversion	329
	D.2	Viscosity of glucose syrups at different Dextrose Equivalents and temperatures. Reproduced by courtesy of The Corn Refiners Association.	334
	D.3	Maize starch Baumé tables. Reproduced by courtesy of The Corn Refiners Association.	348
	D.4	Sucrose Brix table – Brix – % sucrose w/w, specific gravity and Baumé (145 modulus)	349
	D.5	Sucrose Brix – refractive indices at 20°C	351
	D.6	Glucose syrup tables – commercial Baumé, DE, % solids – at 60°C (140°F)	352
	D.7	Sieve specifications	352

Bibliography — 357

Index — 361

Preface

The object of this book is to explain to people new to glucose syrups how these syrups are made, and importantly how, why and when to use a glucose syrup in a particular application. This book is not intended to be an academic tome – just a practical basic reference book, based on over 40 years of experience talking to food technologists, new process and project engineers, sales personnel and buyers. It is also not the intention to tell those skilled in their profession how to do their job, or how they should make their product, but hopefully to help them understand why they are using a particular glucose syrup and to explain how to get the best out of a glucose syrup and how to adapt a recipe to use a glucose syrup. Since this book is about the manufacture and application of glucose syrups, detailed sugar chemistry has been kept to a minimum.

As this book is about the manufacture and applications of starch-derived sweeteners, sweeteners and polyols derived from sugars other than starch-based sugars have been omitted. Also omitted are details about polydextrose and cyclodextrins, because neither are sweeteners.

Included in the book are tables and useful technical information together with a glossary of terms which are used in both the manufacture and applications of glucose syrups. The glossary contains simple explanations of these terms and is not intended to be a definitive dictionary of scientific terms.

The old Haworth structure for the dextrose molecule has been used as opposed to the modern conformational chair structure.

In the chapters covering applications, because glucose syrups are a very versatile product, it has not been possible to discuss every application. Only the major syrup applications have therefore been considered in detail, which collectively make up about 90% of total sweetener sales.

As there is often no one single recipe for a product, and as each producer has its own particular recipe or process, where a recipe is given, it is for guidance only and should be verified by experimentation, and that the end product complies with local legislation – trust but verify.

The information in this book is given in good faith and is accurate to the best of the author's knowledge and makes no warranty or representation of any kind, however, expressed or implied as to the suitability of use. The provision of this information is not to be construed as an authorization to infringe any patent.

A note on nomenclature

Whilst glucose syrups are produced all over the world, the nomenclature and units of measurement used by the industry and its customers are not universal. For example in Europe, some producers use maize to produce glucose syrups whereas in America maize is called corn and glucose syrups are referred to as corn syrups. Because a glucose syrup and a corn syrup are the same type of product, the term 'glucose syrup' has been used throughout this book.

A similar situation exists with regard to high-fructose syrups. In Europe, high-fructose syrups are called either isoglucose or high-fructose glucose syrups (HFGS), but in America, they are known as high-fructose corn syrups (HFCS) or HF syrups. As isoglucose, high-fructose glucose syrups (HFGS), high-fructose corn syrups (HFCS) and HF syrups are all the same type of product, the term 'high-fructose glucose syrup' (HFGS) has been used in the book when referring to these types of products.

In the medical and pharmaceutical industries, the white crystalline powder dextrose monohydrate is referred to as glucose, whilst to the glucose industry, glucose refers to the syrup.

Where a glucose syrup contains predominantly one sugar, for example dextrose, there is a tendency to refer to that syrup as a 'dextrose syrup'. Similarly, if maltose is the predominant sugar, then the syrup will often be referred to as a 'maltose syrup'.

Further differences in nomenclature relating to the glucose industry are covered in Chapter 3: Glucose Syrup Manufacture.

Where the word 'sugar' is used in the text, it is used in its widest sense and covers both sucrose and glucose syrups, unless otherwise stated. Where the word sucrose is used in the text, it will refer specifically to the sugar obtained from either sugar cane or sugar beet – the white crystalline product which is used in a domestic kitchen.

Glucose syrups are sold on their solids content, and 'commercial Baumé' is the scale used to measure density. Other terms used for the solids in a glucose syrup are true solids or refractive index solids (RIS).

The term Brix solids is sometimes incorrectly used when discussing the solids content of a glucose syrup. Brix is used by the sugar industry to measure the solids of a sucrose solution, and is only applicable to sucrose solutions. Further details covering solids can be found in Chapter 4, under Specifications, and in Appendix B, Simple Calculations, Section B.7.

A note on nomenclature

Acknowledgements

It would not have been possible to produce this book without the help of a large number of people, organizations and companies who have been unselfish in giving their time and patience and I thank them for their generosity.

I would like to put on record my appreciation to the following people for their help and expert knowledge: Philip Ashurst, Jenny Clements, Neil Cobbett, Tom Cripps, Bill Crosson, John Croxford, Philip Ecclestone, Charles Francis, Graham Godbold, Chris Goddard, Mike Grainger, Roger Hayward, Peter Hill, John Higinbotham, Peter Holder, Simon Hook, Peter Hooper, Philip Jennings, Mike Jones, Chris Marchant, Carl Moore, Tim Morris, John Regan, Rod Stanbridge, David Thacker, Norman Westley (late), John White, Hilary Young.

I would like to formally thank the following libraries and organizations for their help: British Lending Library; Chatham and Kent Libraries; Oklahoma State University Library, USA; Royal Society of Chemistry Library; University at Medway Library; Association des Amidonniers et Feculiers, Belgium; Bank of England (Education & Museum Group); Banque de France (Public Relations; Central Science Laboratory, York; Corn Refiners Association; H.M. Custom & Excise National Museum, Liverpool; The Embassy of Japan, London, Home Grown Cereal Authority; The Institute of Brewing; Institute Francais, London; National Archives, Kew; the Royal Swedish Academy of Science (Centre for History of Science); The Sugar Bureau, London; Victoria and Albert Museum, London (Ceramics and Glass Collection).

I must also thank the following companies for allowing me to reproduce their information and for supplying photographs: Baker Perkins; Briggs of Burton; Carlton & United Breweries, Australia; Chapman & Hall; Endecotts Ltd.; Royal Pharmaceutical Society of Great Britain (National Formulary); Tate & Lyle Americas Inc.; Tate & Lyle Molasses Ltd., and Vitech Scientific Ltd.

Finally, I would like to thank my family for their forbearance during the writing of this book and to David McDade and the staff at Wiley–Blackwell, Oxford, for their help and guidance.

Chapter 1
History of glucose syrups

Glucose syrups are clear colourless viscous liquids, derived from starch, but, who first produced a glucose syrup, when, where and why, or was it all a lucky discovery?

1.1 Historical developments

The glucose industry was born in Russia, in 1811, when a German chemist **Konstantin Gottlieb Sigismund Kirchoff**, working in Russia, showed to the Russian Imperial Academy of Science in St Petersburg three flasks. According to a report of this meeting (Memoires de L'Academie Imperiale Des Science de St Petersburg, **4**, 27, 1811), one flask contained syrup produced artificially from vegetables (potatoes, wheat and buckwheat). The second contained some 'sugar' obtained from this syrup by drying, and the third contained a syrup made from the 'dried sugar'. But, there was no mention in this report of how Kirchoff had made the syrup. Fortunately, this was revealed in the Philosophical Magazine of 1812 (Vol. 40, July to December).

Evidently, Kirchoff discussed the results of his experiments with an eminent Swedish chemist, **Prof. Jons Jacob Berzelius (1779–1849)** of Stockholm, who then told The Royal Institute, in London, about Kirchoff's experiments, and that the different vegetables had been heated with sulphuric acid to obtain the syrup. At the suggestion of **Sir Humphry Davy (1778–1829)**, Kirchoff's experiments were repeated at the Royal Institute, and produced similar results to those of Kirchoff. It was however, not until 1814, that the Swiss chemist, **Nicolas Theodore de Saussure (1767–1845)** showed that the syrups produced by Kirchoff contained dextrose.

To this day, we still do not know why Kirchoff carried out these experiments, or what he was trying to achieve, although there have been many suggestions. Kirchoff was born on 19 February 1764 in Teterow Germany. In 1792, Kirchoff was working in St Petersburg as an assistant in a chemist's shop, and became a chemist in 1805. He was then elected an Associate of The Russian Academy of Science in 1807, and became an 'Adjunct', or member, of The Academy in 1809. According to the Memoires de L'Academie Imperial Des Science de St Petersburg, **4**, 27, 1811, at that time, Kirchoff was working on the analyses of different geological minerals. This suggests that Kirchoff was not a food chemist, and that his discovery of glucose syrups – a sweet product, was purely a fortuitous accident for the food industry. Because of Kirchoff's interest in minerals, and the fact that St Petersburg had a thriving porcelain industry, there is the possibility that he was looking for an alternative adhesive or flux to use in the gilding of porcelain. Gum Arabic had been used as the adhesive in the early days of the gilding process, but that was replaced by honey in the mid-eighteenth century (private correspondence with The V & A Museum, London).

Basically, gold leaf and honey were ground together to form a paste, which was then painted onto the glazed porcelain. Essentially, the honey was acting as a flux, to improve the adhesion of the gold to the porcelain.

However, in 1811 with Napoleon Bonaparte wagging war all over Europe, both honey and sugar became very scarce, and according to Daniel Green's book, 'CPC Europe – A family of food companies' (1979), the price of sugar in Berlin in 1811 was £45 per hundredweight, which in today's money (2007) would have equated to £14.98 per pound (£33.00 per kilo). With this high price of sugar, it is not surprising that honey, which had previously been used in the porcelain industry as an adhesive for gilding, would now be used as an alternative to sugar for sweetening. Therefore, it is possible that Kirchoff with his knowledge of chemistry and minerals would have had the perfect background to look for an alternative adhesive to honey, so as to keep the St Petersburg porcelain industry in business. The fact that he ended up with a sweet sticky product was, like so many scientific discoveries, a very fortuitous discovery. The immediate impact of Kirchoff's discovery can be gauged from the report in the Philosophical Magazine (1812), which mentions that Kirchoff's work was adopted 'on a large scale in the Russian Empire, as the price of cane sugar had increased so much and the supply was so uncertain'. So the sweet sticky substance which Kirchoff had inadvertently made became a useful alternative to sugar. And that was the start of the glucose industry.

In 1812, Kirchoff was elected an Extraordinary Academician in Chemistry. This encouraged him to continue his research into starch, which resulted in 1816, with the discovery that enzymes were also capable of breaking down starch. Kirchoff died, in his adopted country of Russia, in St Petersburg on 14 February 1833, aged 69.

Some people have suggested that glucose was discovered by an unknown German. Kirchoff was a German, and we know that he discussed the results of his glucose experiments with Prof. Berzelius before they were published. As Kirchoff was a member of several scientific societies, it is quiet possible that he also discussed his results with other interested scientists, hence the origins for the suggestion that an unknown German discovered glucose syrup. This unknown German I would suggest was in fact K.C.S. Kirchoff. It is interesting to note that the initials of Kirchoff vary in different publications. The reason for this is that the Russians spell Konstantin with a 'K', whilst Western Europe spells it with a 'C'.

Coincidentally to Kirchoff discovering how to make glucose syrup, **Napoleon Bonaparte** asked the French government for 100,000 French Francs (equivalent to £162,303 in 2005), as a prize for anybody who could make sucrose from an indigenous French plant. The reason for this request was that the successful British blockade of the French ports, prevented sucrose being imported from France's overseas colonies. Napoleon's reaction to this successful blockade was typical of a politician, he issued an embargo on all merchandise from England and her colonies to give the impression that the sugar shortage in France had nothing to do with the successful blockade! The prize of 100,000 Francs was won by a French analytical chemist, **Louis-Joseph Proust (1754–1826)** in 1810 for producing 'sugar' from grapes (Memoires de L'Academie des sciences, belles-lettres et arts d'Anger, 14, 1997–1998, pp. 225–246.). Napoleon also awarded Proust with the Legion d'Honneur. However, the 'sugar' which Proust had produced was not sucrose, but dextrose.

It would be another three years before sucrose was produced from a French plant. On 2 January 1813, Napoleon was told that **Benjamin Delessert (1773–1847)**, from Plassy, near Paris, had succeeded in making loaves of white sugar (so called because the sugar was moulded into the shape of a loaf). Delessert had produced the sucrose from the root of the sugar beet plant – *beta vulgaris altissima*. This was the start of the sugar beet industry in France and a political time bomb – a development which, some 164 years later, would have a considerable influence upon the future development of the European glucose industry. As for Delessert, he did not receive a 100,000 Francs. However, in recognition of his work, Napoleon gave Delessert his own Legion d'Honneur sash (Sugar and All That ... A History of Tate & Lyle, 1978, p. 21). Delessert name has however been immortalised by giving his name to the 'Dessert' course of restaurant menus.

The UK glucose industry started in 1855, when a Frenchman **Alexandre Mambre (1825–1904)** started to produce solid glucose in Spitalfields, London. In 1876, he moved production to Hammersmith. Whilst there had been a starch industry in the United Kingdom, since about 1600, the starch was used mainly for laundry and textile purposes – there had been no attempt to use the starch for making glucose.

Alexandre Mambre was the son of a French farmer, and was born in 1825, in Valenciennes, Northern France. It is believed that the reason he came to England in 1855 was to secure a patent to produce brewing sugars made from potatoes, and he knew that in England, his patent would be protected, whereas in France, this would not have been the case. Mambre was, however not alone in producing 'brewing sugars', but these other producers were using sucrose as their starting material, and produced an invert-type product for use in brewing, mainly for adding colour to the beer. In 1850, there was a tax on sugar used in beer, therefore, it is possible that by using these products for colouring, tax would have been avoided, and this might also be a reason why British beers at that time had a very characteristic dark colour.

In about 1861, Mambre changed the spelling of his name to Manbre, and died in England, in 1904. The company which he started would eventually become Manbre and Garton, which was subsequently purchased by Tate & Lyle in 1976, who then closed the company, selling off the glucose assets in 1980 to Cargill.

In the United Kingdom, prior to 1962, sugar had been taxed, on and off for several years, because it was an easy tax to collect direct from the producer, and glucose syrups, together with molasses and saccharin, were included within this taxation, which was known as Excise Duty. Therefore, in a typical glucose refinery, there would have been a room, known as 'The Tank Room', in which an excise officer would have been permanently stationed. It was his private kingdom, which nobody was allowed to enter, so as to ensure that there was no interference with the collection of the Excise Duty.

Each Tank Room would contain a minimum of two tanks, each of a known volume. All syrups leaving the refinery would have to go into one of these tanks, and once the tank had been filled, the excise officer would note the volume and temperature of the syrup in the tank. Next, he would measure its specific gravity, using a hydrometer. From these three measurements, the officer would calculate the solids present and hence the amount of duty to pay. Then, and only then, would the syrup be allowed to go forward to the next processing stage, that is evaporation. For ease of handling, the solids of the glucose syrup entering the Tank Room would have been about 40%.

The duty in 1960 was five shillings and four pennies (5s 4d), per hundredweight (112 pounds), based on syrup solids. This would have been equivalent to 27 pennies per hundredweight, or £5.40 per ton in decimal currency. For a syrup containing 80% solids, the duty would have been £4.32 per ton, on a product which was selling at about £40 per ton, excluding duty. This duty would also have applied to dextrose. A monthly return of the daily production would be prepared, and the total duty payable for that month by the glucose manufacturer would then be paid to the local collectors' office. For a refinery producing 100 tons of syrup per day, at 80% solids, 30 days a month, the duty would have been £12,960 per month, or £155,520 per year.

If the syrup was exported, or used in a product, which was exported, then the Excise Duty could be recovered from HM Customs and Excise. Fortunately, for the UK glucose industry, the excise duty on sugar, molasses, glucose syrups and saccharin was abolished in 1962.

1.2 Analytical developments

As Kirchoff's discovery grew into an industry, technology was borrowed from several other different industries, particularly from the sucrose industry, with many different people contributing to the process and to the methods of analyses – both physical and chemical.

Following **de Saussure's** confirmation that Kirchoff's glucose syrup contained dextrose, the German chemist **Herman von Fehling (1812–1885)**, in 1848, worked out a method for determining the total reducing sugars in a syrup, using a solution of alkaline tartrate and copper sulphate – the well-known Fehling's Titration. Since dextrose and the other sugars present in a glucose syrup are reducing sugars, this meant that the infant glucose industry now had a method for determining the total reducing sugars present in a glucose syrup – what we now refer to as the dextrose equivalent (DE) of the syrup. Or, to put it another way, how far the starch had been broken down. The principle behind the Fehling's method is that reducing sugars convert the blue alkaline copper sulphate solution to a red precipitate of cuprous oxide. It is interesting to note that the glucose syrup industry used this method, although slightly modified (see below), for over 100 years to run its process and to characterise its products. And it is still a recognised method for determining reducing sugars.

The Fehling's method had several major drawbacks. One was the determination of the end point of the titration. In 1923, two British chemists, **Joseph Henry Lane (1883–1951)** and **Lewis Eynon (1878–1961)**, both of whom had a background in sugar beet technology, and who ran their own analytical consultancy (established in 1910), suggested the addition of methylene blue as an internal indicator to improve the recognition of the end point. The Fehling's method then became known as the Lane and Eynon method.

Other drawbacks of the Fehling's method are that the titration is carried out using a boiling solution, and is time consuming, taking about forty-five minutes for three titrations. Additionally, the detection of the titration's end point is very dependent upon the skill of the analyst – it is not the easiest of titrations to perform. In 1979, **M.G. Fitton**, who worked for Corn Products, Belgium, showed that the DE could be measured using a cryoscope (freezing point osmometer), with result being available within two minutes, and as it is an automatic determination, it did not rely on the eyesight of the analyst to determine the end point. This method uses the principle of sugars depressing the freezing point of a solution. One drawback of this method is that the other soluble substances present in the syrup, typically sodium chloride resulting from pH adjustments of the syrup, will also depress the freezing point. With the introduction of de-ionised (de-min) syrups, it is necessary to calibrate the cryoscope for both de-ionised and non-deionised syrups. It should be appreciated that the cryoscope will only determine the DE of a syrup – ideal for a lot of routine production analyses, and, like the Fehling's method, it will not determine which sugars make up the DE. For determining the individual sugars which make up the DE, the industry had to wait for the next major analytical development – chromatography. The word 'chromatography' is derived from the Greek, and means 'colour writing'.

In 1850, the German chemist **Friedlie Runge (1794–1867)** was the first person to separate dye colours using paper chromatography, but it was the Russian botanist **Mikhail Tswett (1872–1919)**, who recognised that chromatography could be used as a method for separating different substance, and in 1910 published details describing the process. The method was further developed by **Archer Martin (1910–2002) and Richard Synge (1914–1994)** in the early 1940's. In 1941 Martin then suggested the possibilities of high-pressure liquid chromatography (HPLC). In 1946, **Stanley Partridge (1913–1992)** developed the use of partition paper chromatography for the examination of sugars, and in 1948, so as to separate larger amounts of sugars than was possible using paper chromatography, **Hough, Jones and Wadman** used columns of cellulose. Thirty years later, in 1978, this pioneering work would be used on an industrial scale to produce fructose syrups containing up to 90% fructose, by replacing the cellulose with ion exchange resins.

Whilst both paper and thin layer chromatography, together with gas–liquid chromatography, were used as research tools for sugar analyses, the real breakthrough for the glucose industry came in 1973 when the American chemists **Brobst, Scobell and Steele**, working for A.E. Staley Manufacturing Company, Decatur, IL, U.S.A. (now part of Tate & Lyle Americas Inc), produced a viable automatic HPLC which could quickly determine the sugars present in a glucose syrup. As the system is computer compatible, the DE of the syrup can be calculated, and if a blend of syrups and sucrose have been used, the ratios can also be calculated, together with any process inversion. Besides being quick, this method does not use any expensive or dangerous solvents – just de-ionised water, making it both cheap and safe to operate. This was a major step forward in the analysis of glucose syrups, especially as the industry can use either an acid or a combination of both acid and enzyme to produce glucose syrups, which have the same DE, but have totally different sugar spectra and hence different properties and processing characteristics. By comparing the different sugars present in a syrup, it is possible to get an indication of how the syrup has been produced.

The development of HPLC by **Brobst, Scobell and Steele** was invaluable for the next major development in the glucose industry, namely the commercial development of high-fructose glucose syrups. HPLC also enabled the accurate analyses of different sugars, alcohols, polyhydric alcohols and other products present in syrup blends, or in finished products. It is a very versatile and powerful analytical tool, and invaluable to any applications chemist.

The use of the cryoscope and HPLC within the glucose industry have now almost totally superceded the Fehling's method for characterising a glucose syrup.

In the early days of the glucose industry, to quickly determine the solids present in a syrup, the industry used the Baumé hydrometer. The Baumé hydrometer was invented in 1768 by a French chemist, **Antoine Baumé (1728–1804)**, who worked in the French alkali industry, where he used it to determine the solids of different solutions used in the saltpetre, bleaching and sal ammoniac processes. The Baumé hydrometer is still used today in the production of glucose syrups as a quick method for measuring the solids in starch slurries and glucose syrups, as well as in other process streams. It is a very simple and inexpensive piece of equipment. Because of the way that the Baumé hydrometer was adopted by the glucose industry, there were some inconsistent readings, therefore the glucose industry standardised on a 'modulus' of 145 for its hydrometers, and is the constant used to convert specific gravity into degrees Baumé.

The Baumé hydrometer having originating from the alkali industry is standardised using a solution of common salt. In 1835, to make the hydrometer more suitable for the sugar industry, a German chemist, **Karl Joseph Von Balling (1805–1868)**, recalibrated the Baumé scale so that it could read sucrose solids – to give us degrees Balling. In 1854, an Austrian mathematician **Adolf F. Brix (1798–1870)**, improved the accuracy of the Balling scale to give us the well known Brix Solids. And finally, in 1918, **Dr Fritz Plato**, a German, made further improvements and corrections to the Balling scale to produce the Plato tables, which are now used extensively in the brewing industry. Basically, the Balling, Brix and Plato scales are identical up to the fifth and sixth decimal place. In summary, the glucose industry uses the Baumé scale, whilst the sugar industry uses the Brix scale. The brewing industry, however, uses the Plato scale, possibly due to it being more accurate, which is financially important when dealing with the paying of duty on beer.

Today, for a more accurate determination of glucose syrup solids, the glucose industry uses a refractometer. As well as being more accurate, the refractometer is also quick and easy to use. The instrument works on the principle that a single ray of light is diffracted when it passes from one substance to another. The amount of diffraction can be related to the amount of solids present in the syrup. The most common refractometer used in the industry is the Abbe refractometer, which was developed by a German physicist **Ernst Abbe (1840–1905)**. Originally, he was Professor of Optical Physics at the University of Jena, and directed research at the Carl Zeiss factory from 1866. He became a partner in 1875, and on the death of Carl Ziess in 1888, became the director of Carl Zeiss. The solids of a glucose syrup which are determined using a refractometer are referred to as 'The RI Solids'. The temperature at which the determination is made should always be stated. It is therefore standard practice for the refractometer to be temperature controlled.

1.3 Process developments

Running parallel with developments in analytical methods were improvements in the production of glucose syrups. In Europe, the glucose industry was based on either potato or wheat starch, because both of these starches were readily available within Europe. In North America, however, the glucose industry which also used potato and wheat starch, started to look at the possibility of using maize (corn) of which there is a plentiful supply, and in 1841, **Thomas Kingsford (1796–1869)**, an English immigrant, developed an alkaline process to obtain starch from maize. In the year 1875, an Italian, **M.M. Chiozza** developed a process in which the maize was soaked or steeped in a solution of sulphur dioxide. **Dr Arno Behr** of The Chicago Sugar Refining Company used this sulphur dioxide steeping process to replace the Kingsford alkaline process and is still used by the industry today. At that time, American companies produced either starch or glucose syrups, but in 1883, The Chicago Sugar Refining Company started to produce both glucose syrups and starch, and this is possibly the start of the glucose industry, as we know it today, a totally integrated starch and syrup operation.

One of the advantages of using either maize or wheat as a raw material for making starch is that both can be easily stored and transported before being processed. Potatoes, on the other hand, like sugar beet, have to be harvested in a 'campaign'. Whilst potatoes can be stored for a limited time in a clamp in a field, ideally, they should be processed as soon as they have been harvested, with the resultant starch being dried, and then stored. The dried starch can then be made into a slurry before being converted into a glucose syrup.

With the industry being able to produce syrups, the production of crystalline dextrose, the end product of starch hydrolysis, was proving more difficult. The production problems were eventually solved in 1923 by **W.B. Newkirk**, who worked for Corn Products, Argo, USA, by devising a process of controlled crystallisation. This involved seeding a dextrose rich syrup, and allowing it to cool with continuous agitation with the dextrose crystals being recovered using a centrifuge.

When acid is used to make glucose syrups which have a DE higher than 60, the resulting syrup has an undesirable off taste. The problem of this undesirable off taste was solved in 1938 by **Dale and Langlois**, who worked for A.E. Staley, Decatur, IL, USA. They used 'fungal enzymes' derived mainly from moulds of aspergillus to further change the sugar spectrum of a conventional acid converted 42 DE syrup to produce the very successful 63 DE syrup. This syrup is sweeter than a conventional 42 DE syrup, and enabled glucose syrups to be used in many new and different applications, other than confectionery products. In 1951, **Langlois and Turner** further developed the use of their 'fungal enzyme' for the production of dextrose. This development was a major advance in dextrose production and would mean the end of the conventional acid process for making dextrose. All future dextrose production would use either a combination of acid and enzyme or purely enzymes. The 'active ingredient' in Langlois and Turner's 'fungal enzymes' was glucoamylase, sometimes referred to as amyloglucosidase or AMG for short.

The next major milestone in the industry occurred in 1957, when **Richard Marshall** and **Earl Kooi**, discovered the existence of an enzyme which could be used to convert dextrose into fructose. Since fructose is approximately 50% sweeter than sucrose, this discovery had a major impact on the glucose industry – the glucose industry could at long last, compete with the sucrose industry, on a sweetness for sweetness basis. This discovery had possibly the greatest commercial impact on the glucose industry than any other discovery.

Chapter 2
Fructose containing syrups

2.1 Introduction

From the earliest days, the glucose industry had been striving to have a product which was as sweet and as cheap as sucrose, and fructose was the answer.

Fructose is approximately 1.2–1.7 times more sweet than sucrose, and whilst it had been available for several years, up to the early 1950s, it had been produced by one of three methods:

1. The acid hydrolysis of inulin, found in the roots of dahlias, chicory, Jerusalem artichokes and similar plants.
2. Separating fructose from inverted sucrose.
3. Heating dextrose with an alkali.

Because of the way it was produced, it was both expensive and frequently had an unpleasant after taste, as well as a poor colour. All these factors effectively limited its commercial applications.

The breakthrough came in 1957, when Richard Marshall and Earl Kooi showed that it was possible to convert dextrose into fructose using an enzyme. Unfortunately, it would take another 10 years before this discovery would become a commercial reality.

Marshal and Kooi were working for Corn Products, Argo, in America, when they discovered that the micro-organism *Pseudomonas hydrophila* if grown on a xylan-rich substrate would produced the enzyme xylose isomerase, and this enzyme was able to convert D-xylose into D-xylulose. This enzyme also had an interesting side reaction, which under the correct conditions could convert dextrose into fructose. The Holy Grail of the glucose industry had been achieved. Unfortunately, the celebrations were premature because there were several major technical problems to overcome before this laboratory experiment could be converted into a commercial reality. Besides which, in 1957, sucrose was both cheap and freely available. Corn Products possibly thought that it would be too expensive to solve the technical problems involved, and as there could be no guarantee that they would get a premium for fructose, Corn Products probably felt that they would get a better return on their capital expenditure from other products. Marshall and Kooi's discovery, like so many discoveries, was before its time – a solution waiting for a problem! What neither Corn Products nor the glucose industry realised was that in the 1970's there would be a world sugar shortage – a shortage created by a combination of politics and poor harvests of both sugar cane and sugar beet.

So what were these major technical problems? Basically, they were as follows:

1. For the xylose isomerase enzyme which Marshall and Kooi had obtained from *Pseudomonas hydrophila* to convert appreciable amounts of dextrose to fructose, it required either an arsenic or a fluoride salt as a co-factor, both of which would be highly undesirable in a process, producing an ingredient destined for the food industry.
2. The xylose isomerase enzyme obtained from *Pseudomonas hydrophila* was not heat stable, that is, it was heat labile. Using this enzyme, the conversion of dextrose to fructose would have to be carried out at 40°C or lower. At this temperature, pathogens would also grow, which meant that the use of heat to prevent these pathogens from growing during the process would also inactivate the xylose isomerase enzyme. Whilst the heat stability of the enzyme could be improved by the presence of cobalt salts, these salts would again not be acceptable to the food industry.
3. The glucose industry prefers to use liquid enzymes. The reason for this is that liquids are easier to handle than powders – they can be pumped, making them suitable for an automatic dosing system. Also when handling powders, there is always the health risk from inhaling the dust.
4. The process to produce the enzyme would have to be both cheap and easy to operate to give high enzyme yields, with the enzyme being cost effective in use. From their experience to date, it would appear that xylose isomerase was none of these.

Faced with these sorts of problems, one can understand why Corn Products were reluctant to proceed with Marshall and Kooi's discovery.

Fortunately for the glucose industry, the Japanese government had become aware of Marshall and Kooi's discovery. Japan being an island with very few natural resources, spends a lot of time looking for any technological development, from anywhere in the world, which they believe could be further developed. It also has a long history of using fermentations to produce food, for example sake from fermented rice and monosodium glutamate flavouring from soya beans. It was therefore not long before The Japanese Agency of Industrial Science and Technology started to evaluate Marshall and Kooi's discovery.

In 1965, Prof. Takasaki of the Japanese Fermentation Institute, part of The Japanese Agency of Industrial Science and Technology, discovered that the xylose enzyme could be produced from micro-organisms of the genus *Streptomyces*. The significance of this was that the xylose isomerase produced from the *Streptomyces* genus could withstand heat – typically up to 80°C, and at this temperature, pathogenic bacteria would not survive. Additionally, this xylose isomerase did not require arsenic or fluorides as co-factors, only magnesium. Neither did it require cobalt to improve its heat stability. Laboratory work also showed that the enzyme was stable when in solutions of a high dextrose concentration, and was able to convert about 50% of the dextrose present into fructose. Finally, this enzyme is produced inside the bacterial cell, as opposed to the xylose isomerase produced from *Pseudomonas hydrophila*, which was produced on the surface of the cell wall and could therefore be easily detached and die.

This development became the basis of the Takasaki–Tanabe Enzyme Process, and in 1966, the Japanese used this process for small-scale commercial production of high-fructose glucose syrup (HFGS).

2.2 Commercial development

The next step in the Fructose Syrup Story was to develop the process into a viable commercial process – to obtain maximum yields of fructose, using the minimal amount of enzyme, and to that end the Japanese started to work with the American corn starch and syrup company Clinton Corn Refining. This resulted in Clinton having the sole rights in the United States to use the Japanese technology, and they were also allowed to sub-license the technology to other companies.

In 1967, Clinton used a batch process with a liquid isomerase enzyme to produce its first commercial HFGS, which only contained 15% fructose. Both the syrup and its method of manufacture were far from ideal. It required a long residence time, which resulted in the syrup developing both a poor colour and flavour, and due to the low fructose content, the syrup was not particularly sweet. It was also expensive on enzyme usage. A new approach was required – the enzyme would have to be produced either very cheaply or immobilised so that it could be used more efficiently. But how do you immobilise an enzyme? Should the whole of the bacterial cell containing the enzyme be immobilised, or should the enzyme be extracted and then be immobilised? What method of immobilisation should be used? What type of carrier, if any, should be used? Finally, whichever approach or combination of approaches are used, the process for immobilising the enzyme had to be cost effective, and the resulting immobilised isomerase enzyme must perform efficiently to produce maximum fructose yields at minimal costs. A tall order! Fortunately, by using the isomerase from *Streptomyces*, Clinton had several different options:

1. They could just add whole cells to the substrate, and then filter out the cells, after the required fructose level had been reached. This would be expensive, because the cells could only be used once and a lot of cells would be required.
2. They could extract the enzyme and just add it directly to the substrate. Again this would be expensive, and the enzyme could not be recovered.
3. They could immobilise either the whole cell, or just the enzyme. This would effectively 'trap' the enzyme, and allow it to be used several times.
4. Finally, isomerase from *Stretomyces* could withstand temperatures up to 80°C, which meant that heat could be used to immobilise the enzyme, without adversely affecting the activity of the enzyme. Additionally, the conversion process of dextrose to fructose could be run at up to 80°C, and at this temperature, pathogenic organisms would not survive. This was the original approach used by Takasaki in 1968 to immobilise the isomerase enzyme.

In 1968, Clinton's experiments with immobilising the isomerising enzyme were sufficiently advanced, that they used a combination of both a liquid enzyme and an experimental immobilised enzyme to produce HFGS. The resulting syrup contained 42% fructose. This

42% fructose containing syrup was to become the standard HFGS, and is often referred to as 'first generation HFGS'. Clinton continued to develop the novel concept of an immobilised enzyme, and in 1972 were suitably rewarded, when they were able to produce HFGS using only an immobilised enzyme in a continuous process.

An immobilised enzyme looks like the hundreds and thousands, which can be found spread on the top of chocolate cakes. The immobilised enzyme is placed in a column maintained at 60°C, and a de-ionised syrup containing at least 95% dextrose, with a solids of about 40%, is passed down the column. As the dextrose percolates through the column, some of it is converted into fructose, to give a syrup containing about 42% fructose. By using an immobilised enzyme, the enzyme can be used several times, before it becomes exhausted, making the enzyme use very cost effective. This development was to become one of the earliest and most successful uses of an immobilised enzyme on an industrial scale.

In 1968, so as to avoid any accusations of operating a HFGS monopoly, Clinton sub-licensed the Japanese technology to produce fructose syrups to A.E. Staley Manufacturing Company, another American corn syrup producer. Part of this licencing agreement was that Staley had to use the Clinton immobilised enzyme. This sub-licencing arrangement was to have far reaching effects several years later in Europe, where Staley had an interest in two European syrup producers, namely a 50% interest in Amylum of Belgium, and a 25% interest in their subsidiary company, Tunnel Refineries of the United Kingdom.

Not unnaturally, Staley was far from being happy at having to rely on a competitor for their supply of the critically important isomerising enzyme. Therefore, in 1974, Staley started talking to a Danish enzyme company Novo Industries S/A about the possibility of supplying an isomerising enzyme. Novo produced an enzyme from a strain of *Bacillus coagulans*, which had the advantage that the enzyme could be used in a continuous process, at a temperature of 65°C and by keeping the pH above 8.0, the addition of cobalt would no longer be necessary.

During the period from 1974 to 1976, there was a world sugar shortage due to below average sugar beet harvests in the United Kingdom, Continental Europe and Russia, and bad weather conditions in the Caribbean, which reduced the sugar cane harvest in that area. This meant that the Caribbean countries, which are major sugar suppliers to both the United States and Europe, were unable to meet their export commitments.

At this time, the United States was importing nearly 60% of its sugar (sucrose) requirements. The bulk of the sugar was used by industry, with the soft drink industry being the largest user. Soft drinks frequently contain acids (e.g. citric, phosphoric) and acids 'invert' sugar, that is the sugar is converted into a mixture of dextrose and fructose – the very same sugars present in HFGS. The new syrup therefore had a potential ready made market, and as the United States is the world's largest grower of maize, it made a lot of sense to use the starch from the US Midwest maize belt, and to convert it into HFGS, rather than import scarce and expensive sugar, especially as the sweetness of both sucrose and HFGS are approximately the same. Within a few years, 50% of the sugar used by the American food industry would be replaced by HFGS, a fact which was not lost on the European farming lobby.

The next technological development came in 1978, when industrial chromatography was used to increase the fructose content of HFGS. The theoretical maximum fructose content that can be achieved using the isomerising enzyme is 47%. By using industrial chromatography, it was now possible to separate the fructose from the dextrose of a conventional 42 HFGS, and produce HFGS containing 90% fructose. The remaining dextrose fraction could then be re-circulated back into the 95 DE syrup, and either isomerised to make more 42 HFGS, or be blended with the 90% fructose to make a HFGS syrup which contains 55% fructose.

And then in 1987, Staley started to produce crystalline fructose from the 90% fructose syrup – a truly remarkable 30-year journey for both the glucose industry and for technology.

2.3 Europe and the HFGS (isoglucose) production quota

In Europe unfortunately, HFGS was not being so readily embraced. In 1975, Staley's Belgium partner, Amylum, started to produce HFGS, followed in 1976 by Tunnel Refineries in the United Kingdom, with both companies using Staley's technology and Novo's enzyme. A Dutch company, Koninglije Sholten Honig (KSH), also started to produce HFGS in Holland, but using Clinton's technology. Several other European syrup producers were also interested in producing HFGS. This interest was being driven by the success which HFGS was having in replacing sugar in America, caused by the world sugar shortage, and the resulting high sugar prices.

This rapid increase in the use of HFGS to replace sugar in America and now in Europe sets alarm bells ringing in both the European political and agricultural circles. The reason for this was that during the Second World War and immediately afterwards, there was a food shortage within Europe. To make sure that this did not occur again, one of the objectives of the European Community was to make the community self-sufficient in food, and so the Common Agriculture Policy (CAP) came into being. The CAP was of most benefit to countries which were mainly agricultural with lots of small farms, such as France, as opposed to manufacturing countries, such as the United Kingdom.

The CAP works by using a combination of subsidies to encourage farmers to grow as much as possible on the one hand, and a series of quotas to try to balance supply with demand on the other, and as Europe has to import sugar, the CAP encouraged farmers to increase sugar beet production – Napoleon's dream of a French home sugar industry had come true. However, when it became apparent that HFGS was cheaper to produce, and could be used to replace a lot of European produced beet sugar, the European politicians decided to take drastic action to protect their European sugar beet farmers.

In 1977, such was the pressure of the political lobby in the EEC against HFGS production that the European Commission put a levy equivalent to £56 per tonne on HFGS based on syrup solids. This effectively made it uneconomic to produce, and at least one company KSH stopped building its HFGS plant in the United Kingdom, and because the plant was already half built, the financial penalties which followed helped to bankrupt KSH. Cargill of America eventually purchased this half-built plant in 1978, but they did not go ahead with HFGS production, mainly because of the political uncertainties surrounding the future of HFGS within Europe.

The introduction of this £56 per tonne levy was challenged on the grounds that the EEC had imposed a wholly unfair burden on HFGS producers, contrary to The Treaty of Rome, 1957, Article 177. The European Commission therefore reduced it to £28 per tonne.

In 1978, two European glucose producers, not satisfied with the new reduced levy, took the European Commission to the European Courts of Justice in The Hague, Holland, and charged the Commission with breaching the 1957 Treaty of Rome, Title VI, Article 130g, which covers the exploitation of inventions, claiming that the Commission was preventing the development of HFGS production. The Courts of Justice ruled against the Commission, who then had to remove the £28 per tonne levy. (For the record, the European Commission's production levy on HFGS was based on 'units of account per 100 Kg of dry matter', which in the United Kingdom equated to £56 and £28 per tonne.)

Finally, in 1979, the Commission decided that as HFGS could be used as a substitute for sugar, the production of HFGS should be subject to a production quota, the same as sugar, and therefore HFGS should be included in the Sugar Regime. The quota would be expressed in dry metric tonnes of syrup, containing '42% dry fructose equivalents'. The HFGS production quota for each producer was then based on the amount of HFGS, containing 42% fructose on a dry basis, currently being produced by that company. Additionally, the company would not be allowed to move their quota from one country to another country, but they could, however, trade the HFGS between countries. A further 5% of the total quota would be allocated to companies who were just starting to produce HFGS. Because the Cargill plants at Tilbury, England and Bergen op Zoom, Holland, were not producing HFGS when the quotas were set, they forfeited their right at that time to produce HFGS within the EEC. However, with the enlargement of the EU and company acquisitions, Cargill is now a European HFGS producer.

Where a company which produces HFGS is taken over by another company who does not produce HFGS, then the HFGS quota can pass to the new owners, but the quota has to remain within the original country. It was this situation which enabled Cargill to become an HFGS producer in Germany, when they acquired Cerestar's HFGS facilities.

Similarly in 2007, when Tate & Lyle, who had purchased both Tunnel Refiners and Amylum, sold its HFGS interests in the United Kingdom, Belgium, France, Spain and Italy to Syral (a subsidiary of Tereos France, a French sugar company), it enabled Syral to produce HFGS in these countries. Whilst these plants have a new owner, the quotas remain in their original country.

In 2008, however, Syral decided to close its UK production unit by the end of 2009, and since this was the only HFGS production unit in the United Kingdom, this means that HFGS will no longer be produced in the United Kingdom. So what happens to the UK isoglucose quota? The quota is returned back to the EU.

Production quotas are monitored by the Intervention Board, who take production samples to check the fructose content of the syrup. They will also inspect the company's syrup production records, and if a company is found to have failed to comply with its quota agreement, it has to pay a fine.

To confuse the rest of the world, the EEC also decided to refer to HFGS, as 'isoglucose', and defined the syrup as having a fructose content of between 10% and 80%. Anything outside these limits would not be subject to the production quota. The thinking behind this was that at 10% fructose, the syrup would not be sufficiently sweet to replace sucrose, and above 80%, it would be too expensive to replace sugar.

According to the EEC Council Regulation No. 1293/79, dated 25 June 1979, the production quotas were as follows (Table 2.1):

Table 2.1 Original 1976 EEC isoglucose quotas.

Country	Isoglucose
Belgium	56,667
France	15,887
Germany	28,000
Italy	16,569
United Kingdom	21,696
Total	**138,819**

When a new country is admitted to the EU, it negotiates its own production quota, which would usually be based on its current production. One interesting quota negotiation was that of Spain, who was not a member of the EEC in 1979 when the HFGS quotas were set. In Spain, there was at least one company producing HFGS containing 55% fructose as well as 42% fructose. Therefore, when Spain joined the EEC in 1986, its production quota was based on the combined 42% and the 55% fructose HFGS production. However, once Spain became a member of the EEC, this company stopped producing 55% fructose HFGS, and only produced 42% HFGS. As quotas are based on HFGS containing 42% fructose, this company was able to effectively and legally increase its production of 42% HFGS by a further 13%.

With each new country joining the EU, the total isoglucose production within the EU increased. However, the quotas for each country, once agreed, remained unchanged, within certain limits, until 2005, when it had to be changed – a change brought about by the enlargement of the EU to 25 countries and the need to review the EU's Sugar Regime, so that it complied with the World Trade Organisation's ruling. This meant changing the sugar quotas over a period of several years, and increasing the isoglucose quota by 100,000 tonnes each year for next 3 years.

On 20 February 2006, the EU issued Regulation No. 318/2006. This all-embracing regulation covers the new quotas, which have been summarised in Table 2.2, which includes the new National and Regional Quotas, Additional Quotas for sugar (sucrose) and Supplementary Quotas for isoglucose. All quotas are expressed in tonnes. Where no figure has been included, it indicates that the country does not have a production unit. The sugar quotas for both France and Portugal also includes sugar produced in their overseas dependencies.

It is interesting to compare the 1979 isoglucose quotas of the original five producing countries, with their quotas for 2006. According to The Association des Amidonniers et Feculiers, there are now 20 isoglucose production units within the 25 EU countries.

This regulation also explains what happens to any excess production.

So as to make as much HFGS available to as many users as possible, some European producers have produced blends of HFGS with other glucose syrups. The most common

Table 2.2 2006 EU isoglucose quotas.

Country	Sugar	Isoglucose	Inulin
Austria	405,812	–	–
Belgium	882,301	71,592	215,247
Cyprus	–	–	–
Czech Republic	474,932	–	–
Denmark	452,466	–	–
Estonia	–	–	–
Finland	156,087	11,872	–
France	4,120,687	19,846	24,521
Germany	3,655,466	35,389	–
Greece	327,502	12,893	–
Hungary	411,684	137,627	–
Ireland	209,260	–	–
Italy	1,567,443	80,302	–
Latvia	76,505	–	–
Lithuania	111,995	8,000	–
Luxembourg	–	–	–
Malta	–	–	–
Netherlands	931,435	9,090	80,950
Poland	1,772,477	26,781	–
Portugal	89,671	9,917	–
Slovakia	217,432	42,547	–
Slovenia	62,973	–	–
Spain	1,006,961	82,579	–
Sweden	385,984	35,000	–
United Kingdom (1)	1,221,474	27,237	–
Total	**18,540,537**	**610,680**	**320,718**

Note: (1) The sole UK producer of isoglucose is proposing to stop UK manufacture in 2009.

syrup blend is with a 63 DE syrup. By producing blends of HFGS with standard glucose syrups, this strategy also helps to 'tie in' the sales of conventional glucose syrups to a HFGS producer. One of the advantages of using a fructose blend is that it can replace some of the sucrose, thereby increasing glucose sales. Blends containing less than 10% fructose would be outside the production quota, but since they lack sweetness, they only have a limited application.

A 63 DE syrup is used because it is the sweetest non-readily crystallising glucose syrup. If a 95 DE syrup is used, whilst it is sweeter than a 63 DE, the total dextrose content of the blend would be too high, resulting in a syrup which would crystallise, unless it is kept warm, as is the case for a conventional D.95 syrup. If a 42 DE syrup is used, then the sweetness level would be substantially reduced. Of course, if the application requires the use of either a D.95 syrup or a 42 DE syrup, then there is no reason why they cannot be used, or for that matter any other syrup.

2.4 Inulin

So as to avoid the HFGS production quota restrictions, some European companies started to produce a fructose syrup from chicory. Chicory roots contain inulin – a fructose polymer, which can be extracted from the chicory roots in a process similar to that of extracting sucrose from sugar beet using hot water, lime and carbonation. The liquor is then hydrolysed using the enzyme inulase to produce a syrup containing 80% fructose, which is carbon treated, and evaporated to about 80% solids. The fructose syrup is then blended with a dextrose-rich syrup to make a syrup which has the same sugar spectrum as an HFGS containing 42% fructose.

Unfortunately, this attempt to evade the HFGS production quota restrictions was short lived, because the EEC realised that this blended fructose syrups could also be used to replace sugar. Therefore, they decided to include it in the Sugar Regime, and make it subject to a production quota. Additionally, because the fructose is not derived from starch, and despite this blended HFGS syrup having exactly the same sugar spectrum and properties as a starch-based HFGS, the blended syrup has to be called either a glucose fructose syrup, or a fructose glucose syrup, depending upon which is the major ingredient. The EU is obviously determined to support their beet farmers, and to stifle any innovation which might upset this status quo.

Is it possible to tell if the fructose has come from starch or chicory? Providing that the starch is from a cereal, then the answer is yes, by using carbon isotope ratio spectrometry. In nature, carbon exists in two forms as either the carbon 13 or carbon 12 isotopes. The ratio of these two isotopes can be measured, and from these measurements, it is possible to tell which photosynthesis route the plant has used to fix the atmospheric carbon dioxide. Root crops fix their carbon dioxide using the C3 route, known as the Calvin cycle, whilst cereals use the C4 route, known as the Hatch–Slack cycle. Therefore, chicory, being a root crop, will use the C4 route, and maize and wheat, being cereals, will use the C3 route. The isotopes found in the C3 plant, such as chicory, will be in the range -22 to -33 parts per thousand, and the C4 plants, such as cereals, the range will be -8 to -11 parts per thousand below the reference limestone value.

2.5 Conclusion

There is no shadow of doubt that HFGS is the greatest innovation that the glucose industry has ever experienced. It has totally changed both the technical and commercial approach of the glucose industry, as well as the way sweeteners are used by the world's food industry. Whilst production quotas restrict the availability of HFGS within the EU, outside the EU, HFGS is freely available, with fructose contents available from 42% to 90%, and is being readily embraced by all sectors of the food and drinks industries in many different and diverse applications.

Chapter 3
Glucose syrup manufacture

3.1 Introduction

A glucose syrup is defined as 'A purified and concentrated aqueous solution of nutritive saccharides derived from starch', and having the following characteristics:

1. Dry matter of not less than 70%.
2. A dextrose equivalent (DE), expressed as d-glucose, of not less than 20% based on dry matter.
3. A sulphated ash content of not more than 1% on a dry basis.

Starch has a DE value of zero, whilst dextrose, the final end product of starch hydrolysis, has a DE value of 100. For the lay person, the term dextrose equivalent or DE can be regarded as an indication of how far the conversion process from starch to dextrose has gone. For the scientist, DE is a measure of the total reducing sugars present in a glucose syrup. It does not tell you how much dextrose is present in the syrup. The amount of dextrose in a syrup, together with other sugars, can be determined using HPLC (high-pressure liquid chromatography).

The DE of a glucose syrup can be measured using the Lane and Eynon Fehling's Titration, or by using a cryoscope (a freezing point osmometer). The cryoscope uses the principle that sugars will depress the freezing point of a solution, and this depression of freezing point can then be related to the sugar concentration in the syrup, and hence the DE of the syrup. The DE can also be calculated from the HPLC sugar analysis.

Based on the definition of a glucose syrup, fructose syrups derived from starch can be called a glucose syrup, because (1) they are derived from starch, and (2) since fructose is a reducing sugar, like dextrose, the DE, if measured, would be more than 20%. This means that it can be very difficult to make the correct interpretation of an ingredient label declaration, without the use of an HPLC sugar analysis.

The definition of a glucose syrup also means that fructose syrups derived from inulin (chicory) cannot be called a glucose syrup. They have to be called 'fructose syrup'. Similarly, when a 'fructose syrup' derived from inulin (chicory) is blended with a glucose syrup, the resulting blend has to be called either a 'glucose fructose syrup' or a 'fructose glucose syrup' depending upon whether the glucose syrup or the fructose syrup is the major ingredient.

In the glucose industry, the word 'glucose' is taken to mean a syrup, whilst the word 'dextrose' is taken to mean the white crystalline solid 'dextrose monohydrate'.

When somebody refers to a 'dextrose syrup', or a 'maltose syrup', they are usually referring to a syrup which contains predominately dextrose, such as a D.95, and in the case of a 'maltose syrup', they mean a syrup which contains mainly maltose. However, both of these syrups can also be correctly called a glucose syrup.

'Glucose syrup solids' mean spray-dried glucose syrup.

Products derived from starch which have a DE of less than 20 are called maltodextrins. Sometimes, products with a DE of less than 10 are referred to as 'hydrolysed cereal solids'.

3.2 Reducing sugars

Reducing sugars, such as those present in glucose syrups, are so called because when they are heated with alkaline solutions of certain metallic salts, such as copper or silver, the metal is precipitated out as either the oxide or as the metal. In the case of copper, the reducing sugars produce a red precipitate of cuprous oxide, which is the basis of the Lane and Eynon Fehling's Titration for determining the amount of reducing sugars present in a glucose syrup.

The reason for this reaction is that reducing sugars, such as dextrose, contain an aldehyde group in its molecule, and it is this aldehyde group which reacts with the alkaline metallic salt to reduce it, in the case of copper sulphate, to the characteristic red cuprous oxide. Figure 3.1 is a diagram of the dextrose molecule showing the aldehyde group in the C1 position. (The carbon atoms are numbered in a clockwise direction, with the first carbon atom coming after the oxygen atom, and is referred to as carbon one or C1.)

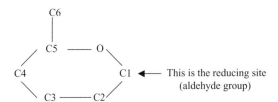

Figure 3.1 Skeleton of the dextrose molecule showing the reducing site.

When two dextrose molecules are joined together, as in maltose, one of the C1 positions will no longer be available – because it is joined to the next molecule at the C4 position. This means that whilst there are two dextrose molecules joined together, only one reducing site is available therefore the reducing power of maltose is only about half that of dextrose. Similarly, where there are three dextrose units joined together, as in maltotriose, the reducing power is only about one-third that of dextrose. Basically, as the number of dextrose molecules in the dextrose chain increases, so the reducing power of the sugar is reduced.

Figure 3.2 shows how the C1 site of the left-hand dextrose molecule is no longer available, because it is joined to the C4 site of the right-hand dextrose molecule, thereby leaving the C1 site of the right-hand molecule as the only available reducing site.

Incidentally because of the reducing properties of glucose syrups, they have a reducing action on certain food colours.

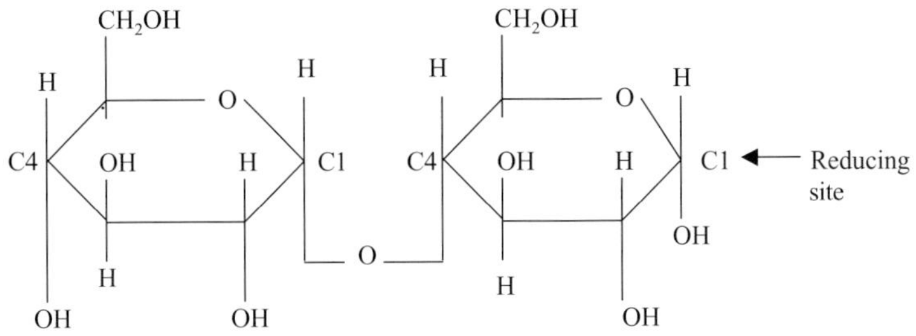

Figure 3.2 Diagram of a maltose molecule.

3.3 Starch

In the above definition of a glucose syrup, there is no mention of the type of starch which can be used. Whilst in theory any starch can be used, in practice, it is a combination of the availability of the starch at a commercially acceptable price and politics, together with the cost and technology involved in converting the starch to a syrup. In the United States, glucose syrups are made from maize, because it is so readily available. In the United Kingdom, American maize used to be the preferred starting material until the United Kingdom joined the EU (Common Market), which put in place tariff barriers making the import of American maize and maize from other countries outside the EU too expensive. This meant that UK glucose producers could either use wheat which grows very well in the United Kingdom and Northern Europe, or source maize from within the EU, typically from South West France and Mediterranean Europe, without incurring import duties. Both Holland and Poland make glucose syrups from potatoes, which again are a major agricultural crop. In Scandinavia, barley is the raw material, whilst in Australia, wheat is the preferred starch, because it grows a lot of wheat. In Asia, cassava (also known as tapioca) and rice are used. Basically, glucose syrups can be made from any starch. Currently, in Northern Europe, wheat is the preferred starting material, because it is a more reliable crop to grow.

The most important points to remember is that starch used for manufacturing glucose syrup must be of the highest possible quality, particularly it must have a low protein content, ideally less than 0.3%, with no metallic or microbial contamination. Both protein and metal ions have been implicated as causes for poor colour and ageing characteristics in syrups. Whilst starch of a lower quality can be used to make glucose syrup, the resulting syrup will have a bad colour, poor storage properties, and possibly off flavours, and it will also be difficult and expensive to improve the quality of the syrup, if at all. Basically, it is more cost effective not to produce colour in the first instance. The starch should also not have any flavour, or colour and be free from any substance which could be harmful. For example, cassava tubers, from which cassava starch is obtained, contains a small amount of hydrogen cyanide (prussic acid). It is therefore important that the hydrogen cyanide does not contaminate the cassava starch and possibly the final syrup.

When extracting starch, several by-products (or co-products as the industry prefers to call them) are produced, and these by-products can then be sold to help off set the cost of producing the starch. A good example is wheat, where the by-product is vital wheat gluten. This is sold to flour millers who use it to increase the protein content of flour, which has been made from low-protein wheat. Fortunately, feed grade wheat can be used to produce starch for syrup production, as well as the more expensive bread making wheat, with the resulting vital wheat gluten, in both cases being perfectly acceptable for use in bread making flour. In the case of maize, both the fibres and the gluten are sold for cattle feed. Incidentally, maize gluten does not react with water, like wheat gluten, which forms a tenacious sticky mass. Maize gluten just sinks in water, like particles of sand. This makes the separation of maize starch from the maize protein relatively easy, and ensures a low protein in the maize starch – important for making syrups with a good colour. Roughly speaking, to process one tonne of maize, requires one tonne of water, whereas one tonne of wheat flour requires four tonnes of water, however, with modern processing techniques, water consumption in both processes is being substantially reduced.

As previously mentioned, starch is a carbohydrate, and there are several other carbohydrates such as cellulose (old newspapers) which could be used to make a 'glucose syrup', but it could not be called a glucose syrup, because it would not have been made from starch. However, if a high DE glucose syrup was made from cellulose, it would be eminently suitable as a fermentation feedstock for the manufacture of industrial ethanol.

Other carbohydrates, for example pectin, and alginates, would not be suitable because they do not have the correct molecular structure.

Whilst starch is considered to be a polymer of dextrose, starch is in fact made up of two distinct fractions – amylose and amylopectin. Depending upon the type of starch, the relative proportions of amylose to amylopectin will vary, but generally will be in the range of 20–25% amylose and 75–80% amylopectin. In the case of waxy maize, the proportion of amylopectin is approaching 100%. Both amylose and amylopectin have a slightly different molecular structure, see figure 3.3.

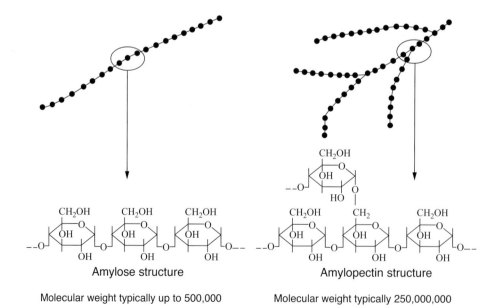

Figure 3.3 Structural differences between amylose and amylopectin.

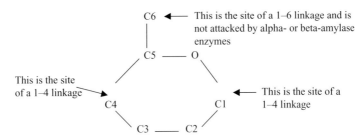

Figure 3.4 Position of 1–4 and 1–6 linkages.

The amylose fraction consists of dextrose molecules joined to each other at the C1 and C4 positions to form a long, spiral chain. The sites where the C1 and C4 are join together are referred to as a 1–4 linkages (see Fig. 3.4). In the amylopectin fraction, there is also a chain of dextrose molecules joined at the C1 and C4 positions, but with an additional straight chain attached at the C6 position (see Fig. 3.4). This site at the C6 position in the amylopectin structure is referred to as a 1–6 linkage. Like the amylose chain, the amylopectin chain is also in the form of a spiral. Finally, both the amylose and amylopectin spiral chains in turn form spiral chains with each other.

Both the 1–4 and 1–6 linkages are very critical when using enzymes for making glucose syrups, because some enzymes can attack these sites, whilst other enzymes cannot. Therefore, the correct choice of enzyme becomes crucial for a successful process.

It should be appreciated that starch is a naturally occurring product and each particular starch has slightly different physical properties. Possibly, the two most important physical properties of a starch to the glucose syrup manufacturer are the gel temperature and the resultant viscosity. First, the starch has to be totally gelled before the enzymes can further convert it to syrup, therefore, knowing the temperature at which the starch gels becomes important. Secondly, if the gelled starch is too viscous, it could be too difficult to process. Table 3.1 illustrates these two properties for three commonly used starches.

Table 3.1 Gel temperature and viscosity for different starches.

	Maize	Wheat	Potato
Gel temperature (°C)	62–74	52–64	56–69
Approximate viscosity (at gel temperature and at 5% starch solids)	14 poises	2 poises	58 poises

3.4 Enzymes

Before discussing the process for making glucose syrups in detail, we should realise that our body is doing exactly the same thing as the glucose syrup manufacturer is doing, namely the breaking down of starch into dextrose, using both acid and enzymes. The main difference is that nature is far more efficient, and does not require high temperatures, pressures and large expensive stainless steel factories. If you chew a small piece of bread in your mouth, after a few minutes you will notice a sweetness develop. This sweetness is the result of the amylase enzyme in your saliva breaking down the starch in the bread into the sugar maltose, and that is what the glucose industry does all the time – it converts starch into the sugars dextrose, maltose, fructose, etc.

What are enzymes? Enzymes are biological catalysts, derived from living organisms, such as yeast, bacteria and moulds, and are very specific in their reactions. In 1816, Kirchoff discovered that enzymes could be used to break down starch into 'sugars', so the use of enzymes are not new to the starch industry. Brewers use the enzymes in barley to convert the barley's starch into sugars, which the yeast then converts into alcohol. The starch industry uses four main types of enzymes to convert starch to syrup, namely α-amylase, β-amylase, glucoamylase and pullulanase, and then there is one particular enzyme, isomerase, which converts dextrose syrup into fructose syrup.

For an enzyme to be effective in converting starch, the starch must be made into a slurry, which is then cooked to form a gel or paste. Next, both the temperature and pH have to be correct for that particular enzyme.

Where liquid enzymes are used in a batch process, which is being continuously stirred, research has shown that continuous stirring is not necessary. Satisfactory results can be obtained if the reaction is stirred for only thirty minutes, every four hours, thereby saving energy and reducing running costs.

3.4.1 α-amylase

This enzyme randomly attacks the gelled starch, at the 1–4 linkages to produce dextrose and maltose, but it is unable to hydrolyse the 1–6 linkages. The 1–6 linkages will therefore become part of the sugars which make up the higher sugars of a syrup.

Most α-amylases work in the pH range of 5.0–7.0, and at a temperature of 65–70°C however, the glucose industry now uses high temperature stable α-amylases. These high temperature stable enzymes can typically withstand 120°C for twenty minutes, at pH 6.0.

3.4.2 β-amylase

Commercially, there are two main sources of this enzyme. The first source is diastatic malt which is obtained from malted barley. Diastatic malt also contains traces of α-amylase, together with other enzymes which together can be beneficial when hydrolysing a poor quality substrate.

The second source of β-amylase is from bacteria, which is a more pure β-amylase. However, this purity is reflected in the enzyme price, making it more expensive than diastatic malt.

β-amylase attacks the 1–4 linkages to produces predominately maltose, with lesser amounts of dextrose, and like α-amylase, it is unable to hydrolyse the 1–6 linkages, which again form part of the higher sugars of a syrup.

The operating pH is 4.5–5.0 and an operating temperature of 55–60°C.

3.4.3 Glucoamylase

Glucoamylase is also known as amyloglucosidase or AMG. This enzyme is capable of hydrolysing both the 1–4 and 1–6 linkages to produce dextrose, and therefore it is sometimes referred to as a 'saccharifying' enzyme.

The operating pH is 4.5–5.0 with an operating temperature of 55–60°C.

3.4.4 Pullulanase

This enzyme is sometimes referred to as a 'debranching' enzyme, because it can hydrolyse the 1–6 linkages, which other enzymes cannot attack, thereby making more of the higher sugars available for hydrolysis to maltose and dextrose. It is for this reason that pullulanase is often used in conjunction with other enzymes. A good example is in the production of maltose syrups where pullulanase is used in conjunction with a β-amylase to produce high yields of maltose, typically 70% or higher, but with only 5% dextrose or less.

Generally, the operating parameters for pullulanase will be the same as the operating parameters of the other enzyme. When used in the production of a maltose syrup with a high maltose content, the operating parameters will be the same as those of a β-amylase.

3.4.5 Isomerase

This enzyme converts dextrose syrups into fructose syrups. The previously mentioned enzymes are relatively cheap and therefore are not reused, being removed from the syrup during the refining stage. They are either liquids or powders and are used to either convert a gelled starch or to alter the sugar spectrum of a syrup.

Isomerase is a totally different enzyme. It is a very expensive enzyme, and is therefore immobilised on a porous carrier. The immobilised enzyme resembles the hundreds and thousands found on the top of chocolate cakes. The syrup passes over the immobilised enzyme, as opposed to being mixed with the syrup, and being an immobilised enzyme, it is being continually used, making it operationally more cost effective.

The operating pH is 7.5–8.0, and a temperature of 60–65°C. The specific conditions of pH and temperature of an isomerising enzyme will depend upon the origin of the enzyme.

Immobilising an enzyme is an added production cost, and as there is also some loss of enzyme activity, immobilising a cheap and readily available enzyme is uneconomic.

3.4.6 Lesser enzymes

Where wheat starch is used for making glucose syrups, filtration problems can occur due to minor amounts of pentosans, hemi-cellulose and β-glucans, which because of their colloidal nature can block the filters. These products can be broken down by enzymes such as pentanase, cellulase and β-gluconase.

3.5 The process

The initial process of converting starch to a glucose syrup is usually referred to as either 'hydrolysis' or 'conversion', prefixed by either 'acid' or 'enzyme', depending on whether acid or enzyme has been used, to give the terms 'acid hydrolysis' or 'enzyme hydrolysis'.

Originally, there was only one type of glucose syrup – 42 DE glucose syrup, also known as 'Confectioners Glucose', and was made by the hydrolysis of a starch slurry with acid, under pressure in a batch process. Unfortunately, the quality of the syrup would vary considerably from one batch to the next, typically with traces of unconverted starch being present in a highly coloured syrup. Nowadays, most syrups are made using a continuous converter which produces a more consistent starch-free syrup with improved colour.

One of the problems when using acid to make glucose syrups with a DE of 50, or higher, and this includes dextrose, is that the end product has an unpleasant bitter taste, due to side reactions, typically the production of hydroxymethylfufural (HMF) and other bitter compounds. Another problem is that the resulting syrups are very coloured. It is for this reason that enzymes are now used for producing high DE syrups.

Today, with the availability of both acids and enzymes, there are now several different techniques at the industry's disposal to convert starch into a glucose syrup, resulting in syrups of any DE or sugar spectrum. As a general rule, acids are cheaper than enzymes, but enzymes are more specific.

- Using only acid is known as acid hydrolysis, and is used to make 35 and 42 DE syrup. See Figure 3.5.
- Using both acid and enzyme is known as acid enzyme hydrolysis, and is used to make 63 DE syrup. See Figure 3.6.
- Using acid to paste the starch, followed by two enzyme treatments, is known as acid enzyme enzyme hydrolysis, and usually abbreviated to PEE (paste enzyme enzyme). See Figure 3.7. This type of hydrolysis is used to make 95 DE or 70% maltose syrups.
- Using two enzymes is known as enzyme enzyme hydrolysis, and can be used to make a 42 DE which has a low dextrose content. *N.B.* The sugar spectrum of an acid produced 42 DE syrup and an enzyme produced 42 DE are frequently different.
- Any combination of the above processes or by the blending of different syrups.

As mentioned previously, starch is a polymer of dextrose units, all joined together. The hydrolysis process is essentially one of breaking down the chain of starch into dextrose units, either individual units or several units joined together. An easy way of understanding what happens to the starch during this first stage is to compare it to the demolition of a brick wall. A brick wall can be demolished by either hitting it with a ball and chain, which produces a selection of bricks, individual bricks, or several bricks still joined together. The other way is for a workman to remove each brick individually with a pick. In the case of starch, think of the bricks of the wall as being dextrose units, and the ball and chain demolition approach is how acid breaks down the starch, ending up with individual dextrose units and several dextrose units still joined together – two dextrose units making maltose, three making maltotriose, and so on. The workman approach is how enzymes work. They detach one dextrose unit at a time, working along the dextrose chain.

Before considering how the different methods of starch hydrolysis are used to make different types of glucose syrups, it should be appreciated that different manufacturers will produce their syrups their own way, depending upon the available equipment and raw materials – there are no golden rules, just hydrolysis, clarification, colour removal and evaporation. What is very important is that there is no detectable starch in the syrup, after the hydrolysis stage – that means no blue colour reaction with iodine.

The following process descriptions, pH values and temperatures are only indications and must be considered only as a guide, and not exact process parameters.

3.6 Acid hydrolysis

Acid hydrolysis is used to make a 35 or 42 DE syrup. It is a relatively simple process, which is also one of its limitations.

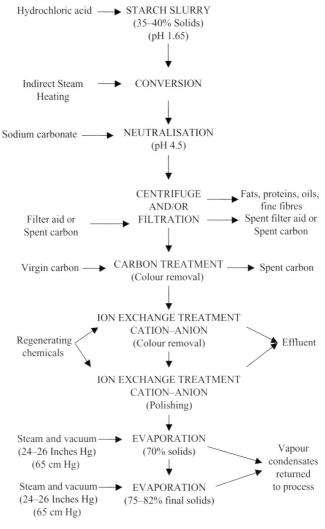

Figure 3.5 Production of de-ionised acid converted 35 or 42 DE glucose syrup.

(1) Starch slurry

The starch is made into a slurry with water, preferably soft water, at a temperature of about 20–30°C. The solids of the starch slurry should be between 30% and 40% and should contain about 200 ppm of sulphur dioxide.

   a. Soft water is preferred because hard water contains calcium salts, and most calcium salts are insoluble. These insoluble salts could therefore form a scale on the surface of the equipment or a haze in the finished syrup.
   b. Starch appears to disperse more easily in warm water (20–30°C). This pre-warming of the starch slurry also helps with the gelatinisation of the starch, when it goes through the converter. The water is heated with waste heat, thereby improving the overall thermal efficiency of the process. If the temperature of the water is too high, there is the risk that the starch will gelatinise before it reaches the converter.
   c. The solids of the starch slurry will depend upon the type of conversion process. The solids for an acid conversion process can be about 40%, but for an enzyme conversion process, the solids tend to be lower at about 35%.

   The solids of the starch slurry are as high as possible, so that less energy will be required in further process evaporation stages, but such are the physical properties of starch, that if the solids exceed about 45%, the starch will no longer exist as a slurry, but as a moist solid, totally unsuitable for pumping.

   A further consideration for the solids of the starch slurry is the viscosity of the starch when it is gelled. If the viscosity of the gelled starch is too high, then there is the risk that the converter will become blocked.

   The possibility of using higher starch solids and an extruder to produce a glucose syrup has been investigated, but did not prove viable.
   d. The reason for adding sulphur dioxide to the starch slurry are twofold. First, the sulphur dioxide will minimise any microbiological spoilage of the slurry. Microbes contain protein, and protein causes colour formation. The second reason is that the sulphur dioxide will reduce colour formation during the conversion process. During the conversion process, reducing sugars are formed, and reducing sugars will react with proteins to produce colour – the well known Maillard reaction, but by having sulphur dioxide present in the conversion process, it will block this reaction, thereby reducing colour formation.

(2) Acid addition

Acid is now added to the starch slurry to reduce its pH to 1.65 for acid conversion. As a double check, some companies will also measure the conductivity of the acidified starch slurry. Whilst any acid can be used, the preferred acid is food grade hydrochloric. The reasons for this are as follows:

   a. Hydrochloric acid is readily available, and has a greater hydrolytic power than sulphuric, nitric or phosphoric acids therefore less acid is required.
   b. Most salts of hydrochloric acid (chlorides) are very soluble, therefore there is less chance of any scale or haze being formed.

c. The molecular weight of chlorides is less than salts of other acids, therefore there will be less ash in the final syrup.
   d. Because less acid has been used, less salts will be formed, which means that there will be less salts for the ion exchange resins to remove when producing de-ionised syrups.
   e. If sulphuric acid is used, because its hydrolytic power is less, more acid would have to be used, resulting in more ash being produced. Since the molecular weight of sulphates are higher than those of chlorides, there would be a greater loading on the ion exchange plant, which would have to be a lot larger and hence more expensive. Additionally, as most sulphates are insoluble, there is the slight risk that they might cause a haze in the final syrup.

(3) Conversion

The next stage is known as the conversion or hydrolysis stage. This is the stage where the individual starch granules are gelatinised and broken down by the acid, and turned into glucose syrup, and involves using a combination of high temperatures and pressures. Modern processes typically use a two-stage plate and frame heat exchanger to obtain the required residence time for the hydrolysis reaction.

The acidified starch slurry is pumped into the first heat exchanger, which is heated, indirectly to a temperature of 143°C. At this temperature, the starch slurry is instantly gelled to form a very viscous paste. As the paste leaves the heat exchanger, about 90% of it is recycled and mixed with about 10% of the ungelled acidified starch slurry which is feeding the heat exchanger, and is again heated to 143°C. The reason for recycling part of the gelled starch is to ensure that the starch is properly gelled, and to reduce the initial viscosity of the freshly gelled starch to a pumpable paste. It also helps to keep the ungelled starch granules in suspension during the initial conversion stage. The gelled starch now passes to the second heat exchanger, which has a temperature of 135°C. The residence time at 135°C is about twenty minutes and this ensures that all of the starch has been fully gelled and is converted into a 42 DE acid converted syrup.

   a. The final DE can, within certain limits, be changed by altering the processing conditions such as time, temperature or pH.
   b. When producing a very low DE syrup, it is important to be able to re-circulate the converted starch through the heat exchanger to increase the conversion time. The reason for increasing the conversion time is because minimal quantities of acid are used to prevent over-converting the starch, which would result in a slower conversion. By increasing the conversion time compensates for the reduction in acid and ensures that the product is starch-free.
   c. Plate and frame heat exchangers are cheaper, more energy efficient, and easier to maintain than the older shell converters, which consisted of a series of tubes to give the required residence time.

(4) Neutralisation

As well as converting the starch into a glucose syrup, these harsh conditions also solubilise the small amount of protein present in the starch. The pH therefore is adjusted to 4.5, with sodium carbonate, which stops the conversion, but pH 4.5 is also the isoelectric point for the proteins present in the starch, which are now precipitated out of solution. Typically, the original starch would have a protein content of less than 0.3%.

(5) Centrifuge and filtration

The object of this stage is to clean up the syrup by removing all the precipitated protein resulting from the neutralisation, together with any fine fibres, fats and oil from the original starch, before the carbon or ion exchange resins remove the colour from the syrup. Failure to remove these impurities would result in the surface of the carbon or ion exchange resins being coated with an impervious layer, thereby preventing colour removal from the syrup.

This clean up is in two stages. The first stage is to pass the syrup through a continuous centrifuge which removes the bulk of the insoluble material. The next stage is to filter the syrup to make sure that everything has been removed. This ensures that the surface of the activated virgin carbon or ion exchange resins is not coated with any insoluble material.

a. The filter aid used at this stage is 'spent powdered carbon' – that is carbon powder which has already been used to remove colour. By using spent carbon as a filter aid at this stage, it helps to reduce processing costs by not having to buy filter aid. Where spent powdered carbon is not available, that is where granulated carbon is used, a suitable filter aid is Perlite, an expanded volcanic rock.
b. One of the disadvantages of using a centrifuge is their high maintenance, therefore, the alternative to a centrifuge is an advancing knife rotary vacuum filter. With this type of equipment, the syrup is mixed with filter aid or spent carbon. The fibres, fats and oils adhere to the filter aid on the rotating drum due to the applied vacuum, but because the drum is rotating, when this layer reaches the knife blade, it is scrapped off, thereby producing a fresh surface of filter aid. In some cases, the disposal of spent carbon can be a problem.
c. Some starches such as wheat contain colloid-type carbohydrates which block filters, and in some processes, this fouling of the filter cloths can be minimised by the addition of β-gluconase to the syrup, prior to filtration.

(6) Carbon treatment

The object of the carbon treatment is to improve the colour of the syrup. This colour removal is achieved by using either activated carbon (virgin carbon) or ion exchange resins, or a combination of both. As a general rule, activated carbon is used to remove the 'organic based impurities', such as proteins, and other colour precussors. Ion exchange resins are used to remove metallic ions, that is salts and minerals.

Some companies believe that for the maximum 'removal' of organic compounds, the powdered activated carbon should be composed of a 50/50 mixture of wood and peat-based activated carbons. The downside of using activated carbon derived from peat is that it can add minerals back into the syrup.

Where activated carbon is used for decolourisation, the syrup is kept at 65°C, and virgin carbon is added to the coloured syrup at an addition rate of about 1.0% or less, based on the syrup solids. After a contact time of about thirty minutes, the carbon is filtered off, and is now referred to as 'spent carbon'. This spent carbon is then used as the filter aid for the previous filtration stage.

An alternative to powdered activated carbon is granulated carbon, which is used in a continuous, countercurrent process. This is achieved by placing the carbon in a column, and the syrup passes down the column, whilst the granulated carbon travels to the top. When the carbon reaches the top of the column, it is spent, and is regenerated, by heating it in a furnace to about 900°C. After regeneration, the carbon is returned to the bottom of the column. The flow rate of the syrup is controlled so as the carbon is in contact with the syrup for about thirty minutes. Since granulated carbon has a larger particle size than powdered carbon, and is regenerated, it is not suitable as a filter aid.

a. Colour is formed in the syrup during the conversion stage because of two reactions – caramelisation and the well known Maillard reaction. When the starch is converted into syrup, the temperature is 120°C, resulting in caramelisation of the syrup. During the conversion of the starch into syrup, the residual protein in the starch is solubilised, which then reacts with the reducing sugars of the syrup to give Maillard browning.
b. The addition of sulphur dioxide to the starch slurry will block the reactive sites of the reducing sugars – that is the aldehyde groups, thereby reducing colour formation due to the Maillard reaction – proteins and reducing sugars.

Typically, the total protein content of the original starch should be less than 0.3%, and the total protein content in the syrup after carbon treatment will be about 0.03% and 0.01% or less if the syrup has been treated with ion exchange resins.

(7) Ion exchange treatment

Ion exchange treatment can be used either in conjunction with activated carbon to remove colour, or it can be used on its own. The normal temperature of the syrup for ion exchange treatment is usually about 50°C. Whilst activated carbon will remove colour, it will not remove salts. Ion exchange will remove both colour and salts, however ion exchange resins have to be regenerated, and the chemicals used for their regeneration and the disposal of the chemical effluent can be expensive. This is one of the reasons that some companies use both activated carbon and ion exchange for colour removal. Ion exchange resins do not last forever. Over a period of use, the ion exchange resins start to break up and become ineffective, and have to be replaced.

Where ion exchange resins are used for colour removal, there are generally three pairs of cation and anion resins units, in series and operated on a carousel principle. The first pair removes the bulk of the colour and minerals. The second pair acts as a 'polishing unit', whilst the third pair is on standby. When the first pair is no longer able to remove any colour, or the conductivity of the syrup increases, the pair are taken out of service and regenerated. The second pair now removes the colour and minerals, with the third pair acting as the 'polishing unit'. The procedure is then repeated when the second pair are exhausted, and so on. When to regenerate an ion exchange unit is usually based on the conductivity of the syrup leaving the unit, and in a modern installation, the regeneration cycle would be started automatically by a computer once the conductivity has exceeded the pre-set limits.

a. Dilute acids such as hydrochloric acid are normally used to regenerate the cation resins, whilst dilute sodium hydroxide and ammonia are used for regenerating the anion resins. The normal procedure is to use ammonia for about nine regenerations, followed by one regeneration with sodium hydroxide. The reason for this combination of sodium hydroxide and ammonia is to reduce the amount of wash water used. Whilst sodium hydroxide is excellent for regenerating the resin, it requires a lot of water to remove the last traces from the resin before they can be used again to treat syrup. Ammonia, on the other hand, requires less wash water than sodium hydroxide, but it does not regenerate the resin as efficiently. When using ammonia, there can be a problem due to the extra nitrogen in the wash water which can be detrimental to the working of an effluent treatment plant. Resins with a high pH will cause high colours in a syrup, therefore it is important that ion exchange resins are thoroughly washed after regeneration.
b. Ideally, to prevent large swings in the pH of the effluent, the spent regenerating chemicals and wash waters are mixed together, thereby neutralising each other. Large swings in effluent pH can stop an effluent plant from working.

(8) Evaporation

The clean syrup is now evaporated under vacuum, of about 25 to 26 inches (65 cm) of mercury, to about 70% solids. The solids of the original starch slurry would have been about 30–40%, and it is still at 30–40% solids after being refined. Unfortunately, these solids are ideal for microbial growth, so the syrup must be evaporated up to 70% solids to prevent microbial growth, and either stored at these solids, prior to further processing, or evaporated up to a final solids of about 80–82%, depending upon the syrup. The exact final solids will depend upon the DE of the syrup – its viscosity and osmotic pressure. As a general rule, the lower the DE, the lower the solids of the syrup. This is because low DE syrups are viscous, and at higher solids, they would be impossible to handle. This is one of the reasons why low DE syrups and maltodextrins are usually spray dried. The other reason for spray drying is that low DE syrups, because of their high molecular weight, have a low osmotic pressure, and hence are more susceptible to microbiological growth. However, high DE syrups, such as 95 DE, can also be a problem not due to osmotic pressure, but to dextrose crystallisation. Typical solids for a 95 DE syrup would be about 75%. In the case of HFGS, the final solids are generally 71%. These solids are the same as those of liquid sugar, which makes it easy for direct replacement.

Also at this stage, it is possible to blend different syrups, change the pH or sulphur dioxide content, or whatever other changes the customer requires!

a. The reason why syrups are evaporated under vacuum is because the evaporation temperature of the syrup is lower, than at room temperature, which means that there is less risk of colour formation due to excessive heat. Additionally, by boiling at as lower a temperature as possible, there is a cost saving, because a lower steam pressure can be used, and therefore less energy is required.

3.7 Acid enzyme hydrolysis

A good example of a syrup produced by acid enzyme hydrolysis would be a 63 DE syrup. This process was developed so as to make a syrup which was more sweet than a 42 DE syrup, but not as sweet as a 95 DE, and did not have the characteristic unpleasant flavour associated with acid converted syrups of 50 DE and higher.

The starting material is a conventional acid produced 42 DE syrup.

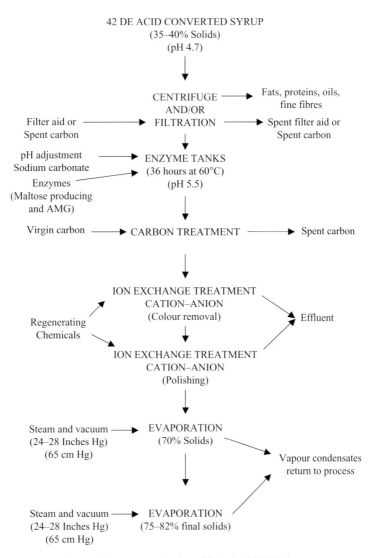

Figure 3.6 Acid enzyme production of de-ionised 63 DE glucose syrup.

1. After the acid produced 42 DE syrup has been neutralised and passed through the centrifuge and/or filtered, the syrup is diverted to an enzyme tank.
2. The pH is adjusted to 5.5 and the temperature adjusted to 60°C.

3. Only after the pH and temperature have been adjusted are the two enzymes added – a maltose producing enzyme, typically a diastatic malt or a bacterial β-amylase, and the dextrose producing enzyme, AMG. One advantage of using a cereal β-amylase (i.e. diastatic malt) as the source of a β-amylase is that it contains other enzymes which can be beneficial when the substrate is difficult to convert. The disadvantage is that its activity is less compared to a highly purified bacterial β-amylase. This means that a larger amount has to be used, and being a protein, it can cause a slight increase in colour. By using a bacterial β-amylase, because they are more potent and more highly purified, less enzyme is required, which results in less colour being formed, however, the disadvantage is that they are generally more expensive.
4. The DE and sugar spectrum are monitored, and when the correct sugar spectrum has been reached, the syrup then goes forward for carbon treatment, and so on. A typical residence time for the enzyme reaction would be about 36 hours, depending upon the amount of enzyme added, to give a final DE of 61–64.
   (i) The enzymes continue the conversion process started by the action of the acid, by breaking down the higher sugars into maltose and dextrose.
   (ii) If the enzymes are added before either the pH and temperature has been corrected, the enzymes will be inactivated.
   (iii) The enzyme tanks are run on a continuous batch process – which means that one tank is filled after another, with the first tank to be filled being the first tank to be emptied. This ensures that there is a continuous supply of syrup to go forward to the next stage, and if one particular enzyme tank, for whatever reason, fails to produce the correct sugar spectrum, then that tank can be isolated and subsequently blended with syrup from other tanks.
   (iv) Some starches such as wheat starch contain colloid carbohydrates which foul the filters, but the addition of β-gluconase to the enzyme tanks can reduce this fouling of the filter cloths. Therefore in the flow diagram, figure 3.6, there would be an additional filtration stage, normally a filter press after the liquor leaves the enzyme tanks and before carbon treatment. This approach of adding enzymes to improve filtration is easy to apply to any starch conversion process where there are enzyme tanks, that is where AMG or β-amylase is added.
   (v) One problem in adding enzymes, which are proteins, to a syrup containing reducing sugars being held at 60°C for 36 hours is colour formation due to the well-known Maillard reaction – reducing sugars reacting with proteins. It is for this reason that the carbon dose rate is about 2.0%, based on syrup solids for an acid enzyme hydrolysis, as opposed to 1.0% for a straight acid hydrolysis.
   (vi) The advantage of using β-amylase is that it is a more pure enzyme, so less is required, resulting in less colour being formed.
   (vii) The removal of the colour is expensive, because extra activated carbon has to be used. It is best to avoid making colour in the first place.

3.8   Paste Enzyme Enzyme Hydrolysis (PEE)

In the PEE process, the acid converts the starch into a low DE paste, which the enzymes then further convert. One of the advantages of the PEE process is that it is possible to obtain a lower starch-free initial DE than with a conventional acid conversion, and this low initial DE allows higher final sugar yields to be obtained, which is why the PEE process is used to make high-dextrose syrups, for example D.95, and high maltose syrups, for example HM 70. The low DE obtained with the PEE process also means that the PEE process can also be used to make maltodextrins.

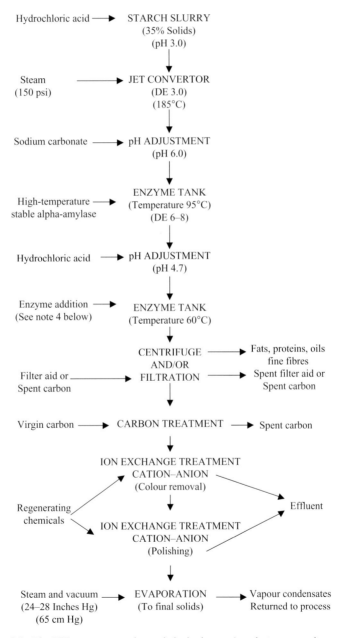

Figure 3.7 The PEE process to produce a de-ionised syrup (e.g. dextrose or maltose syrups).

1. To make a high DE syrup or a high maltose syrup, using the PEE process, the pH of the starch slurry is adjusted to 3.0 with hydrochloric acid.

2. The acidified starch slurry at about 35% solids is pumped into a stream of high-pressure steam (150 psi to give a temperature of 185°C). At this pH, temperature and pressure, the starch slurry is immediately gelled and sheared to produce a paste. The successful production of a homogeneous paste is very dependent upon the design of the jet inside the converter, as it is essential that the starch granules are not only completely gelled, but also fragmented and totally dispersed. After this initial acid pasting, the DE will be about 3.0.
3. The pH is now adjusted to 6.0 and a high-temperature stable α-amylase enzyme is added. Typically, this type of enzyme will withstand 120°C for twenty minutes. The DE at this stage will be about 6.0–8.0, and starch-free.
4. The pH is now lowered to 4.7 by the addition of hydrochloric acid, and the temperature is lowered to 60°C, before the next enzyme is added, which will be either AMG for a dextrose syrup or a maltose producing enzyme for a high maltose syrup. Where a high maltose syrup is being produced, the debranching enzyme pulluanase is also added.
5. Once the required sugar spectrum has been reached, the syrup is then centrifuged and filtered, prior to carbon treatment, ion exchange and evaporation.
   (i) The enzyme tank where the high-temperature α-amylase is added consists of three tanks joined in series. This arrangement allows three things to happen. First, it allows the process to be run continuously. Secondly, it gives the enzyme sufficient time to break down the pasted starch into a 6.0–8.0 DE syrup, and thirdly, it allows the very low DE syrup to cool. The pasted starch will have entered the first tank at about 105°C, and will leave the last tank at about 95°C.
   (ii) Where a high maltose syrup is being produced, the maltose enzyme tanks, that is where the maltose producing enzyme is added, are often run as a continuous batch process, but where a dextrose syrup is being produced, the dextrose enzyme tanks, that is where the AMG enzyme is added, are usually run as a continuous process. In both cases, the residence time in the enzyme tanks is about 36 hours.

Ideally, after the pH has been adjusted to 4.7, prior to the enzymes being added, the syrup should be filtered, but because of the low DE, the syrup is far too viscous, therefore the centrifuging and filtering stages occur after the syrup has left the enzyme tanks.

Maltose syrups are evaporated up to 80% solids. In the case of a dextrose syrup, after it has been refined, it is evaporated up to 75% solids, and the 95 DE syrup is either sold as syrup or is used to make crystalline dextrose or it may be spray dried to make a 'total sugar'. It can also be used for the production of high-fructose glucose syrups. The reason why dextrose syrups are only evaporated up to 75% solids is because at higher solids, the dextrose would crystallise. To prevent the syrup from crystallising, the syrup has to be stored at 50–60°C.

3.9 Crystalline dextrose production

To make crystalline dextrose, D.95 syrup is pumped to a series of horizontal crystallisers. See figure 3.8. There are generally three or four crystallisers in series, one on top of the other, with the syrup flowing from the top crystalliser to the one below, and then to the last crystalliser. Each crystalliser has a water jacket, cooling the syrup to about 20°C. Inside each crystalliser is a spiral ribbon agitator, which slowly rotates. Because the D.95 syrup is a supersaturated solution of dextrose, as the syrup cools, the dextrose crystals come out of solution to form a white mass similar to a coarse or gritty fondant. This mixture of dextrose crystals and syrup is referred to as 'magma'. The slowly rotating agitator has two functions. First, it keeps the mixture of dextrose crystals and syrup continuously moving, thereby preventing the formation of a solid block of dextrose. The second function is to help in the cooling process by allowing warm syrup to come in contact with the cold water jacket.

It is important that the operating conditions of the crystallisers are not changed. This means that the temperature regime of the crystallisers, the syrup composition, that is solids and dextrose content, feeding the crystallisers and the residence time in the crystallisers does not vary. Any variation to any of these parameters can result in the production of dextrose crystals with a variable size. This variation in crystal size will produce variations in the bulk density of the dextrose, and ultimately will mean changing the size of the dextrose bags.

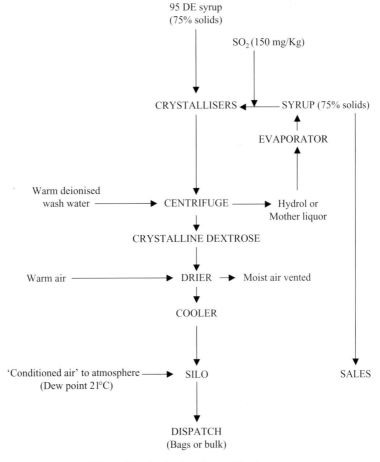

Figure 3.8 Production of crystalline dextrose.

When the magma reaches the last crystalliser, it passes into a centrifuge, which separates the dextrose crystals from the syrup (or mother liquor). The crystals are washed in the centrifuge to remove excess syrup, and then dried in a fluid bed dryer, then cooled before being pneumatically conveyed, in a current of 'conditioned air' to a silo before being packed. The 'conditioned air' used to convey the dextrose to the storage silo should have a dew point of 21°C or less, so as to prevent the dextrose crystals clumping together, and forming a solid lump.

The reason why dextrose forms lumps in storage is because of the air trapped in between the crystals. If the air contains too much moisture, then over a period of time, with changes in temperature, this moisture will condense to form a dilute film of syrup on the surface of the dextrose crystals. When the moist crystal faces come in contact with adjoining crystals, for example in a bag at the bottom of a pallet, the dilute film of syrup on the two surfaces acts like a cement, joining the crystals together, to form a solid lump. The smaller the crystals, the greater the surface area and points of contact, so greater will be the lump formation. Therefore, to avoid lump formation, the air in between the crystals must have a dew point of 21°C or less. Above 21°C, it will be a disaster.

Typically, it will take 3–5 days from the time the D.95 syrup enters the first crystalliser to when crystalline dextrose leaves the dryer.

The mother liquor (also known as greens or hyrol) contains about 85% or more dextrose and is evaporated up to 75% solids, and sold to the fermentation industry as an alternative to molasses. In some companies, because the syrup still contains a high proportion of dextrose, the syrup is returned to the process for a second crystallisation.

Some companies supply dextrose as a solution, particularly to manufacturers of sorbitol. The reason for this requirement is that it saves companies the inconvenience of having to handle crystalline dextrose, and making it into a solution. The dextrose solution is delivered directly to the point of use.

To make a dextrose solution, the crystalline dextrose is added to sterile de-ionised hot water. Hot water is necessary, because dextrose has a negative heat of solution, that is as the dextrose dissolves, it absorbs heat from the water, thereby cooling the water, and as the water cools, the solubility of the dextrose is reduced. Therefore, unless heat is supplied, it would be impossible to obtain a solution with a high dextrose concentration. The solubility of dextrose at 20°C is only 48%, but at 55°C, it increases to 73%. To obtain a concentration greater than 73%, it is necessary to evaporate the syrup, as with a standard syrup. Before the dextrose solution is evaporated, it is filtered using ultrafiltration, and then evaporated, up to 75% solids.

It must be remembered that dextrose solutions are very fermentable, therefore good hygiene is essential at all stages of the process.

3.10 Total sugar production

Total sugars differ from dextrose because they are made by a different process.

1. Total sugars are made by spray drying a 95 DE syrup. Therefore, all the lesser sugars present in the original 95 DE syrup are still present in the spray-dried end product.

2. Crystalline dextrose is made by crystallising the dextrose out of a 95 DE syrup. Crystallisation is essentially a purification process separating dextrose from the other sugars present in the original 95 DE syrup to give a product containing 99.9% dextrose.
3. A further difference is the moisture content and crystal form. Dextrose contains about 8–10% water, which is chemically bound, and the crystals are in the α-form. With a total sugar, the moisture content is about 0.5%, with the dextrose crystals being in the β-form. The low moisture content of a total sugar means that it can be used in certain applications, such as chocolate confectionery, where dextrose because of its higher moisture content would be totally unsuitable.

The spray drying of a 95 DE syrup requires a specially designed drier, which has a product re-circulation facility. The reason for the product re-circulation facility is because dextrose, unlike sucrose, will not instantly crystallise or form a solid. Therefore, to encourage the dextrose to solidify, some of the spray dried product is returned to the drier, where it acts as a carrier and seed for the fresh syrup. Depending upon the type of drier and the quality of the syrup feeding the drier, a typical recycle ratio would be 7 parts of dry product to 3 parts of syrup.

Spray driers use a lot of energy, and with energy prices increasing, the future of spray-dried 95 DE syrup is unclear.

3.11 Enzyme enzyme hydrolysis (E/E)

In an enzyme enzyme conversion process (E/E), a high-temperature stable α-amylase enzyme replaces the acid which is usually used for the initial pasting of the starch in a PEE hydrolysis. This acid replacement with an enzyme requires the pH of the starch slurry to be adjusted to 6.0 before being pumped into a stream of high pressure steam. Some companies add calcium ions, usually as calcium chloride, to the starch slurry, because some α-amylase enzymes require calcium ions to act as a stabiliser/activator. However, as explained when discussing the acid hydrolysis process, calcium salts could cause a haze in the final syrup. Additionally, if the enzyme/enzyme produced syrup is to be used to make HFGS, then the presences of calcium ions will adversely affect the isomerisation process unless they are removed.

The solids of the starch slurry should be about 30–35%. The reason for the starch slurry having lower solids than for acid hydrolysis is because the initial viscosity of the starch paste is very high, therefore by using lower starch solids, the viscosity of the paste is reduced to more manageable proportions. If the viscosity of the starch paste is reduced by shearing, then there is the possibility that the enzyme could be destroyed. Thereafter, the conversion process follows the normal procedure – adjusting pH to 4.7, further enzyme treatment, carbon treatment, ion exchange treatment, and finally evaporation to final solids.

The way in which acids hydrolyse starch is relatively crude enzymes on the other hand are very selective, which makes the enzyme/enzyme process of interest where a very specific sugar spectrum is required. It is also claimed that the colour development when using an enzyme/enzyme hydrolysis is less compared to an acid hydrolysis, because the temperatures used are less, despite longer reaction times.

3.12 Isomerisation

Isomerisation is the name given to the process of converting a dextrose syrup into a fructose syrup, for example HFGS. See figure 3.9. In the conversion of starch to a dextrose syrup, liquid enzymes are used, but in the conversion of a dextrose syrup to a high fructose glucose syrup, an immobilised enzyme is used – the first large-scale commercial use of an immobilised enzyme. Another major technological achievement is the use of ion exchange resins on an industrial scale to produce an enriched fructose syrup by separating fructose from dextrose using a continuous process.

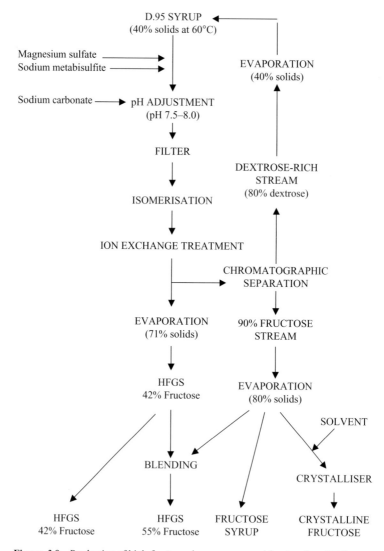

Figure 3.9 Production of high-fructose glucose syrups and fructose from D.95 syrup.

To convert a dextrose-rich syrup into a high fructose glucose syrup, the process is as follows:

1. The starting syrup is a de-mineralised D.95 syrup, with a conductivity of 10 micro-Siemens or less (conductivity is a measure of how well the syrup has been de-ionised).

   To produce a D.95 syrup, some companies will add calcium salts to the starch slurry, because the calcium ion stabilises the amylase enzyme, which is used to make a D.95 syrup. Unfortunately, calcium ions in the D.95 feedstock would prevent the isomerising enzyme from working, because calcium ions will displace the magnesium ions, and it is the magnesium ions which are essential for activating the isomerising enzyme therefore, it is essential that the calcium content of the D.95 syrup is less than 1 mg/Kg.

2. The solids of the syrup should be about 40%, with a dextrose content of at least 95%, but preferably higher. The temperature of the syrup should be 60°C. The operating temperature of the isomerising process is a balance of maximum fructose production with minimal colour formation in the syrup together with maximum enzyme life. Fructose is more susceptible to heat colour formation than dextrose, because it is more chemically reactive than dextrose.

3. A solution of magnesium sulphate is added to give a concentration of about 45 mg/Kg of magnesium. The addition of magnesium is necessary because magnesium acts as an activator for the immobilised enzyme – no magnesium, no conversion of dextrose to fructose. If there is an increase of calcium salts in the syrup for whatever reason, then it is essential to increase the magnesium sulphate so as to prevent the isomerase enzyme being deactivated.

4. Sodium metabisulphite solution is added to give a sulphur dioxide content of 50–100 mg/Kg in the feed syrup. This acts as an oxygen scavenger. The exact addition will depend upon how much dissolved oxygen is present in the syrup.

5. Sodium carbonate solution is now added to adjust the pH of the syrup to about 7.0, depending upon the origin of the isomerase enzyme and its specific requirements.

6. The syrup is filtered to ensure that no particulate matter will foul the surface of the immobilised enzyme.

7. The syrup then passes to the isomerisation columns. The isomerase enzyme looks very similar to the hundreds and thousands found on chocolate cakes. The enzyme is packed into a series of vertical columns. The syrup percolates down the column, and as it passes over the immobilised enzyme, some of the dextrose is converted into fructose. Typically, each column will be approximately 10–12 feet tall, with a diameter of about 3 feet. The important design feature of a column is to ensure an even syrup flow and distribution through the column, with no compaction of the immobilised enzyme. Over the years, as the techniques for immobilising the enzymes have improved, to produce a mechanically stronger product, so the height to diameter ratio of the columns has become less critical.

   As the immobilised enzyme is expensive compared to the other enzymes which the glucose industry uses, it is important to obtain the maximum amount of fructose from each column. Unfortunately, over a period of use, the immobilised enzyme loses its ability to convert dextrose into fructose, which can result in variations in the fructose content of the syrup, as it leaves the isomerising column. To overcome this problem, most companies operate three or four enzyme columns on a carousel principle, with three columns producing syrup, and the fourth column on standby, similar to the way in which an ion exchange process operates. Whilst theoretically, the maximum fructose content that can be produced is about 47%, operationally, it is nearer to 45%, and will subsequently drop to about 38% over the operational life of the column. Therefore, the skill in operating the process is to control the syrup flow rates and temperatures for each column, so as to continuously produce a syrup containing approximately 42% fructose. The syrup flow from the three columns is continuously blended to ensure a syrup containing 42% fructose is produced. When the fructose content of a column starts to drop off, to keep the fructose level as near to 42% as possible, the flow rate can be reduced, and/or the temperature of the syrup feeding the column can be increased, or both. Eventually, as the fructose level falls below 38%, the immobilised enzyme is replaced. At this point, the column which had been on standby now becomes a production column. The spent enzyme is then replaced and the recharged column now becomes the standby column. And so the process goes on. The life of a column will vary depending upon throughput but with careful management, the life will be about 100–150 days.

8. After isomerisation, the HFGS 42 is de-ionised to remove any minerals and to improve the colour which might have developed during processing. Frequently, there is an additional ion exchange treatment which is designed to remove any off flavours that might have been formed during the process.

9. The HFGS is then either evaporated to 71% solids and sold or sent to the chromatographic enrichment process to produce a fructose syrup containing about 80% fructose.
    (i) The reason why HFGS is evaporated up to 71% solids is because at those solids, HFGS 42 can be used to replace liquid sugar on an equivalent weight for weight basis.
    (ii) Additionally, the osmotic pressure at 71% solids is sufficient to kill any yeast or bacteria which might have contaminated the syrup.

10. The chromatographic enrichment process is similar to a conventional ion exchange process. Basically, the syrup passes over an ion exchange resin, which preferentially retains the fructose, leaving the dextrose in the syrup, to produce a dextrose rich syrup. This dextrose rich syrup is sometimes referred to as 'raffinate'. This dextrose rich syrup will typically contain more than 80% dextrose (together with lesser amounts of other sugars, e.g. maltose, maltotriose, etc.) compared to the 52% dextrose which was present in the D.95 syrup feeding the chromatographic enrichment process.

11. The dextrose rich syrup is now evaporated up to about 40% solids, and then added back to the D.95 syrup which is feeding the isomerisation columns. With this continual re-circulation of the dextrose rich stream, there is a slow build-up of the other sugars,

and eventually the dextrose rich stream has to be returned to the dextrose process, so that the higher sugars can be converted into dextrose.

12. The fructose is washed off the ion-exchanged resin to give a fructose rich syrup which is evaporated to about 80% solids.
    (i) It can be sold as a fructose rich syrup.
    (ii) It could be blended with HFGS 42 to make HFGS 55.
    (iii) It could be converted into crystalline fructose. Because fructose is a very soluble sugar, (80% at 25°C), it cannot be crystallised out of solution like dextrose. Therefore, according to USP 4724006, the fructose-rich syrup, at 92% solids and containing 95% fructose is mixed with solvents, which precipitates the fructose out of solution. The precipitated fructose crystals are then separated from the solvent, washed, dried and packaged. The solvent is recovered, purified and then reused in the process.

It is interesting to consider the relationship between 95 DE syrup and crystalline fructose. A 95 DE syrup can be used to make HFGS and ultimately fructose. A 95 DE syrup can also be used as a feedstock to produce ethanol. With the solubility of fructose in 95% ethanol being less than 5%, a company which is producing 95 DE could use it to produce both ethanol and a fructose-rich syrup. The ethanol could be used to precipitate the fructose out of solution to produce crystalline fructose, whilst the ethanol could be recovered and reused. Since ethanol is a 'bonded spirit', it is necessary to denature the spirit for two reasons. By denaturing the spirit, there is no longer any requirement to pay duty on the spirit, and the second is to deter it from being illegally consumed. Ethanol can be denatured by the addition of methyl alcohol.

3.13 Syrups for particular applications

During the 1990s, the possibility of producing glucose syrups which had very specific molecular weights for use in urology treatments was investigated. The work involved passing a glucose syrup through semipermeable or selectively permeable membranes which allowed only certain size molecules to pass through and to reject others. Problems were however experienced due to the suitability of the membranes and the pressures involved. Additionally, as the market for such specialised syrups was very limited, the project was terminated.

3.14 Summary of typical sugar spectra produced by different processes

Because of the many processing permutations available to make glucose syrups together with the possibility of the blending of different syrups, it is now possible to produce a 'standard syrup' such as a 63 DE syrup by blending, as well as syrups with different DEs, different sugar spectra, different properties, and hence suitable for different applications. Below are the sugar spectra of six typical syrups:

Table 3.2 Sugar spectra of six typical glucose syrups.

	45 DE acid	45 DE enzyme/enzyme	63 DE acid/enzyme	95 DE PEE	HFGS PEE	Maltose PEE
% Dextrose	21	3	36	94	52	2
% Maltose	16	52	33	4	4	70
% Maltotriose	13	20	8	1	1	20
% Higher sugars	50	25	23	1	1	8
% Fructose	–	–	–	–	42	–

As can be seen from Table 3.2, as the DE increases, the percentage of higher sugars becomes less. This reduction in higher sugars results in a corresponding reduction in viscosity. The higher the DE of a syrup, the lower its viscosity. However, this increase in DE will also increase the risk of dextrose crystallisation.

Table 3.3 Relationship between syrup viscosity and sugar spectra of glucose syrups.

Syrup	True solids	Viscosity (in centipoises at 50°C)	Dextrose	Maltose	Maltotriose	Higher sugars
28 DE	80.0%	10,000	4%	11%	15%	70%
35 DE	80.0%	10,000	15%	12%	11%	62%
42 DE	80.0%	6,500	19%	14%	11%	56%
50 DE	80.0%	5,500	29%	23%	10%	38%
63 DE	80.0%	1,000	34%	33%	10%	23%
95 DE	75.0%	100	94%	4%	1%	1%
HM 50.	80.0%	7,000	5%	50%	20%	27%
HM 70.	80.0%	1,500	2%	70%	20%	8%

Table 3.3 illustrates how the viscosity of a glucose syrup is reduced with the reduction in the level of higher sugars.

N.B. The reason why the solids of the 95 DE syrup is only 75%, compared to the 80% solids for the other syrups, is because of the high dextrose content, and the risk of dextrose crystallisation.

Chapter 4
Explanation of glucose syrup specifications

4.1 Introduction

It is very important for both the supplier and the customer to talk through the syrup specification when contracts are being agreed. This should then ensure that the customer knows what is being supplied, and the significance of the tests, whilst the manufacturer knows exactly what the customer requires, thereby ensuring a successful commercial relationship.

With continual changes in the way in which syrups are produced, together with changes in both legislation and the requirements of the food industry, specifications will also change. A typical glucose syrup specification would contain most of the following details, however there could be additional information if requested by a customer. Also specifications could be 'tailored' to meet the requirements of a specific industry, for example the information required for the confectionery industry would be totally different to the information required by a brewer. Some specifications will also state the analytical methods which should be used. With the production of ion exchanged, or de-ionised syrups, and the use of HPLC sugar analysis, some people will say that not all of the information included in a specification is now relevant. It must also be stressed that a specification is *not* the same as a typical analysis – the two complement each other! A specification will show the range of a measurement, for example DE 40.0–44.0, whereas a typical analysis will be more specific, for example DE 42.1.

4.2 What specification details mean

Table 4.1 Glucose syrup specifications.

Identification	This could be a name, or a manufacturer's code.
Appearance and form	A brief description such as, 'A clear water white, bright viscous liquid'.
Colour	This is can be measured using a Lovibond Tintometer. In this instrument, the colour of the syrup is compared with a series of coloured glass slides. An alternative instrument could be a spectrophotometer.

(Continued)

Table 4.1 (*Continued*)

Starch test	With modern starch conversion technology, residual starch should no longer be a major problem. To test for starch, add a few drops of iodine. A blue colour indicates the presence of starch. The presence of starch in a syrup would suggest that the syrup has not been correctly produced. Any residual starch could cause a haze in high-boiled sweets, and in brewing could cause a haze in beer, as well as slowing down the rate of beer filtration.
Optical density (transmission)	This is another method for measuring colour, using a spectrophotometer. A beam of light of a known wavelength, for example, 380 nanometers (nm) is passed through a sample of syrup, and the percentage of transmittance is measured. This percentage is the optical density of transmission.
Ageing colour	This is a measure of the change in colour of the syrup, when it is held at a certain temperature, typically 2 hours at 100°C. This is sometimes referred to as the 'heat punishment colour'. Like syrup colour, it is measured using a Lovibond Tintometer, or some similar instrument. Ageing colour is an indication of the likely change in syrup colour during processing, particularly when making confectionery products such as boiled sweets. It could also give an indication of how well the syrup has been made. The addition of sulphur dioxide will improve the ageing colour, but it should be added to the syrup, prior to evaporation to the final solids, *not* after evaporation. With the use of de-ionisation, ageing colours have greatly improved.
Dextrose Equivalent	Sometimes abbreviated as DE. It is a measure of total reducing sugars in the syrup. Where a DE is quoted in a specification, it is usually the mid-point of a DE range, for example a 42 DE syrup will have a DE range of 40–44 DE. Traditionally, the DE has been measured using Fehling's solution (sometimes referred to as the Lane and Eynon's titration). The more modern method is to use a cryoscope or osmometer. It can also be determined by calculation from the HPLC sugar analysis of the syrup. With new types of syrup being available, DE is not always appropriate. Instead, the syrup should be characterised by its sugar spectrum determined using HPLC.
pH	The pH of a syrup will usually be in the region of 4.5–5.5. Dextrose is most stable in acidic conditions. With the introduction of de-ionised syrups, pH is less relevant. However, the pH range of a de-ionised syrup is often 4.0–6.0. This wide range is because of the lack of minerals to buffer the pH to a smaller range. Where pH is very critical, the pH of the syrup can be adjusted using an organic acid, for example citric acid.

Table 4.1 (*Continued*)

Acidity	This is measured by titration. Some confectionery companies depend upon the acidity of a glucose syrup to invert the sucrose, when making hard-boiled confectionery. The invert reduces the risk of the sucrose crystallising in the end product.
Refractive Index	This is a quick way of determining the total solids in a syrup using a refractometer. From the refractive index, the solids can be determined using tables, if the instrument is not already calibrated to measure solids. The temperature at which the determination is made should be stated. Normally, the refractometer will be temperature controlled, for example 20°C.
True solids	This tells you the % solids present in the syrup, and is usually determined by heating a sample of syrup to drive off the water. True solids can also be calculated from the RI solids, using tables.
RI sugar solids	A quick way of determining the % solids in a syrup based on the refractive index of the syrup. The temperature at which the determination is made should always be stated, for example 20°C. For absolute accuracy, as different sugars have different refractive indices, correction tables are required to correlate RI sugar solids and syrup DE with true solids. These correction figures will change with temperature, DE, sugar spectrum and the method of hydrolysis used to produce the syrup (i.e. acid or enzyme, etc.) and if the syrup has been de-ionised or not. The refractometer is usually temperature controlled and calibrated to give a direct solids reading, otherwise the solids figure can be obtained by reference to tables. Alternatively, The Corn Refiners Association have produced a computer program from which the solids can be calculated. For syrups at commercial solids, typical corrections would be as follows: to convert an RI solids reading of a 26 DE syrup at 20°C into true solids, 3.3 must be subtracted from the RI solids reading. For a 42 DE syrup, 2.2 must be subtracted. For a 63 DE syrup, 0.8 must be subtracted, but for a 95 DE syrup, it is necessary to add 1.5. Similarly, for a 42 HFGS, it is necessary to add 1.5. For a 55 HFGS, it is necessary to add 1.65, and for a 90% FX, add 1.69. For a 45% maltose containing syrup, it is necessary to subtract 2.0, and for a 70% maltose containing syrup, it is necessary to subtract 1.9.
Sulphur dioxide	This is usually added to a syrup, in parts per million, to improve the colour of the syrup during manufacture and in storage. The maximum addition is usually 400 mg/Kg, but generally a lot less. The sulphur

(*Continued*)

Table 4.1 (*Continued*)

	dioxide blocks the aldehyde reactive sites in the syrup, thereby preventing the proteins, etc., from forming coloured compounds. With de-ionisation now being used in the syrup refining process, the addition of sulphur dioxide to improve colour is no longer required, because proteins and other impurities will have been removed.
Foam index	Some companies require the addition of anti-foam to prevent the syrup foaming in their process. However, with modern refining procedures (e.g. de-ionising), foaming is no longer the problem which it was in the past.
	Some companies will not accept syrups which contain anti-foam (e.g. brewers), because the anti-foam might affect their end product. In the case of brewers, the foam or 'head' of the beer could be adversely affected.
	Companies might have their own foam test, for example the Bickerman method, where air is sucked through a sample, or zinc and acid might be added to the syrup, and the bubbles of gas produced will create a foam. The height of the foam can then be measured.
Metals	With the introduction of de-ionised syrups, metal contamination is often now in the region of parts per million. Typical metals reported would be iron, lead, copper and arsenic, usually at a level of less than 1 part per million. All have a detrimental effect on either the syrup, or the consumer.
	Other metals would be sodium 25 mg/Kg, potassium 1 mg/Kg, calcium 5 mg/Kg, and magnesium 5 mg/Kg.
Chloride	This is important because of its corrosive effect on certain grades of stainless steel. The chloride comes from the acid used in the conversion process. With de-ionised syrups and improved grades of stainless steel, together with the use of enzymes for the starch conversion, chlorides are no longer a big problem.
	The amount of chlorides in a syrup can now be quickly measured using a selective ion electrode. Typical chloride value would be 30 mg/Kg maximum.
Other minerals	Sulphate, typically 5 mg/Kg maximum for a de-ionised syrup.
Sulphated ash	Because most syrups are made using hydrochloric acid, a lot of the ash will be chlorides, but chlorides are volatile during ashing, therefore by converting the ash into the more stable sulphate, it will not be lost. Additionally, sulphates weigh more than chlorides, thereby improving the accuracy of the ashing procedure. To convert a sulphated ash figure into chloride, multiply the sulphated ash by 0.37.

Table 4.1 (*Continued*)

Conductivity	With the introduction of de-ionised syrups, the mineral content is so low that conductivity is a more accurate indication of impurities present due to metals or salts.
	Conductivity is expressed in micro-Siemens. Typical conductivity should be less than 30 micro-Siemens, ideally less than 10 micro-Siemens.
Acetaldehyde	Acetaldehyde gives an unpleasant taste to a syrup. It is a waste product of micro-organisms. Therefore, if detected in a syrup, it suggests that the syrup has been made in a dirty plant. Acceptable levels would be less than 10 parts per billion.
Sugar spectrum	With the use of many different methods and combinations of hydrolysis, the sugar spectrum of a syrup is now a more accurate method of characterising a syrup, and is determined by HPLC sugar analysis. Sugars usually reported are dextrose, fructose, maltose, maltotriose and higher sugars.
Fermentable sugars	This is of interest to brewers and other fermentation industries.
	Readily fermentables are dextrose, fructose and maltose.
	Slowly fermentables are maltotriose.
	Non-fermentables are the remaining sugars.
	Some industries will base their purchasing price on the fermentability of the syrup.
Brewer's Extract	This is only included in specifications for syrups to be used in brewing. It is another way of expressing the syrup solids. The units used are 'litre degrees per kilo'.
EBC colour	Some brewers require their syrup to be coloured with caramel colouring. The colour of the resulting syrup is reported in European Brewing Convention Units, and is determined by comparing the colour of the syrup with a series of coloured glass standards. With the demand for light-coloured beers such as lagers, the requirement for the addition of caramel has declined.
Specific gravity	Another way of determining the solids in a syrup is using a hydrometer calibrated to read specific gravity. As syrups are very viscous at room temperature, the determination is usually carried out at a specific elevated temperature, usually 60°C (140°F).
	It is used to convert volumes of syrup into a weight of syrup, or vice versa.

(*Continued*)

Table 4.1 (*Continued*)

Commercial Baumé	Glucose syrups are sold on their solids content, and commercial Baumé is the figure which is used.
	Baumé is a quick way of determining the solids of a syrup by using a hydrometer, which is calibrated in 'degrees Baumé'. Because of the viscosity of a glucose syrup, Baumé readings are carried out at a specific elevated temperature, usually 60°C (140°F), with the readings being adjusted back to 15.6°C (60°F). Readings are usually expressed as 'Degrees Bé'. Tables can then be used to convert the Baumé reading into % solids. For example, '41.0 Commercial Baumé syrup' means that the solids of the syrup are 77.8% – but it does not tell you anything about the sugars present in the syrup.
	When measuring and comparing Baumé readings, it is important that the modulus of the hydrometer which is used is declared. The glucose syrup industry uses the '145 modulus'. The term 'modulus' refers to a constant used to convert specific gravity into degrees Baumé.
Microbiological	This will show that the syrup has been manufactured and stored under good hygienic conditions and that it does not contain any harmful microbes. Ideally, the glucose syrup should meet the American Bottlers Standard. Typical values could be as follows:
	Coliforms, Staph. aureus and Salmonella – absent
	Mesophilic Bacteria – 100 per gram maximum
	Yeasts – 20 per gram maximum
	Osmophilic yeasts – 10 per gram maximum
	Moulds – 10 per gram maximum
	The combination of high syrup solids and the molecular weight of the sugars in a syrup creates a high osmotic pressure within the syrup, which will destroy most microbiological organisms such as yeast. In the case of fructose syrups, this osmotic pressure effect makes the syrup 'self sterilising'. It should however be appreciated that whilst the syrup might be sterile, the waste products of microbiological activity could still be present in the syrup, for example acetaldehyde.
Mycotoxins	Some specifications will include mycotoxins, which are naturally occurring toxic compounds produced by fungi. The presence of mycotoxins indicates that the starting material was contaminated by fungi, usually due to incorrect storage, and should therefore have not been used to make a glucose syrup.
	Typical mycotoxin producing fungi are *Fusarium, Aspergillus* and *Penicillum*. Of these, *Fusarium* fungi are probably the most prevalent toxin producing fungi on cereals grown in the northern temperature regions of America, Europe and Asia.

Table 4.1 (*Continued*)

	Trichothecenes is a group of about 150 related compounds which are produced by *Fusarium*. Toxins T-2, HT-2, deoxynivalenol (DON) and nivalenol are probably the most important. The toxins Zearalenone and Fumonisins are also produced by *Fusarium*.
	Aflatoxins consist of a group of approximately 20 related fungal metabolites such as aflatoxin B1, B2, G1 and G2. Of these, aflatoxin B1 is considered to be the most toxic, and is produced by *Aspergillus flavus* and *A. parasiticus*. The latter also produces aflatoxins G1 and G2.
	Ochratoxin A is produced by several different fungi, but particularly by *Penicillium verrucosum*.
Ergoline alkaloids	These compounds are present in ergot-contaminated wheat, and is produced by the fungi *Claviceps purpurea*.
Viscosity	Viscosity is measured at different temperatures, and results are expressed in centipoises, for example 160 centipoises at 15.6°C (60°F). Generally, the viscosity of a syrup will be greatly reduced by heating the syrup, making it easier to handle or pump but be careful of colour development. Low DE syrups are more viscous than high DE syrups, because low DE syrups contain lots of higher sugars, and higher sugars mean high viscosity. The following are approximate viscosities for different syrups at 15.6°C (60°F):
	35 DE (80% solids) = 2,000,000 Centipoises
	42 DE (80% solids) = 500,000 "
	52 DE (80% solids) = 400,000 "
	63 DE (80% solids) = 200,000 "
	95 DE (75% solids) = 1,000 "
Recommended storage temperature	Syrups are generally stored at a certain temperature for ease of handling and to prevent crystallisation. The storage temperature will vary for each syrup depending upon the DE of the syrup and the solids. The following are typical recommended storage temperatures:
	28 DE = 60°C; 35 DE = 60°C; 42 DE = 60°C; 50 DE = 60°C; 63 DE = 50°C; 95 DE = 55°C.
	HFGS = 30°C; Maltose syrups = 50°C.
Labelling	For ingredient declaration purposes in the final end product, it should be labelled as glucose syrup. If the syrup contains fructose, which has been made from substances other than starch, for example from inulin, then the syrup has to be labelled as either glucose fructose syrup, or fructose glucose syrup, depending on which is the major sugar.

(*Continued*)

Table 4.1 (*Continued*)

	For nutritional labelling purposes, dextrose, maltose and maltotriose can be called 'sugars', whilst the higher sugars of a glucose syrup can be called 'starch'. Therefore, by replacing a high-dextrose syrup with a low DE syrup which contains predominantly higher sugars, it is possible to claim a reduction in the 'sugar' content of the product! However, the calorie contribution from the syrups would remain unchanged, providing that the replacement was on a one-for-one dry basis.
Availability and packaging	Tankers or drums. This will depend upon the type of syrup. Generally, all syrups are available in both tankers and drums, with the exception of HFGS and 95 DE syrups, because both of these syrups have a high dextrose content which will crystallise if the syrups are not stored at the correct temperature. Twenty-five kilograms multiply sacks or bulk road tanker for dextrose and maltodextrins.
Legislation	It should be mentioned that the glucose syrup conforms to current and relevant food legislation, together with Health and Safety requirements. See also EU Directive No. 1881/2006, which covers levels of contaminants in foodstuffs.
Toxicity	Syrups and dextrose are considered to be non-toxic. However, for people suffering from diabetes, care should be taken not to ingest as they can affect blood sugar levels. In the case of powders such as fructose, dextrose, spray dried syrups and maltodextrins, there could be a dust nuisance when handling.
Corrosion properties	Glucose syrups have a pH in the range of 4.0–6.0. Syrups which have not been de-ionised will contain chloride, which can attack certain metals.
Fire risk	Syrups are considered to be non-flammable. Both dextrose and maltodextrins will burn if subjected to direct heat.
Explosion risks	Syrups do not have an inherent explosion risk. Dextrose and maltodextrins are a Class 1 risk.
Handling procedures	Bulk syrups are hot when delivered, therefore suitable protection to hands and face should be employed. Drums of syrup will be at ambient temperature. For drums and bags of dextrose and maltodextrins, suitable lifting techniques should be observed.
Protective clothing	Since syrups can be hot, hands and faces should be suitably protected. Face mask should be worn with prolonged exposure to dust when handling dextrose and maltodextrins.

Table 4.1 (*Continued*)

First aid requirements	Syrup, dextrose and maltodextrins can be removed from hands and faces with warm water. Any product in the eye can be removed with sterile saline solution (0.9% w/v NaCl). If hot product has entered the eye, medical attention should be sought.
Disposal of spillage and waste	Syrups can be removed with hot water. Where a large quantity of syrup has been spilt and enters the local sewer, the Water Authority should be notified, due to a possible increase in the COD (Chemical Oxygen Demand) loading on the local treatment plant. Dextrose and maltodextrins should be swept up, bagged and disposed off in an appropriate manner, for example waste skip.

4.3 Dry products

The specification for dry products such as dextrose monohydrate, crystalline fructose and maltodextrins would be similar to that given in Table 4.1, where appropriate, and would include bulk density – both loose and compacted, also mesh size or particle size distribution. In the case of dextrose, there would be the additional specification for optical rotation, which is another indication of its purity. If other sugars are present, then the optical rotation will be affected.

Some specifications might also include the caloric value and melting point of the product.

With starch, there would also be information about the gelling temperature, and viscosity – usually determined using a Brabender, whilst the microbiological specification would be slightly different from the syrup microbiological specifications. The starch microbiological specification would probably be based on the National Canners Association suggestions, and would be equally applicable to dry products which are to be used in canned products, for example maltodextrin, dextrose.

Total thermophilic spore count. Of five samples, none shall contain more than 150 spores per 10 grams, with an average for all samples not exceeding 125 spores per 10 grams.

Flat sour spores. Of five samples, none shall contain more than 75 spores per 100 grams, with an average for all samples not exceeding 50 spores per 10 grams.

Thermophilic anaerobe spores. Not more than three of the five samples (60%) shall contain spores, and in any one sample, not more than four (65%) of the six tubes are to be positive.

Sulphide spoilage spores. Not more than two of the five samples (40%) shall contain these spores, and any one sample shall contain not more than five colonies per 10 grams.

Thermophilic anaerobes, for example *clostridium*, are also known as 'hydrogen swells', because they produce hydrogen which would distend the container.

Flat sours are acid producing facultative anaerobes, for example *Bacillus macerans*. The significance of these bacillus is that they produce acid, but no gas. The acid which they produce will corrode the metal container, unless the surface is protected by lacquer and can distort the can.

Sulphide spoilage spores, for example *Desulphotomaculum nigrificans*, are also known as 'sulphide stinkers', because they produce hydrogen sulphide. The hydrogen sulphide, with its characteristic smell of bad eggs would render the end product inedible. They also produce acids which would corrode the inside of unprotected metal containers.

Products in bags, such as dextrose, fructose, maltodextrins or starch, can pick up odours from their surroundings or from the pallet on which they are stored. To prevent the bags becoming contaminated from the pallet or during storage, a plastic sheet should cover the pallet before it is loaded with bags. Once the pallet is loaded, the pallet should be either covered with a plastic shroud or the pallet should be shrink wrapped. Shrink wrapping also helps to hold the bags together when the pallets are being moved around the warehouse and when in transit. Finally, the pallets should be stored in a dry, odour, insect and rodent-free area, and minimum of 3 feet (1 m) away from a wall. By keeping pallets away from walls makes it easier to monitor and to control pests.

4.4 Syrup problems and their possible causes

On rare occasions, the glucose syrup might not reach the exacting standards which both the industry and customers expect. Sometimes the syrup has been incorrectly produced or stored by the customer. Here are a few of the problems together with possible causes. Whenever there is a problem with a syrup, always check the 'problem sample' against a retained sample, and if the problem relates to a hazy syrup or to black specs in the syrup, first look at the sample under a microscope, and then, if possible, take photomicrographs of the haze or specs. These microphotographs can then be used when discussing the problem with either the customer or production. These photomicrographs can also be used to build up a 'Rouges Gallery' for future reference as well as an aid to monitoring problem areas in the process.

If the problem relates to incorrect solids, pH, DE or sugar spectrum, take more than one sample.

- If the sample has solidified, melt the entire sample before carrying out any analyses.
- Calibrate the refractometer and check solids of the syrup on more than one refractometer. Low solids can be due to condensation, if the sample is taken from the top of the tanker. If the sample is taken from the tanker discharge pump, low solids can be due to tanker wash water remaining in the pump.
- Calibrate the pH meter, and check the pH of the sample on more than one pH meter.
- Check the DE on more than one cryoscope.
- Check sugar spectrum using different HPLC equipment, against standards. Inject the sample three times, and take the average of the three readings. Compare the results with the specification for that particular syrup.

Problems can be experienced with bags of dextrose monohydrate going solid during storage. The reason for this is due to moist air being trapped with the dextrose during the filling of the bag and subsequent storage. A more detailed explanation is given in Chapter 3, Section 3.9.

Table 4.2 Syrup problems and their possible causes.

Symptom	Possible cause	Confirmatory test	Comment
Hazy syrup	1. Yeast	Smell of alcohol? Microscopic examination to confirm yeast. Check syrup solids.	Yeasts grow in syrups which have a low solids.
	2. Starch	Add dilute iodine solution. A blue colour confirms starch.	Starch is due to poor conversion. Common in low DE syrups.
	3. Filter aid	When the syrup is diluted and centrifuged, is there a white deposit at the bottom of the tube? Filter aid. Microscopic confirmation.	Filter aid is passing through the safety press. Check filter cloths for damage.
Black specs	1. Rust	1. Dilute sample and filter. Dry filter paper. Are the specs attracted by a magnet? If yes, particles are rust.	Check for corrosion.
	2. Carbon black	2. If particles are non-magnetic, particles could be carbon. Confirm with microscopic examination.	Check filters for damage.
	3. Dead yeast or mould	3. Microscopic examination. Are there cell structures?	Carry out microbiological examination.
	4. Tank or pipe lining	4. Microscopic examination. Do the particles have a rugged structure?	Check tank and pipe linings for damage.

(*Continued*)

Table 4.2 (*Continued*)

	5. Broken ion exchange resins	5. Microscopic examination. Are the particles semi – transparent? (Amber colour?)	Check condition of ion exchange resins.
	6. Burnt syrup	6. Microscopic examination. Do the particles have a brown colour? Do the specs have a burnt or caramel smell?	Heating coils not turned off, when level of syrup is low.
White crystals	Dextrose crystals	Do the crystals disappear when the syrup is warmed in a hot water bath? If so, the crystals are dextrose. *N.B.* Melt all the crystals before carrying out any analyses.	Syrup has been stored at too low temperature. Check the sugar spectrum of the syrup. Dextrose content too high.
Syrup is solid	Dextrose crystallised. If the sample has solidified, melt the entire sample before carrying out any analyses	As for dextrose crystals.	Common problem with high DE syrups, which have been stored at too low temperature.
Brown-coloured syrup	1. Poor refining	Treat with activated carbon or ion exchange resins. Does the colour improve?	Check carbon dose rates, and ion exchange operation.
	2. Syrup stored at too high temperature	Does the syrup have a caramel smell? Is the colour still present after carbon treatment?	Check temperature of syrup in storage tank.

Table 4.2 (*Continued*)

	3. Metal contamination	Check for metal ions. Typically iron, zinc and copper.	Presence of metals suggests possible corrosion. If both iron and zinc are present, this suggests galvanised metal.
Blue green colour	Carbon	Dilute syrup, and filter. Examine black residue on filter paper.	Carbon passing through safety press. Check filter cloths for any damage
Pink colour	Ion exchange resins	Is there a fishy/amine smell? Test for amines.	The resin was insufficiently washed before being used.
Fishy or amine smell	Ion exchange resin	Test for amines.	As above.
Chlorine or iodine smell	Chemical in place cleaning solutions	Test for chlorine and iodine.	More thorough washing required.
Smell of petroleum or oil	Dirty tanker	Test for hydrocarbons.	More thorough washing.
Alcohol	Yeast	As for hazy syrup.	As for hazy syrup.
Smell in bagged products, for example dextrose, maltodextrins	The bags have been stored on dirty pallets, or incorrectly near to smelly products	Examine the surface of the pallet. Has the wood of the pallet been treated with preservative? What was the previous load on the pallet? Check warehouse conditions. Is there a smell?	The top of the pallet should be covered with a thick plastic sheet before putting on the bags onto the pallet. A plastic shroud should be placed over the bags, or the pallet should be shrink wrapped. Store pallets away from smelly products.

4.5 Bulk tank installation

Before installing a bulk syrup storage tank, there are several basic questions which should be considered.

1. Does the syrup usage justify the capital investment?

    It is generally considered that the minimum syrup usage to justify a bulk tank installation is 500 tonnes per year, or 10 tonnes per week. The advantages of bulk syrup deliveries are cost savings – syrup in bulk is cheaper than syrup in drums, and there are less labour costs involved, because there are no drums to handle. The syrup can be delivered by a pipe directly from the storage tank to where it is to be used.

    The useful economic life of a bulk tank made of steel is about 10–20 years.

2. Is there sufficient space on the site for a bulk tank?

    Bulk tanks can be either vertical or horizontal. Vertical tanks are preferred for two reasons: (a) vertical tanks take up less ground space, providing there is sufficient available headroom. (b) Vertical tanks have a smaller surface area. This is important, because if there is condensation inside the tank, it will form a dilute syrup layer, in which yeast and other microbes can grow. By keeping the surface area as small as possible, any contamination is kept to a minimum. With a horizontal tank, there is a larger surface area for microbial contamination. Additionally, it is difficult to re-circulate syrup in a horizontal tank.

3. Are the tanks to be built inside or outside the building?

    For ease of handling, glucose syrups should be kept warm, which means that in Europe, the tanks would have to be heated. If the tanks are to be situated outside a building, then the tanks will have to be insulated to reduce heat loss and the insulation will have to be weather proof. If the tanks are to be place on a concrete base, then the bottom of the tank should be insulated to prevent any heat loss to the ground and corrosion from ground water.

    If the tanks are to be situated inside a building, then they will require less insulation, or possibly no insulation if the tank is inside an insulated hot room.

4. How far is the storage tank away from where the syrup is to be delivered and from where it is to be used?

    Ideally, the pipe runs should be as short as possible, self-draining and trace heated, with temperature-controlled electrical heating tape.

5. Will there be a sufficiently clear area for a road tanker to deliver the syrup? Will the storage tank be sufficiently close to the tanker discharge point, and will the ground be sufficiently solid to carry the weight of a loaded road tanker?

    The average road tanker requires a minimum turning circle of 22 metres, and the total weight of a road tanker when loaded could be 44 tonnes.

4.6 Bulk tank design

Having checked the above, the next consideration is the design of the bulk storage tank. See Figure 4.1. Most glucose suppliers will provide help and guidance on a bulk tank installation. The following should be considered only as a checklist:

1. What should be the capacity of the proposed tank?

    If the syrup usage is 10 tonnes per week, then the capacity should be 15 tonnes. This extra capacity will give more flexibility with syrup deliveries. Instead of waiting until

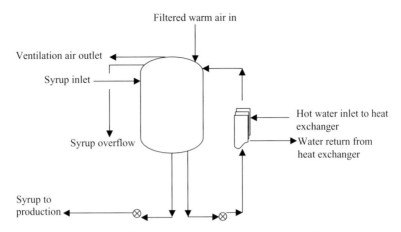

Figure 4.1 Schematic of a typical glucose bulk tank installation.

the tank is completely empty before accepting the next delivery, a new delivery can be made when the tank still contains some syrup, thereby ensuring that production is not stopped due to lack of syrup. Always build in extra storage capacity – it is cheaper to do so at the design stage, rather than alter an existing installation at a later date.

2. The foundations for the tank.

    Glucose syrups are approximately 50% more dense than a corresponding volume of water. Therefore, the foundations for the tank should reflect this increase loading requirement.

3. The tank should be vertical and circular, made of 316 stainless steel, finished to dairy or brewing standards. Circular tanks are easier to clean – there are no corners.

4. The tank should be supported off the ground, and placed on load cells, or incorporate some measuring instrument to show how full is the tank, with a duplicate read out in the production control room.

5. The bottom of the tank should be dished shape, with an off take point at its lowest point. It is from this off take point that the syrup will be pumped to the production area. This off take point, if required, could also act as a drainage point for the tank. If the take off point is also to be used as a drainage point, then there should be an interlock system, so that a full tank of syrup cannot be accidentally discharged down the drain. There should also be a small sample tap on the side of the tank, near the bottom for taking microbiological or other samples.

6. At the top of the tank, there should be an inlet point, where syrup deliveries would enter the tank. At this point, there should be a high-level warning system, which will give both an audible and visual alarm at the tanker discharge point and in the control room. This will help to prevent the tank from being over-filled. There should also be an easily visible overflow pipe for when the tank is over-filled, with a coarse filter over the exit so as to prevent anything entering the tank. See also (8) below.

7. To keep the syrup at the correct temperature, the syrup should be circulated from the bottom of the tank, through an external, water-heated, heat exchanger with the syrup being returned to the top of the tank. By using hot water in the heat exchanger, and controlling the temperature of the water with thermostatic valves, there is less chance

of the syrup over-heating with subsequent colour development in the syrup. If heating coils are used at the bottom of the tank, there is always a risk that the level of the syrup will fall below the level of the heating coils, and any syrup left on the surface of the heating coils could become caramelised, and could result in black specs in the syrup.

8. It is important to prevent condensation forming in the headspace of the tank. Any condensation in the headspace will condense to form a dilute syrup layer on the top of the syrup. This layer of dilute syrup is an ideal breeding ground for yeast and other microbes. The problem of condensation can be avoided by having forced air ventilation. Air from outside is filtered, warmed and then blown into the headspace. The volume of air required to prevent condensation should be approximately five times the vessel volume per hour. The air is then vented to the atmosphere through the overflow pipe. To prevent condensation and reduce the possibility of insects and other airborne contamination entering the headspace, a positive air pressure should be maintained at all times.

Some people suggest the use of ultraviolet (UV) lights inside the tank headspace. The problems with this approach are that over a period of time, the lamps become covered with syrup from splashing, and whilst UV light can penetrate the surface of water, UV light cannot penetrate the more dense glucose syrup.

9. Providing a bulk storage tank is in continuous use, with the same syrup, cleaning is not generally required. In fact, more problems are created by cleaning tanks, than by not cleaning them. The reason for this is that glucose syrups are viscous liquids, and require a lot of very hot water to remove them. Any dilute syrup left in the bulk tank or in any of the pipe work after 'cleaning' is a potential breeding ground for yeast and other microbes, which will then contaminate the next syrup delivery.

Do not clean a tank unless it is really necessary. If a tank has to be cleaned, check the refractive solids of the water leaving the tank, or use a test paper sensitive to dextrose. A suitable test paper is one used by diabetics to test for sugar in their urine. Wash water leaving the cleaned tank should contain zero solids.

10. During the designing of the tank, it is advisable to include a manhole inspection plate. This should be situated near the bottom of the vertical side to allow the inside of the tank to be inspected, if that is ever required.

11. A bund wall should be built around the base of the tank, with the floor of the bund area sloping towards the centre, where a drain should be situated. This bund area serves two main purposes. The first is to retain any syrup spillage which might occur due to the tank being over-filled or through leaks, and the second reason is to collect the water which is used to lubricate the syrup pumps, which are delivering the syrup to the production area, and the pump circulating the syrup through the heat exchanger. The pumps should be mounted off the ground, so that they are not submerged if there is a syrup spillage.

Chapter 5
Application properties of glucose syrups

5.1 Introduction

To be successful at using glucose syrups, it is important to understand their properties. Because of the advances in glucose syrup production, there are now several different types of syrup, with each syrup having its own particular characteristics, and hence its own particular application niche. Before undertaking any work involving a glucose syrup, ask the questions – Why use a glucose syrup, and will it achieve what is required? Unfortunately, most people will think of 'glucose syrups' as being an alternative sweetener for sucrose. How wrong can they be! Whilst, glucose syrups are sweet, this is just one of their many functional properties.

Traditionally, glucose syrups have always been described by their dextrose equivalent (DE). This is a leftover from the days before HPLC sugar analysis, and the problem it has created is that you can have two syrups with the same DE, but they will have totally different sugar spectra and these differences in sugar spectrum will, in some instances, affect how the syrup performs. It is therefore advisable when describing a glucose syrup to quote both the DE and the sugar spectrum. For example, consider two '42 DE syrups'. One made using only acid – a conventional confectioners glucose syrup and, the other, a maltose syrup, made using both acid and enzyme.

Table 5.1 Table showing the differences in the sugar spectra of syrups with the same DE but produced by different methods of hydrolysis.

Process	DE	Dextrose	Maltose	Maltotriose	Higher sugars
Acid	42.0	19.0%	14.0%	11.0%	56.0%
Acid enzyme	42.0	6.0%	44.0%	13.0%	37.0%

Because the acid produced 42 DE syrup contains more dextrose than the acid–enzyme produced syrup, it will be more hygroscopic, and produce darker toffees, than the acid–enzyme 42 DE syrup. These differences are entirely due to the higher proportion of dextrose in the acid produced syrup. Whilst both dextrose and maltose are reducing sugars, it is the dextrose which is the more chemically reactive of the two sugars.

Another difference is in the viscosity of the two syrups due to the difference in the higher sugars. The acid enzyme converted 42 DE syrup contains less higher sugars, and therefore has less viscosity than the straight acid converted 42 DE. This difference in viscosity means that the two syrups are likely to behave differently on different pieces of production machinery.

In 1963, W.J. Hoover produced an excellent chart which correlated the DE of syrups to their functional properties (Symposium Proceedings, Products of the wet milling industry in food, published by Corn Refiners Association, Inc., 1971). Since 1963, there have been many advances in glucose syrup production, notably the introduction of maltose and fructose syrups, and so these two syrups have been added to the chart. Fructose syrup – HFGS – has been positioned on the extreme right, with the 95 DE syrup and dextrose because they both have similar properties, except, of course, sweetness. The positioning of maltose syrups within this chart is more difficult because the maltose content can typically vary from 28% to 70% or even as high as 80%. Additionally, a lot will depend upon the particular application. As an approximate guide, maltose syrups with a low maltose content will have similar but not exactly the same properties as a 40 DE syrup, whereas a syrup containing 70% maltose will have properties similar to a 70 DE.

Another development has been the introduction of de-ionised syrups (also known as demineralised syrup or ion exchange treated syrups) as a routine production process. The terms de-ionised, demineralised or ion exchange treated syrup all refer to the same process, namely the removal of salts, minerals, proteins, colour and colour precursors, etc., from a syrup.

Whilst some de-ionised syrups were available before, they were produced in limited quantities and consequently sold at a premium, with their main use being in health promoting medicinal soft drinks. The technology used to produce these early de-ionised syrups was based on the then current water treatment technology, using the same mixed bed resins. With the production of HFGS and the need for de-ionisation, syrup producers opted for individual cation and anion columns, filled with resins specific for treating syrups as opposed to treating water. The expertise gained in the de-ionising of syrup for high fructose glucose syrup production, syrup producers started to use this new technology to de-ionise other syrups, so as to improve their colour and storage properties, however de-ionisation also gave the syrups several other advantages. So what are these advantages?

Traditionally, the glucose syrup industry has used sulphur dioxide to block the Maillard colour reaction between the reducing sugars and the residual protein colour precursors in the syrup. By de-ionising the syrup, these colour precursors have been removed, thereby reducing colour formation, and hence giving the syrup exceptionally good storage properties. Additionally, legislation in some countries prohibit the use of sulphur dioxide in certain products, therefore de-ionised syrups can be used, if applicable, in products for those countries.

The presence of salts and minerals can have an effect on taste, flavour and perceived sweetness. With these impurities removed, the background flavour from the syrup disappears, allowing the true flavour of the syrup to come through, and because of this flavour enhancement, it is now possible to reduce the quantity of the expensive flavours required, thereby offering a cost saving. Similarly, with the minerals having been removed, the buffering capacity of the syrup is reduced, allowing a reduction in acid additions, for example citric acid.

Microbiological stability of the syrup is improved because with de-ionised syrups there are no minerals present to sustain microbiological growth. The removal of the minerals also lowers the ERH of the syrup, which could be beneficial in improving the shelf life of a product.

One of the main salts in a glucose syrup is sodium chloride, resulting from the acid hydrolysis. Sodium chloride will corrode some metals, for example 314 stainless steel, but with a de-ionised syrup, the sodium chloride, together with other salts, is virtually zero, as shown by their low conductivity, typically less than 30 micro-Siemens.

Finally, due to the negligible protein content of a de-ionised syrup, there is minimal foaming. This can be used to advantage when de-ionised syrups are used in the manufacture of hard-boiled confectionery or in other processes where foaming can be a problem.

All these advances make W.J. Hoover's chart more relevant when deciding which syrups to use to develop a particular functional property. Understanding this chart and remembering how de-ionisation can affect syrup properties is the key to successful glucose syrup application technology.

5.2 Summary of properties

Table 5.2 Properties and functional uses of syrups.[a]

Property or functional use	Type of syrup
	28 DE — 42 DE — 63 DE — 95 DE — HFGS
Bodying agent	←———————————
Browning reaction	——————————→
Cohesiveness	←———————————
Fermentability	——————————→
Flavour enhancement	——————————→
Flavour transfer medium	——————————→
Foam stabiliser	←———————————
Freezing point depression	——————————→
Humectancy	←——————————→
Hygroscopicity	——————————→
Nutritive solids	←——————————→

(*Continued*)

Table 5.2 (*Continued*)

Property or functional use	Type of syrup
	28 DE 42 DE 63 DE 95 DE HFGS
Osmotic pressure	→→→→→ (increasing to HFGS)
Prevention of sucrose crystallisation	←←←←← (increasing to 28 DE)
Prevention of coarse ice crystal formation	←←← (favours low DE)
Sheen producer	↔ (mid-range)
Sweetness	→→→→ (increasing to HFGS)
Viscosity	←←←←← (increasing to 28 DE)

^aReproduced by kind permission of Corn Refiners Association.

When referring to Table 5.2, remember the following:

1. Low DE syrups contain lots of higher sugars, and lots of higher sugars mean high viscosity – 10,000 centipoises at 50°C.
2. High DE syrups, on the other hand, contain less higher sugars, and are therefore less viscous – 100 centipoises at 50°C.
3. High DE syrups contain lots of reducing sugars, and reducing sugars mean both sweetness and colour reactions.

In deciding which syrup to use, you might have to make a compromise.

5.3 Bodying agent

The term 'body' is usually used when talking about a liquid product. A typical example would be when a low DE syrup is used in a soft drink. The higher sugars in the low DE syrup increase the viscosity of the drink. This increase in viscosity improves the mouthfeel of the drink by giving the drink body. This is of particular importance when using high intensity sweeteners, which lack any bodying capability.

5.4 Browning reaction

The browning reaction involving glucose syrups can be one of two types. The first is the browning reaction which occurs when glucose syrups are heated with proteins – the well known Maillard Reaction, the reaction of reducing sugars with proteins. Therefore, high DE syrups, which contain lots of reducing sugars, particularly dextrose and fructose, will produce lots of colour when heated with a protein. As a general rule, fructose will produce more colour than dextrose, which in turn will produce more colour than maltose. An example of Maillard browning is the brown colour formed when making toffee – the reducing sugars of the syrup reacting with the proteins in the milk. When toffees are made with a maltose syrup which contains less dextrose, the resulting toffees do not have either the same dark colour or strong flavour, compared to toffees made with a conventional 42 DE acid produced syrup. This suggests that the dextrose is the more relevant sugar, rather than the 'total reducing sugars', which DE measures.

Sucrose is not a reducing sugar, and therefore does not form Maillard colour with proteins unless it is inverted, that is converted into dextrose and fructose.

The second colour reaction is caramelisation, that is the colour formed when sugars are heated or burnt.

5.5 Cohesiveness

Because low DE syrups have lots of higher sugars, they are viscous and sticky, and it is this combination of properties which is the reason for their cohesiveness. This makes these types of syrup ideal for use as a binder as in muesli bars, where the syrup holds the various ingredients together to form a solid bar.

5.6 Fermentability

Glucose syrups are composed of a mixture of sugars which micro-organisms such as yeast and bacteria can break down and use as an energy source. How far the glucose syrup can be broken down is known as 'the fermentability'. As a general rule, yeast and bacteria can easily use the simple sugars, such as dextrose, fructose and maltose – these sugars are therefore referred to as readily fermentables. Some yeasts and bacteria can slowly make use of maltotriose – which is referred to as a slowly fermentable sugar. Sugars higher than maltotriose which are not broken down are referred to as non-fermentable sugars.

Glucose syrups such as a 95 DE syrup, HFGS and maltose syrups which contain a high percentage of easily fermented sugars are ideal for applications involving fermentation. Low DE syrups such as a 28 DE with little dextrose or maltose would provide very little fermentability.

5.7 Flavour enhancement

The sugar spectra of a glucose syrup will often enhance or change the perception of a flavour. Generally, high DE syrups with their low viscosity have a greater flavour enhancement than low DE syrups.

Both high DE and fructose syrups will enhance fruit flavours, particularly citrus flavours. Maltose syrups will enhance cereal flavours, whilst the cooling effect of crystalline dextrose adds to the cooling effect of a mint flavour.

The reason why some sugars enhance some flavours but not others is possibly due to the molecular structure of the flavour and the sugar having a joint occupancy of several taste receptor sites involved in the perception of sweet, sour, salt and bitter sensations. This synergistic effect can be compared to having the correct key to open a lock – it either fits or does not.

Low DE syrups are not so effective at enhancing flavours possibly due to their higher sugars enveloping the receptor sites of the flavour, so that the sites are no longer freely available. This ability to envelop the receptor sites of a flavour can be used to advantage to mask unpleasant or bitter flavours.

Finally, because flavours are expensive, these subtle flavour changes can often result in a cost saving because less flavour is required.

5.8 Flavour transfer medium

As the taste receptors on our tongues are situated in depressions on the surface, viscous syrups such as low DE syrups are unable to enter these pores, because the large molecules of the higher sugars are too large to reach the receptor sites and therefore low DE syrups will either mask or reduce a flavour. On the other hand, less viscous syrups such as high DE syrups, which contain less higher sugars, are able to enter the pores, and are therefore more able to carry the flavours to the taste receptors making them ideal syrups for transferring flavours.

A low DE syrup such as 40 DE is used in chewing gum as a flavour carrier, but because 40 DE syrup is viscous, the flavour is only slowly released during the chewing process. This flavour release is helped by the saliva in the mouth, which mixes with the 40 DE syrup, thereby reducing its viscosity and allowing the flavour to reach the flavour receptors.

If a 60 DE syrup was used, because it has a lower viscosity, the flavour would have been released far too quickly, resulting in the chewing of a tasteless piece of gum.

5.9 Foam stabilisers

Foam is a dispersion of air bubbles in a liquid. If the liquid involved lacks viscosity, the walls of the air bubbles will be very thin, and delicate, and will easily collapse. To improve the stability of the air bubble and therefore the stability of the foam, it is necessary to increase the viscosity of the liquid. Low DE syrups are very viscous, so when added to a liquid, they will increase the viscosity of the liquid. This increase in viscosity will make the walls of the bubble thicker and stronger, resulting in a stable foam. It is for this reason why low DE syrups are ideal in foamed goods, such as marshmallows, ice cream and other aerated products.

5.10 Freezing point depression

When a substance is dissolved in a liquid such as water, the temperature at which the water now freezes is lowered. This lowering of the freezing point is referred to as 'The Freezing Point Depression', and is one of the colligative properties of a substance. Other colligative properties are boiling point elevation, osmotic pressure and vapour pressure, and they all depend upon the concentration of molecules in a substance, rather than on the weight of the molecule.

How far the freezing point is depressed will depend upon the concentration of molecules in the liquid. The lower their molecular weight, the lower will be the freezing point because there will be more molecules present. Therefore, dextrose and fructose, with a molecular weight of 180, will on a weight for weight basis lower the freezing point more than say a 42 DE syrup, which has a molecular weight of about 430 – because there will be more of the smaller dextrose or fructose molecules present for a given weight than of the larger 42 DE molecules (Table 5.3).

Table 5.3 Relationship between average molecular weight and freezing point depression.

Sweetener	Average molecular weight	Freezing point depression
Sucrose	342	1.0
28 DE Glucose Syrup	643	0.5
42 DE Glucose Syrup	429	0.8
Dextrose	180	1.9
Fructose	180	1.9

The ability to lower the freezing point of a substance is used to advantage when formulating frozen desserts, and in particular ice creams. In their most simple form, ice creams are an aerated frozen emulsion of milk solids, fats and sugars. If the sugar used is sucrose, then the ice cream will be hard. If the sucrose is replaced with either dextrose or fructose, then the ice cream becomes soft. The reason for this is that the molecular weight of sucrose is 342, but for dextrose and fructose, the molecular weight is lower at 180, and the lower the molecular weight, the lower the freezing point. By using a combination of different syrups, it is possible to tailor the freezing point to suit the final product.

Frequently, freezing point is not the only consideration – texture and sweetness might also be a consideration, in which case a compromise between syrup and which properties will have to be made.

One advantage of using glucose syrups for depressing freezing points in foods is that glucose syrups are nutritious and virtually tasteless. Whilst both glycerine and sorbitol are more effective freezing point depressants on a weight for weight basis, they tend to be laxative, have an after taste, have little nutritious value and are more expensive.

An alternative term to freezing point depression is the freezing point factor, which is defined as the molecular weight of sucrose divided by the molecular weight of the sweetener.

5.11 Humectancy

Humectants are substances which are used to prevent products from losing their moisture and drying out. Medium and high DE glucose syrups are good at retaining moisture. However, commercially, a 60 DE syrup is probably the best. Whilst higher DE syrups are better, because of the increased dextrose content, there is always the risk of crystallisation, resulting in a grainy or gritty end product, whereas a 60 DE will not crystallise. Glucose syrups can be used in bakery icing or fondant so as to prevent the product from drying out as well as preventing the sucrose from crystallising and forming a hard icing.

5.12 Hygroscopicity

A substance is said to be hygroscopic when it picks up moisture from the atmosphere, and in some cases, the moisture pick up can be so great that a liquid will results. The rate at which moisture is picked up will depend upon both the surrounding relative humidity and temperature, or more correctly, the dew point of the surrounding air.

Basically, all products made from starch hydrolysates – maltodextrins, glucose syrups, spray dried syrups, crystalline dextrose and fructose, all have the potential to be hygroscopic, but the lower the DE, the slower will be the rate of moisture pickup. It is for this reason that when handling dextrose, the dew point of the air must be below 21°C, otherwise it will pick up moisture and form hard lumps, and why maltodextrins are ideal for use in powdered products.

5.13 Nutritive solids

By definition, all glucose syrups are nutritious – 'nutritive solids derived from starch'. All glucose syrups have the same caloric value, regardless of their DE, because they are all composed of sugars, and all sugars on a dry basis, have a caloric value of 17 kJ/g (4 kcal/g) and this includes sucrose. With glucose syrups being soluble and easily assimilated, they are an ideal source of nutritive solids and, hence, energy.

Whilst all glucose syrups are nutritious, because they are made up of several different sugars, their potential energy is not always readily available. The reason for this is that different sugars are absorbed by the body at different rates. Low DE syrups are slowly absorbed, whilst high DE syrups are absorbed more quickly. This enables the application chemist to formulate different products which can supply energy over a long or short period of time.

It must also be appreciated that most green plants make their own carbohydrates from carbon dioxide and water by photosynthesis. Plants which do not have this capability, such as yeast, moulds and bacteria, have to rely on soluble carbohydrates from an external source, such as glucose syrups for their nutritive requirements, hence, the use of glucose syrups in fermentations.

5.14 Osmotic pressure

Osmotic pressure, like freezing point depression, boiling point elevation, and vapour pressure, is one of the colligative properties of a solution, and, like the other three, depends upon the concentration of molecules present.

Osmotic pressure occurs when a concentrated solution is separated from a dilute solution by a semipermeable membrane. The concentrated solution will pull water through a semipermeable membrane from the dilute solution, thereby becoming progressively more dilute. The osmotic pressure is the pressure required to stop this dilution effect and there are two ways of reporting osmotic pressure:

1. Atmospheres, Bar, Pascals, etc. These units tend to be used within industry.
2. Milliosmols or molar concentrations are generally used for medical or health-related topics. Osmolarity refers to the concentration in moles per litre of solution. Osmolality refers to moles per kilogram of solution.

Because glucose syrups are concentrated solutions of sugars, they will exert an osmotic pull on dilute solutions. Therefore, if a yeast or other type of micro-organism is in contact with a glucose syrup, because their cell wall is semipermeable, the syrup will pull out the liquid from inside the micro-organism, effectively dehydrate the organism, which will then die. This ability to kill off micro-organisms makes glucose syrups very effective preserving agents. However, there are other factors to be taken into consideration.

The osmotic pressure of a glucose syrup is related to its molecular weight – the lower the molecular weight, the greater will be the osmotic pressure; therefore, dextrose and fructose, with a molecular weight of 180, will exert a greater osmotic pressure than say a 42 DE syrup, with a molecular weight of about 429. Table 5.4 illustrates the relationship between the molecular weight of a sugar and the osmotic pressure of a 10% solution at 27°C.

Table 5.4 Approximate osmotic pressure and osmolality of different syrups with different molecular weights.

Sugar	Molecular weight	Osmotic pressure in atmospheres (a)	Osmolality in mOsm (b)
Sucrose	342	7.2	29.2
15 DE Maltodextrin	1200	2.0	8.3
28 DE Glucose Syrup	643	3.8	15.6
42 DE Glucose Syrup	429	5.7	23.3
55 DE Glucose Syrup	327	7.5	30.6
63 DE Glucose Syrup	286	8.6	35.0
Dextrose	180	13.7	55.6
Invert syrup	180	13.7	55.6
Maltose syrup – 50% MT	448	5.5	22.3
HFGS – 90% FX	182	13.5	54.9
HFGS – 55% FX	185	13.3	54.1
HFGS – 42% FX	190	12.9	52.6

Notes:
(a) Based on a 10% molar solution.
(b) Based on a 10% molal solution.

Other considerations to bear in mind when choosing a syrup are the concentrations of the syrups, the solubility of the sugars, the sweetness and the possibility of crystallisation.

Fructose is very soluble, but more expensive than dextrose, whilst dextrose is less soluble than fructose, and cheaper. Because HFGS is a mixture of both dextrose and fructose, it has both a low molecular weight and good solubility, which makes it a virtually self-sterilising syrup.

When a syrup is being used to preserve a fruit, for example when making glace cherries, the lower molecular weight sugars, which exert the higher osmotic pressure, will penetrate the fruit quicker than the higher molecular weight sugars.

One reason for using a glucose syrup, as opposed to sucrose, when preserving fruits, is that some fruits are acidic, with the possibility that the sucrose will be inverted, that is it will be converted into dextrose and fructose. A 10% sucrose solution will have an osmotic pressure of 7.2 atmospheres, but when the sucrose is inverted, then the solution will contain 5% dextrose and 5% fructose. Now the osmotic pressure of a 5% dextrose solution is 7.0 atmospheres, and 7.0 atmospheres for the fructose solution, to give a total osmotic pressure of 14.0 atmospheres – twice that of the 10% sucrose solution. Fortunately, glucose syrups do not invert – their sugar spectrum remains constant. By blending different glucose syrups, it is possible to have a syrup blend which will have both the same osmotic pressure and sweetness as a sucrose solution.

The osmotic pressure which glucose syrups can exert must also be taken into account when using glucose syrups in fermentations. If the syrup concentration is too high, there is the risk that the organism which you are trying to grow will be killed by the high osmotic pressure.

5.15 Prevention of sucrose crystallisation

Sucrose is used in many types of food products, but in some products, it can crystallise, giving the product a coarse, gritty texture. This gritty texture can be controlled by incorporating a low DE glucose syrup such as a 35 or 42 DE syrup in the formulation. Both of these syrups contain lots of higher sugars, which will increase the viscosity of the product. This increase in viscosity will slow down the movement of the sucrose crystals, thereby preventing the small sucrose crystals coming together and forming larger crystals. Basically, the glucose syrup acts as an insulation around the individual sucrose crystals. This coming together of crystals to form larger crystals is the basis of Ostwald ripening.

This property is used extensively when making fondant, where a smooth white texture of very fine sucrose crystals is required. This texture is achieved by using the glucose syrup to keep the sucrose crystals in suspension, which in turn gives the fondant its white opaque lustre. In the case of a hard-boiled sweet, the sucrose is in a supersaturated state, and the glucose syrup is used to prevent the sucrose from crystallising, and making the hard-boiled sweet opaque.

5.16 Prevention of coarse ice crystal formation

The formation of coarse ice crystals can be reduced by increasing the viscosity of the water by the addition of a low DE syrup. Low DE syrups are very viscous, due to their higher sugars. This increase in viscosity prevents the aggregation of the ice crystals. Additionally, the glucose syrup will lower the freezing point, which will encourage the formation of smaller ice crystals. This property is very useful when formulating frozen ice lollies and similar frozen desserts.

5.17 Sheen producer

The bright surface sheen or gloss of a jam or fruit tart increases its appeal to the buyer. It makes the product look both fresh and appetising. This sheen can be further enhanced by the use of a glucose syrup in the formulation of the filling. If the filling is made only from sucrose, there is the risk that it will crystallise, and give the filling an opaque, dull appearance, as well as a gritty texture, due to the sucrose crystals. Similarly, fillings made with some types of starch can have a dull opaque appearance. By using a glucose syrup such as a 42 or 63 DE to replace part of the sucrose, this problem can be overcome as well as improving the product's appearance.

5.18 Sweetness

Sweetness is a complex subject. Unfortunately, there is no chemical test or instrument for measuring sweetness, therefore sweetness is very subjective. It is dependent upon concentration, temperature, pH, acidity, viscosity and any possible synergy with other ingredients. It is for these reasons that all quoted sweetness values should only be considered as an approximate guide.

With the exception of fructose and HFGS syrups, no conventional glucose syrup has the same sweetness as sucrose. In simple terms, as the DE of the syrup increases so does the sweetness, but they are never as sweet as sucrose. Fructose is approximately 50% more sweet than sucrose, so by blending fructose or HFGS with conventional glucose syrups of different DEs, the sweetness of a conventional glucose syrup can be substantially increased which makes glucose syrup blends a very versatile ingredient. Taking the sweetness of sucrose as being 100, then the perceived sweetness of fructose will vary from 120 to 170, depending upon temperature, with the sweetness being greatest at lower temperatures. This sweetness temperature dependency makes fructose and fructose containing syrups useful sweeteners in frozen desserts.

Typical sweetness values of four glucose syrups are as follows:

Sucrose	100
42 DE Glucose Syrup	50
63 DE Glucose Syrup	70
Dextrose	80
HFGS (42% fructose)	95 (90–100)
Fructose	150 (120–170)

Not only do different sugars have different sweetness values but their sweetness is also perceived in a different way. See figure 5.1.

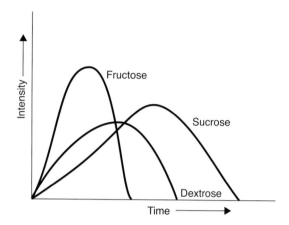

Figure 5.1 Profile of sweetness response.

1. Fructose has a very intense sweetness which only lasts for a short period of time.
2. Sucrose, on the other hand, has a lower perceived sweetness than fructose, but it lasts for a longer time.
3. Dextrose is less sweet than either fructose or sucrose. However, the perceived sweetness lasts for a longer length of time than for fructose, but less than that of sucrose.

If a fructose syrup is blended with say a 63 DE syrup, then the initial intense sweetness of the fructose syrup is reduced but the sweetness profile becomes similar to that of sucrose. The reason for this is that some of the fructose molecules become entrapped in the more viscous higher sugars of the 63 DE syrup. This entrapment of the fructose molecules has two effects. The first is that there is less fructose available to produce the initial intense sweetness and the second is that the fructose which has been entrapped in the higher sugars is slowly released to give a 'sweetness drag' effect, resulting in the blend having a sweetness response profile similar to that of sucrose.

Because acids affect sweetness perception, by changing the type and amount of acid present, the shape of the sweetness response curves can be changed.

This ability to change the perceived sweetness makes glucose syrups ideal for controlling and balancing the sweetness of a product, which in turn can affect how the flavour in a product is also perceived.

5.19 Viscosity

Viscosity is an important property influencing taste, flavour, mouthfeel and how a product can be processed. The viscosity of a glucose syrup is due to the higher sugars which it contains. Low DE syrups contain lots of higher sugars, and are therefore very viscous. High DE syrups contain very few higher sugars, and therefore have less viscosity, for example a 95 DE syrup has a viscosity similar to a sucrose solution at the same solids.

The viscosity of glucose syrups will vary not only with DE but also with solids and temperature. Table 5.5 illustrates the differences in viscosity due to changes both in DE and temperature – a slight increase in temperature will produce a considerable reduction in viscosity.

Table 5.5 Effect of temperature on the viscosity of a glucose syrup.

	Viscosity in centipoises			
Temperature (°C)	42 DE at 81% solids	55 DE at 82% solids	63 DE at 82% solids	95 DE at 75% solids
20	250,000	200,000	90,000	1000
30	60,000	50,000	25,000	300
40	18,000	15,000	7,000	200
50	6,000	5,000	2,500	100
60	2,500	2,000	1,000	50
70	1,100	900	450	30
80	550	450	250	20

The ability to reduce the viscosity of a syrup by slightly changing its temperature makes the processing, handling and storage of syrups more manageable. For example, when a low DE syrup is used as a binder in a muesli-type bar, the initial high viscosity would prevent the syrup from coating all the ingredients, which would result in a crumbling mess. However, if the syrup is first heated to reduce its viscosity, the syrup will cover all of the ingredients, so that when the mix is formed into a bar and cooled, the viscosity returns and the syrup now binds all the ingredients together to form a firm product.

As mentioned previously, the taste receptors on the tongue are not on the surface, but in a depression, therefore if the product is too viscous, it will not reach the taste receptors, and so the product will taste bland. By reducing the viscosity of the syrup, the flavours are able to reach the receptors.

By changing the viscosity of the syrup, both the texture and mouthfeel of the end product can be changed. For example, if the above-mentioned muesli bar is stored in cold conditions, when eaten, it will be hard, due to the viscosity of the syrup increasing. Additionally, the product will have a bland taste because the syrup is unable to carry the flavour to the taste receptors. If, on the other hand, the muesli bar is stored in warm conditions, the viscosity of the syrup is less, resulting in a softer bar and improved flavour release.

Viscosity plays an important part in the texture of aerated products where the viscosity of the syrup increases the strength of the wall of a bubble.

5.20 Summary of properties

Summarising the properties of glucose syrups, a low DE syrup will provide the following – body, cohesiveness, foam stability, prevention of sucrose crystallisation and viscosity. A high DE syrup will provide the following – browning, fermentability, enhanced flavour, lowering of freezing point, increase osmotic pressure and sweetness.

If you are looking for some properties from both a low DE syrup and a high DE syrup, then the choice of which syrup to use must be a compromise syrup, such as a 60 DE syrup.

5.21 Differences between glucose syrups and sucrose

There are several fundamental differences between glucose syrups and sucrose.

Glucose syrups are made up of reducing sugars, which will react with proteins to form a brown colour – the well known Maillard reaction. Sucrose, on the other hand, is not a reducing sugar, and it will not form a brown colour with proteins.

The sugar spectrum of a glucose syrup will not change during storage or during processing, unlike sucrose which can invert in acid conditions – that is change into a mixture of dextrose and fructose.

Sucrose is a disaccharide sugar, which means it is composed of two sugars chemically bound together, in this case dextrose and fructose. When sucrose is heated in the presence of an acid, this bond is broken to produce invert, which is a mixture of dextrose and fructose. The conversion of sucrose into dextrose and fructose during processing is known as 'process inversion' as shown in Fig. 5.2.

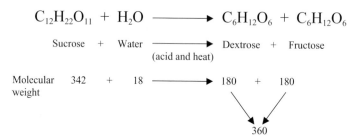

Figure 5.2 Inversion of sucrose.

Looking at the above reaction, it is interesting to note that whilst we started with 342 parts of sucrose, we actually end up with 360 parts of dextrose and fructose – a gain of just over 5%, due to the inclusion of a molecule of water. This type of gain is referred to as the 'hydrolysis gain'.

The enzyme invertase is also capable of splitting sucrose into dextrose and fructose. This use of an enzyme to change sucrose into invert is used by the confectionery industry to change the texture of sweets. An example is when making some types of chocolate-enrobed confectionery, which have a fondant like filling. This filling is made by enzymatically inverting sucrose in the fondant cream centre.

Whilst an ingredient declaration will state that sucrose has been used, because of process inversion, there might not be any sucrose found in the end product – just dextrose and fructose. Therefore, the changes due to the process inversion of sucrose into dextrose and fructose should be born in mind when considering possible applications for high fructose glucose syrups, which are essentially a mixture of dextrose and fructose. Unfortunately, the solubilities of sucrose, dextrose and fructose are very different. Fructose is very soluble, but dextrose is less soluble. This means that a sucrose syrup can be crystal clear and stable, but once converted into an invert syrup, that is a mixture of dextrose and fructose, the syrup will become cloudy during storage due to the dextrose crystallising out, leaving the fructose in solution (Table 5.6).

Table 5.6 Solubility of sugars at different temperatures.

Sugar	20°C	55°C
Sucrose	67%	73%
Dextrose (as anhydrous)	48%	73%
Fructose	79%	88%

Despite the conversion of sucrose into equal parts of dextrose and fructose during process inversion, the sweetness, on a weight for weight dry basis, remains unchanged – there is no increase in sweetness.

If, however, the invert is not fully converted, as in a medium invert, that is a mixture of 25% dextrose, 25% fructose and 50% sucrose, then due to the synergy between the sucrose and fructose, the sweetness of the mixture appears to be slightly greater than that for 100% sucrose.

Chapter 6
Syrup applications: an overview

6.1 Introduction

The object of this chapter is to give an indication of typical applications of different syrups, and as there are now many different types of syrups being produced, the following must only be considered as an approximate guide. As the products of the glucose industry are very versatile, it is impossible to mention every application, especially as new applications are continually being found. Further experimentation will no doubt be required in order to establish the correct ingredient balance for the desired characteristics of the end product.

6.2 42 DE Glucose Syrup

A typical analysis of a 42 DE acid converted syrup would be as follows:

> Dextrose = 19%
> Maltose = 14%
> Maltotriose = 11%
> Higher Sugars = 56%

42 DE glucose syrup was the original glucose syrup made by the glucose industry, using acid, and since it was mainly used by confectioners, it was usually referred to as Regular or Confectioners Glucose.

This syrup is a clear, colourless, viscous syrup containing dextrose, maltose and higher sugars. Because of the relatively low dextrose content together with a large percentage of higher sugars, 42 DE syrups are totally stable under normal conditions and will not crystallise. The syrup will give body and viscosity to a product, making it an essential ingredient for the confectioner to control sucrose crystallisation. Whilst this syrup is considered to be viscous, slight heating will greatly reduce the viscosity.

The main applications of a 42 DE syrup is in confectionery.

Caramels and toffees

In these types of products, a 42 DE syrup reduces graining, making the product more chewy, whilst the dextrose reacts with the proteins of the milk to produce both colour and flavour.

Chewing gum

Because the manufacture of chewing gum does not involve a heating stage to remove excess water, a very high solids 42 DE syrup is required. The role of the syrup is to aid the slow release of the flavour.

Fudge and fondants

In these types of products, the syrup controls the graining of the sucrose.

Gums and jellies

Whilst the gelling agent will have a major effect on the structure, 42 DE syrup is used in the recipe to prevent sucrose crystallising and the higher sugars of the syrup will add body to the product, making it more chewy.

High boilings

A 42 DE syrup is used in high boilings to control sucrose graining.

Muesli bars (cereals, vine fruits and nuts)

A 42 DE syrup can be used in these types of products as a binder, holding the bar together and making the bar more chewy.

Nougat

A 42 DE syrup is used in this type of confectionery to reduce graining and to give texture and structure.

Other applications include the following.

Adhesives

Because a 42 DE syrup does not crystallise, and has relatively low-humectant properties, it is ideal as a plasticiser for water-based adhesives such as those based on animal glue and dextrin.

Bakery glaze

A 42 DE syrup can be part of a bakery glaze formulation with sucrose, a gelling agent such as pectin, an acid regulator and citric acid.

Caramel colour

A 42 DE glucose syrup can be used as a carrier for caramel colouring, where it can also be used to adjust the colour or tinctorial power of the final caramel colour.

Glaze

A 42 DE syrup can be used as a glaze on crystallised fruits and dates.

Ice cream

A 42 DE syrup is used to control the crystallisation of the lactose in the milk, to give a smooth texture, to improve mouthfeel and body, and to supply nutritious solids.

Toys

A 42 DE syrup because of its viscosity and non-crystallising properties has been used as a centre filling in children's flexible bendy toys.

6.3 28 and 35 DE Glucose Syrup

These syrups are characterised by having a high viscosity, and are therefore generally sold at a lower solids than a 42 DE glucose syrup. The 35 DE syrup is usually produced by acid hydrolysis, whereas enzyme/enzyme hydrolysis is used to produce the 28 DE syrup.

A typical sugar analysis for these syrups would be as follows:

	28 DE Syrup	35 DE Syrup
Dextrose	2%	11%
Maltose	10%	12%
Maltotriose	16%	14%
Higher sugars	72%	63%

These viscous syrups have several useful properties, particularly where the product has to be spray dried. The higher sugars which are responsible for the inherent high viscosity of these syrups results in good droplet formation, producing stable droplets which dry as opposed to crystallising. Additionally, when being used in a co-spray dried product, such as a coffee whitener, where there is an oil, the high viscosity of these syrups acts as an excellent emulsion stabiliser.

On a dry basis, a typical coffee whitener will contain about 57% 28 DE glucose syrup, 34% vegetable oil with the remainder being stabilisers, caseinate, emulsifier, free-flow agent and colouring.

These syrups contain less dextrose than a 42 DE syrup, and therefore will pick up less moisture. They can therefore be used to replace a 42 DE syrup, in confectionery production in places with high humidity where moisture pick up is a problem.

Low DE syrups can be used in meat products, particularly in ham where they help in water retention and hence increase the weight of the ham. They also give a deeper red colour to the meat by reducing the nitrate to nitrite in the curing of the meat.

6.4 Glucose syrup solids

These are spray dried low DE glucose syrups, usually covering a range of 20 to 42 DE. The resulting products are stable free-flowing powders, which have exactly the same sugar spectrum as the original syrup, and therefore when dissolved, will produce a syrup which will have the same properties as the syrup from which they were made. Typical spray dried syrups are 20, 25, 35 and 42 DE.

Whilst these products are usually more expensive than a conventional syrup, they nevertheless have certain advantages for small syrup users. Being dry powders, they are easy to use and store.

They can be used in the same applications as conventional glucose syrups. Being dry powders, they can be used in formulations where moisture is at a premium, such as in bakery products, or where it is necessary to increase the solids of a formulation. They can also be used in dry mixes, with dextrose or maltodextrins.

Glucose syrup solids are frequently the major ingredients in foods for babies who suffer from lactose intolerance.

6.5 Maltose and high maltose syrups

These syrups can be used in either confectionery or brewing, and are characterised by having less than 10% dextrose and a maltose content which can vary from 28% to 70% or higher.

Maltose syrups with a maltose content of 28–49% are usually referred to as a 'maltose syrup', whilst syrups with a higher maltose content are referred to as either 'HM syrups' or 'VHM syrups'. The exact sugar spectrum of a maltose syrup will vary from one producer to another and the method used for its manufacture.

Typical sugar spectra for maltose syrups would be as follows (Table 6.1).

Table 6.1 Typical sugar spectra of different maltose syrups.

Sugar	Maltose syrup	Maltose syrup	High maltose syrup	Very high maltose syrup
Dextrose	5%	8%	5%	2%
Maltose	28%	41%	50%	70%
Maltotriose	25%	15%	20%	19%
Higher sugars	42%	36%	25%	9%

Confectionery

The advantages of using maltose syrups in confectionery are as follows:

1. They have a lower viscosity than a 42 DE syrup, which makes for easier processing.
    (i) The viscosity of the hot sugar mass is less.
    (ii) The reduced viscosity means that less air will be entrapped.

(iii) There is less tendency for a crust to form when the sugar mass is being worked.
(iv) There is reduced tailing during depositing, therefore the depositor can run at a higher speed, which means increased throughput.

2. Less browning, due to the reduced dextrose content. With less browning, there will also be less flavour development, because browning and flavour are both part of the Maillard reaction. Where browning is required in a product, this can be achieved by leaving about one-third of a mix behind in a cooker and adding the next mix to it. This means that one-third is effectively cooked twice.
3. Higher sucrose replacement levels can be achieved compared to a conventional 42 DE glucose syrup. This is because of the lower viscosity, compared to a 42 DE, and hence more maltose syrup can be incorporated into a blend before the increase in viscosity makes it unworkable.
4. Because maltose syrups have less dextrose compared to a 42 DE glucose syrup, the moisture pickup is less, which makes them ideal for confectionery production in hot humid conditions.

Brewing

The sugar spectrum of maltose syrups gives them a balanced fermentability, making them ideal for brewing where an easily controlled fermentation is required.

	50% Maltose Syrup	70% Maltose Syrup
Readily fermentables	55%	72%
Slowly fermentables	20%	19%
Non-fermentables	25%	9%

The 50% maltose syrups have a sugar profile similar to a malt wort, with the non-fermentables providing body and mouthfeel to the final beer.

The 70% maltose syrup being highly fermentable lends itself to beers with a high alcohol content.

Maltose syrups, like other glucose syrups, can be added to the wort kettle, where it can offer the following advantages to the brewer:

1. With fluctuations in the price of malt and sugar, syrups can be a cost effective viable alternative.
2. It enables the brewer to extend the brewhouse capacity by increasing throughput without additional capital expenditure. By adding the syrup directly into the kettle (copper), brewhouse capacity is increased, as well as enabling the production of high gravity brewing, together with greater flexibility for other brews.
3. Where the quality of the malt is variable, the use of a maltose syrup will dilute excessive colour and any non-starch components such as polyphenols and proteins, which can cause haze in the final beer.

A typical inclusion rate of 10–30% would be possible. Some brewers have even used a 50% inclusion.

6.6 63 DE Glucose Syrup

A typical 63 DE glucose syrup is an acid enzyme produced syrup and would have the following sugar spectrum:

Dextrose	34%
Maltose	33%
Maltotriose	10%
Higher sugars	23%

A 63 DE syrup with this type of sugar spectrum was first available in the United Kingdom from about the middle of the 1950's, and was made by adding a maltase enzyme from barley to a 42 DE syrup. With the introduction of HPLC sugar analysis, and the availability of several different syrups, it is now possible to match this sugar spectrum by the blending of different syrups.

The main characteristic of a 63 DE syrup is that it has a balanced sugar spectrum and a medium viscosity, making it ideal for use in the production of jams, sauces, brewing and soft drinks, also in soft sugar confectionery and baking. This syrup is an ideal compromise between a 42 DE with its high viscosity and a 95 DE with its high dextrose content and increased sweetness.

Because of the balanced sugar spectrum, this syrup will supply sweetness from the dextrose, but will not crystallise when stored at room temperature. Crystallisation will occur when the syrup is stored at low temperatures. For ease of use, the syrup should be stored at about 50°C.

Baking

A 63 DE syrup can be used as a humectant in fruit cakes, producing a moist cake, with an increased shelf life. Additionally, it will contribute to the sweetness, and improve the brown colour of the crust. A typical inclusion rate would be 5%, based on the dry solids. If the inclusion rate is too high, a brown colouration will develop at the base of the cake, which will only become evident when the cake is cut. This can be a disadvantage especially in a light-coloured sponge.

In a bakery marshmallow, there are two conflicting requirements for the sweetener – high viscosity to entrap the air and sweetness. For high viscosity, the choice of syrup should be a low DE syrup containing lots of higher sugars. For sweetness, the choice would be a high DE syrup, or possibly a fructose containing syrup. Unfortunately, a low DE syrup will lack sweetness, and the dextrose present in the high DE syrup could crystallise out in the product, and a fructose containing syrup would be too hygroscopic. Therefore, a 63 DE syrup is a good compromise syrup, offering an acceptable viscosity, with reasonable sweetness and minimal risk of graining, due to dextrose crystallisation.

A typical application would be 20% sucrose and 80% 63 DE syrup on a dry basis.

Brewing

The fermentability of a 63 DE syrup is similar to the fermentability of a brewer's wort, and it was because of this similarity that 63 DE syrups were the first glucose syrups to be used on a commercial scale by UK brewers.

Typically, 70% of the sugars in a 63 DE syrup can be fermented by the yeast, with the remaining 30% being unfermented during the brewing process. These non-fermentable sugars will, however, provide body and mouthfeel to the end product. Typically, up to 30% of the extract can be derived from a 63 DE glucose syrup.

Generally, the glucose syrup can be added directly to the wort kettle, where it can offer the same advantages to the brewer as the use of an HM glucose syrup.

Fermented drinks and liqueurs

The balanced combination of fermentable and non-fermentable sugars in a 63 DE glucose syrup makes it ideal for use in fermented fruit drinks such as the Russian drink KVAS. The fermentable sugars are easily converted into alcohol, whilst the higher sugars impart body and mouthfeel. This combination of properties gives an enhanced flavour profile to the drink, and highlights the subtle notes of the fruit. A total sucrose replacement is possible.

A 63 DE glucose syrup can also be used in the manufacture of liqueurs. Basically, a liqueur is a distilled spirit, which lacks both body and sweetness. The addition of a 63 DE syrup will provide just sufficient sweetness to mask the harshness of the spirit, thereby enhancing the flavour, whilst the higher sugars will increase the viscosity, adding both body and mouthfeel, so as to produce a rich smooth drink. Depending upon the alcohol content, an initial inclusion rate could be about 20%.

Jams

There are several advantages to be gained by using a 63 DE syrup to replace 50% of the sucrose in a jam.

A 63 DE syrup, because of its molecular weight, has a higher osmotic pressure than an equivalent weight of sucrose, and will therefore make a greater contribution to the micrbiological stability of the jam than if sucrose had been used.

The use of a 63 DE syrup in a jam will help to reduce sucrose crystallisation. During the boiling process when making jam, some of the sucrose will be inverted – this is known as process inversion. As a result of this process inversion, the sucrose is broken down into dextrose and fructose, and it is this additional dextrose which can crystallise out during storage to give a gritty texture to the jam. By replacing some of the sucrose with a 63 DE syrup, this crystallisation due to the process inversion of the sucrose can be reduced.

Whilst the sweetness of a 63 DE syrup is less than that of sucrose, the inclusion of the syrup will reduce the apparent sweetness, but this reduction in sweetness will allow the flavour of the fruit to come through.

Additionally, a 63 DE syrup will enhance the sheen of the jam, making it a more visually appealing product.

Sauces, ketchups and mayonnaise

In these types of products, a 63 DE can be used to replace some of the sucrose. The higher sugars present in the 63 DE syrup will improve the texture and gloss of the product, whilst the slight reduction in sweetness will allow the subtle flavours of the spices to come through.

Soft confectionery

When a 63 DE syrup is used in a pectin jelly, a softer chew will result. The reason for this is the reduced level of higher sugars in this type of syrup compared to a 42 DE syrup.

Soft drinks

Whilst a 63 DE is less sweet than sucrose, it can still play an important part in the formulation of a soft drink, especially when used with a high intensity sweetener.

With high intensity sweeteners, the higher sugars of a 63 DE syrup give both body and mouthfeel something which high intensity sweeteners are unable to provide. Additionally, the syrup will mask the often harsh flavour notes so often associated with high intensity sweeteners, especially saccharin. This masking effect is due to the higher sugars in a 63 DE syrup. The result is to give the overall sweetness a more round character, by increasing the sweetness drag, thereby making it more similar to that of sucrose.

Different sugars are absorbed at different rates by the body, and a 63 DE syrup being made up of dextrose, maltose and higher sugars will therefore be slowly absorbed by the body over a period of time, which makes a 63 DE syrup ideal for use in health drinks. In this sort of application, the 63 DE syrup will provide a slow but steady release of energy.

For a drink which has to be diluted with water, then a 50% replacement of sucrose is possible, especially if artificial sweeteners are being used. In a carbonated drink, a typical inclusion rate would be up to 30%.

Tobacco

A 63 DE syrup can be used as a humectant for tobacco.

6.7 95 DE Glucose Syrup

As this syrup contains mainly dextrose, the syrup is normally used for dextrose production, or as the starting point for making fructose syrups, such as HFGS. Alternatively, it might be spray dried to produce a free flowing powder, which is called a 'total sugar'. When the syrup is spray dried, the dry powder will still contain all the lesser sugars present in the original 95 DE syrup. In several applications, spray dried 95 DE syrup and dextrose are interchangeable. Below are typical analyses for a 95 DE syrup, and spray dried 95 DE syrup.

	95 DE Syrup	Spray-dried 95 DE
% Solids	75.0	99.0
% Moisture content	25.0	1.0
DE	95.0	95.0

Sugar composition	95 DE Syrup	Spray-dried 95 DE
% Dextrose	94.0	93.0
% Maltose	4.0	5.0
% Maltotriose	1.0	1.0
% Higher sugars	1.0	1.0

For spray dried 95 DE, the specification could include bulk density and particle size distribution details. Typically, there could be three grades based on different particle size. Some companies might also express the particle size distribution according to the mesh size.

Coarse	Particles greater than 355 microns
Fine	Particles less than 355 microns
Extra fine	Particles less than 160 microns

The particle size of the extra fine grade would match the particle size of icing sugar.

Whether you use a 95 DE syrup or the spray dried product will depend largely upon the application and the volumes involved. As a general rule, if the annual usage is less than 500 tonnes per annum, then the dry product would be the preferred option. Additionally, 95 DE syrups have to be stored at 60°C, to prevent dextrose crystallisation. If heated storage is not available, then the dry product should be used and made into a syrup before use.

The advantage of using liquid 95 DE syrup is when large volumes are involved, it avoids having to dissolve the spray dried product before it can be used in the process. A typical application would be industrial fermentations, where bulk tanker deliveries can reduce the expensive manhandling of bags, and allow the process to be automated. Bulk syrup deliveries also offer a cost reduction on a tonne for tonne delivery, as well as freeing up expensive warehouse space and the labour cost involved with its handling.

Antibiotics

Being totally fermentable, 95 DE syrup is the ideal carbohydrate source for the production of antibiotics, and because of the high purity of 95 DE syrups, there is a minimal risk of contamination.

Baking

Both spray dried 95 DE and 95 DE syrup can be used in baking. Being totally fermentable, they can be used in bread making and other fermented doughs, where they can replace sucrose as an energy source for the yeast. With the high percentage of reducing sugars present in 95 DE syrups, any residual sugars after the fermentation will react with the proteins of the flour to produce the brown crust, and the distinctive aroma so characteristic of freshly baked bread.

Spray dried 95 DE is ideal for use where there is a requirement for minimal moisture in a recipe.

The extra fine grades of spray dried 95 DE syrup can be used as a partial replacement of icing sugar.

Brewing

A 95 DE syrup can be used in brewing, but there are limitations. Because the syrup is highly fermentable, the fermentation could be difficult to control. As 95 DE syrups have a low molecular weight, if the concentration is too high, the osmotic pressure of the fermentation could be sufficiently high to inactivate the yeast. Additionally, there is the risk that unwanted by-products, such as isobutanol, could be formed, which would adversely affect the flavour of the beer.

Ten percent of the malt could be replaced with a 95 DE syrup, on an extract for extract basis.

Building industry

Spray dried D.95 can be used in additives for retarding the setting of cement and concrete. It is also used for a similar purpose in wall and ceiling plasters.

Caramel colour

D.95 glucose syrups can be used to make caramel colouring, whilst a 42 DE glucose syrup can be used to adjust the colour or tinctorial power of the final product.

Carrier for soluble flavours and colours

Spray dried 95 DE syrup is a free flowing, water soluble powder, and because it has been spray dried, it has the structure of a sponge, giving it a large surface area. This sponge like structure, together with its large surface area, makes it an ideal carrier for water soluble flavours and colours.

Cattle lick blocks

D.95 syrups, because of the high dextrose concentration, will crystallise if the syrup is not kept heated. However, by blending 70 parts of a D.95 syrup with 30 parts of a 42 DE syrup (on a dry basis), adding vitamins and minerals (available as a combine pre-mix) and adjusting the final solids to 85–90% solids, it is possible to produce a nutritious block which cattle will use as a source of vitamins and minerals.

Cider

A 95 DE syrup is used in cider production. Cider is made by fermenting apple juice. Whilst apple juice contains sugars, the amount will vary, depending upon the variety of apple, and when and where it was grown. Therefore, 95 DE syrup is added to the apple juice so as to be sure that the correct amount of fermentable sugars are present at the beginning of the fermentation, and hence the correct amount of alcohol being present in the final product.

Confectionery

Because of the water present in 95 DE syrup, it cannot be mixed with chocolate, but with spray dried 95 DE with a moisture content of less than 1%, there are no problems of incompatibility, allowing for a 10% sucrose replacement. Dextrose monohydrate, on the other hand, with a moisture content of about 8.5%, is totally unsuitable for mixing with chocolate.

Spray dried 95 DE can also be used in French paste work to replace part of the icing sugar.

Feedstock

D.95 syrup is used to produce HFGS by isomerisation.

Fruit preparations

Spray dried 95 DE can be used as a partial replacement of sucrose in the preparation of prepared fruits for use in yogurts, although care must be taken to avoid dextrose crystallisation, especially during storage of the fruit preparation.

Ice cream

Both 95 DE syrup and spray dried syrup can be used in soft scoop ice creams, where it will provide nutritious solids and can be used to replace glycerol to depress the freezing point. The advantage over glycerol is that 95 DE does not have an after taste.

Industrial fermentation

Unlike molasses, 95 DE syrup is highly refined, which makes it suitable for the production of citric acid, lactic acid, xanthan gum and similar fermentation products. Because of the good colour of the syrup, there is minimal post fermentation clean up of both the product and the effluent. A 95 DE syrup can also be used in the production of antibiotics.

It can also be used as a feedstock for growing meat-free mycoprotein products, which are used in vegetarian 'meat analogues'.

Preservative

Because of the low molecular weight of 95 DE products, they can be used to increase the osmotic pressure of solutions, thereby reducing microbiological spoilage. Dextrose can also be used in the bottling of fruits and vegetables, where the low molecular weight of the dextrose allows it to quickly and easily pass through the cell walls. This results in less shrinkage and greater penetration, thereby improving the appearance of the product.

Well drilling

Spray dried 95 DE can be used to retard the setting of grouting used in the drilling of oil wells. Because of the heat developed during drilling, the grouting can dry out, rather than set. This drying of the grout will result in the walls of the well not being correctly sealed.

6.8 Dextrose monohydrate

Dextrose monohydrate is the ultimate end product of starch hydrolysis, and is produced by crystallising dextrose from 95 DE syrup. The process of crystallisation allows only the dextrose to crystallise, leaving the other sugars still dissolved in the 'mother liquor'. The dextrose crystals are recovered and washed using a centrifuge and dried to give a very pure product. A typical analysis for dextrose monohydrate would be as follows:

Moisture	7.0–9.5%
Dextrose content	90.5–93.0%
Dextrose equivalent	99.5%
Fermentable extract	99.5%
Optical rotation	+52.5 to +53.0

Typical particle size analysis is as follows:

Greater than 300 microns	15%
Greater than 150 microns	50%
Greater than 74 microns	85%
Less than 74 microns	15%

Like sucrose, dextrose is also available in several grades based on different particle size distribution.

As a general guide, dextrose monohydrate can be used in all the same applications as 95 DE syrup. Dextrose monohydrate can also be used in the same applications as spray dried 95 DE, with the exception of its use in chocolate. The real advantages for using dextrose monohydrate is in its purity, although for some applications, even further purification is required. Dextrose is less sweet than sucrose, which can be advantageous where a reduced sweetness is required. In some products, the sweetness of sucrose can overpower subtle flavours, such as in savoury foods and sauces. One interesting property of dextrose is that it has a negative heat of solution, that is when it dissolves, there is a cooling effect, which makes it very compatible with mint flavours.

There are several differences between dextrose monohydrate and spray dried 95 DE syrup. Dextrose is pure dextrose; it has a moisture content of about 8.5%, and the crystals are in the α-form. Spray-dried 95 DE syrup contains all the sugars present in the original syrup (e.g. maltose, maltotriose, etc.); the moisture content is less than 1%, and the crystals are in the β-form. These differences can have a marked effect in some applications.

Antibiotics

Being totally fermentable, dextrose is the ideal carbohydrate source for antibiotic production. As dextrose is very pure, there is minimal post fermentation clean up of both the product and the effluent.

Bakery

Like 95 DE syrup and spray dried 95 DE syrup, dextrose can be used in similar baking applications. Being totally fermentable, it can be used in making bread and other fermentable doughs, where it can replace sucrose as an energy source for the yeast.

Dextrose can be used to sweeten whipped dairy cream which is to be used in bakery products. During the manufacture of a whipping cream, a stabiliser such as an alginate is added to stabilise the foam. This stabiliser addition can reduce the perceived sweetness of the whipped cream, therefore dextrose is added to compensate for this loss of sweetness without making the cream excessively sweet. The addition of dextrose will also increase the microbiological stability of the cream, by increasing the osmotic press within the water phase of the cream.

Browning agent

Dextrose can be used to encourage the browning of foods. One example is its use in the preparation of fried foods, particularly chips and French fries.

Carrier

Dextrose can act as a carrier for flavours. A typical example could be an instant tea, where it makes up 80% of the recipe, and carries 7% of tea soluble solids, with the remainder being citric acid, acidic regulator, artificial sweetener and a free flow agent.

Feedstock

Dextrose is the feedstock to produce the sugar trehalose and the polyol erythritol, both by fermentation.

Foundry work

Dextrose can be used to act as a binder in foundry sand, where it increases the green strength of the sand moulds.

Horticulture

Dextrose can be used as a feed for cut flowers.

Industrial fermentation

Being totally fermentable and very clean, dextrose is the ideal feedstock for most fermentations.

Intravenous drips

The sugar circulating in the human bloodstream is dextrose, and it is the energy source for the body to function. In critical situations where the body requires a steady supply of dextrose for its survival, a dextrose solution will be administered by an intravenous drip. The dextrose for this application must pass the pharmaceutical standards of purity, with the dextrose solution being both sterile and pyrogen-free.

Meat preparation

Dextrose can be used in brines or pickling solutions for curing hams. Historically, ham brines contain a mixture of nitrates and nitrites, but for the "curing" activity, nitrite is required, which is produced by bacterial action on the nitrate to produce nitrite. The bacteria need a substrate for growth, so sugar or dextrose is added. Dextrose is preferred when a less sweet cure is required, or if the particular bacterial strain used prefers dextrose as the substrate. Dextrose is also used in paté to increase the shelf life.

Oral rehydration

This application is for people suffering from dehydration due to diarrhoea. Since this condition frequently arises in third world countries, the dextrose is dry mixed with sodium chloride, trisodium citrate and potassium chloride, and then packaged into sachets. Each sachet is then dissolved in a litre of potable water, prior to being drunk by the dehydrated person.

Plastics

Dextrose is the raw material for the production of methyl glucoside, which in turn can be used to make rigid urethane foam.

Sorbitol and ascorbic acid production

Sorbitol is produced by the catalytic hydrogenation of a dextrose solution. For maximum plant efficiency and minimal production costs, it is important that the dextrose solution contains 100% dextrose, and nothing which will poison the catalyst. Therefore, to reach this high standard of purity, the crystalline dextrose will be washed with de-ionised water, and if it is to be supplied as a syrup, it will be dissolved in de-ionised water.

Ascorbic acid – vitamin C, is produced by the fermentation of sorbitol to sorbose, followed by chemical oxidation and acid treatment to produce ascorbic acid.

Sweetness / flavour modulator

Because dextrose is less sweet than sucrose, it can be used to balance the sweetness and flavour in a product.

Tabletting

Dextrose is used in several different types of tablets, from the confectionery thirst quenchers and energy tablets to pharmaceutical tablets. In these applications, the dextrose acts as a sweet bulking agent and a soluble carrier for flavours or the active pharmaceutical ingredient. Additionally, the cooling effect of a dextrose tablet will enhance the flavours, especially mint.

Tanning

Dextrose can be used to replace sodium sulphite as the reducing agent in the tanning of leather, when using chrome alum.

6.9 HFGS and fructose syrups

Table 6.2 Typical analyses of fructose products.

Sugar	42% HFGS	55% HFGS	80% Fructose Syrup	Crystalline Fructose
Fructose	42%	55%	80%	99.5%
Dextrose	52%	41%	18%	0.5%
Maltose	4%	2%	1%	–
Maltotriose	1%	1%	0.5%	–
Higher sugars	1%	1%	0.5%	–

HFGS syrups are clear, colourless syrups, having a very clean taste and a viscosity similar to liquid sugar. The sweetness of these syrups increases as the fructose content increases. With the increase in fructose content, it is possible to increase the solids of the syrup. The solids of a 42% HFGS are normally 71%, but the solids of 55% HFGS and an 80% fructose syrup are generally 77%. The reason why the solids can be increased is because as the fructose increases, the dextrose is reduced, and fructose is more soluble than dextrose. This reduction in dextrose reduces the risk of dextrose crystallisation.

Due to EU regulations within Europe, only HFGS containing 42% fructose is available, together with an 80% fructose syrup, derived from inulin. For the rest of the world, there is a 55% HFGS, however by blending it is possible to have any fructose level. The governing factor is price. As the fructose content increases, the syrups tend to become more expensive, and therefore become less competitive to sucrose.

Sweetness is probably the most important property of a fructose syrup, and whether it is derived from starch or inulin, its overall properties remain the same, and can compete with sucrose when it is used in most liquid applications. Unlike sucrose, the sugar profile of HFGS will not change during processing, and since the perception of flavour can be influenced by the sugar profile, HFGS will ensure a constant flavour.

Table 6.3 Approximate sweetness values of fructose containing syrups.

Sugar	Approximate sweetness
Sucrose	100
Fructose	120–170 (150)
HFGS 42	90–100 (95)
HFGS 55	100–110 (105)
80% Fructose syrup	110–150 (130)

Because fructose is approximately 50% more sweet than sucrose, less fructose can be used to obtain an equivalent sucrose sweetness. This means that in a diabetic food, there could be less 'sugars' for the same sweetness. Another advantage is that fructose, unlike dextrose, cannot be utilised directly by the body. It has to go to the liver, where it is converted to dextrose, which means that there is a delay, before it affects the 'blood sugar levels'.

Whilst sweetness is a common property of both HFGS syrups and sucrose, they are fundamentally different.

Sucrose is a disaccharide sugar, made up of one molecule of dextrose and one molecule of fructose, which are chemically joined together, and because of the way in which they are joined together, sucrose is a non-reducing sugar. This means that it will not react with proteins. Because sucrose is made up of one molecule of dextrose and fructose, the molecule weight of sucrose will be 360 (180 for dextrose and 180 for fructose).

HFGS syrups are composed of separate dextrose and fructose molecules, both of which are reducing sugars. This means that they will react with proteins to give the characteristic Maillard brown colour. In fact, fructose is chemically more reactive than dextrose, and so the reaction is both quicker and darker. As mentioned previously, the molecular weight of both fructose and dextrose is approximately half that of sucrose, at 180. This lower molecular weight means that both HFGS and fructose will have a greater effect on lowering the freezing point of a formulation, compared to sucrose. They will also exert a higher osmotic pressure, thereby acting as an excellent preservative. Finally, fructose is a very soluble sugar. Because of its high solubility, fructose crystallisation is not normally a problem, although dextrose can crystallise out from 42 HFGS with time during storage especially if the temperature falls below 30°C (Table 6.4).

Table 6.4 Solubility of three sugars at different temperatures.

Sugar	20°C	55°C
Sucrose	67%	73%
Dextrose	48%	73%
Fructose	79%	88%

When considering possible HFGS and fructose applications, think of the following:

1. Sucrose replacement, but remember the limitations outlined above. Any replacement should be on a solid for solid basis.
2. Invert replacement, but consider possible volumes involved. Often only a small amount of invert might be used, which is delivered in drums. HFGS is normally delivered in road tankers.
3. Any application where sucrose is used in a product which contains acid. Carry out an HPLC sugar analysis to confirm that process inversion of the sucrose has taken place in the product.
4. If a glucose syrup is present in the existing recipe, consider the possibility of a blend containing HFGS and the glucose syrup, providing that there is a sufficient volume.
5. If there is only minimal amount of water in a recipe, then the use of HFGS might not be appropriate. Whilst a blend of crystalline dextrose and crystalline fructose might be suitable, the cost of the blend could make it too expensive.
6. Where sucrose is being used for its crystallising properties, then it is unlikely that an HFGS would be suitable.
7. There could be a problem of dextrose crystallising in a product if the dextrose from the HFGS and from other ingredients exceeds about 40%.
8. Fructose is more expensive than other nutritional sweeteners, therefore price might limit its applications to value added products such as special dietary foods and drinks.

Typical applications could be:

Bakery products

Most bakery formulations contain minimal amounts of water, so the use of any syrup is not always practical. However, with bread and other yeast-raised goods, HFGS can be a good source of easily fermentable sugars for the yeast.

HFGS, because of its sugar spectrum, can also be used as an effective replacement for invert.

Breakfast cereals

When blended with a low DE syrup for use as a cereal coating, the coating will darken more quickly when the cereal is heated, because fructose is more reactive than either dextrose or sucrose. HFGS will also impart sweetness to the product.

Confectionery

Because HFGS is very hygroscopic, it has limited use in confectionery products. The two main applications are as a replacement for invert and in the formulation of chocolate enrobed soft confectionery. In this case, the outer coating of chocolate acts as a barrier, thereby preventing the HFGS from picking up moisture from the outside atmosphere.

Diabetic foods and drinks

As mentioned previously, because fructose is approximately 50% more sweet than sucrose, less fructose is required to obtain the same sweetness in a diabetic product. Because fructose has to be converted into dextrose by the liver before it can be utilised by the body, there is a time lag between when the fructose is consumed and when it becomes available to the body. This delay helps to reduce the rate and peak of sugar concentration in the blood. However, it must not be considered as a total sucrose replacement, because ultimately it will enter the bloodstream, and affect the body's blood sugar levels. It is therefore not advisable to use fructose to control diabetes.

Fruit preparations

These are fruit containing preparations, such as the fruit added to yogurts, and the fruit sauces found in frozen desserts, etc. The use of HFGS and fructose in these types of preparations offers four advantages. (1) The sweetness of fructose appears to increase at lower temperatures. (2) Because of their low molecular weight, these products will reduce the freezing point of a mix, thereby allowing it to remain liquid or soft in the end product. (3) The low molecular weight will also increase the osmotic pressure in the product, which will improve the prevention of microbial growth. (4) Fructose containing products will have enhanced fruit flavours.

Ice creams and frozen desserts

In these types of applications, HFGS is often used with a low DE syrup or maltodextrin, so that the HFGS can provide sweetness, and the higher sugars of the low DE syrup or maltodextrin counteract the freezing point depression effect of the HFGS.

Jams, preserves, fruit fillings and topping syrups

HFGS when blended with other glucose syrups can be used to replace part or all the sucrose in these types of products.

Sauces, ketchups, mayonnaise and pickles

These types of products usually contain acid, typically acetic acid (vinegar), which will invert any sucrose into dextrose and fructose. Therefore, HFGS can be used to replace all the sucrose in these types of products, on a solids-for-solids basis.

Soft drinks

This is the major use of fructose syrups, for 42 and 55 HFGS, where a 100% sucrose replacement is achievable, with the additional advantage that fructose containing syrups will enhance the fruit flavours, particularly citrus. The reason for this is possibly due to the difference in the sweetness profile of sucrose, compared to that of fructose.

Fructose is ideal for use in reducing the calories of soft drinks. Because fructose is approximately 50% more sweet than sucrose, less of it is required to match an equivalent sucrose sweetness. This difference in sweetness levels means that a drink can be formulated with the same level of sucrose sweetness, but will contain 50% less calories.

6.10 Maltodextrins

Maltodextrins are characterised by having a DE of less than 20, and are normally sold to a DE specification, typically 1, 5, 10, 15, and 18. See Table 6.5. The maltodextrins with a DE of less than 10 are sometimes referred to as 'hydrolysed cereal solids'.

Maltodextrins are usually available as a spray dried, free flowing white powder, which have a bland taste, with very little sweetness. Sometimes, the free flowing powder is

Table 6.5 Typical analyses for maltodextrins.

Percentage of sugar distribution	Dextrose equivalent (DE)				
	1.0	5.0	10.0	15.0	18.0
Monosaccharide – dextrose (DP1)	0.3	0.9	0.6	1.3	1.5
Disaccharide – maltose (DP2)	0.1	0.9	2.8	4.1	6.0
Trisaccharide – maltotriose (DP3)	0.2	1.0	4.4	6.0	8.3
Tetrasaccharide (DP4)		1.1	3.5	4.6	5.8
Pentasaccharide (DP5)		1.3	3.8	5.2	6.9
Hexasaccharide (DP6)		1.4	5.7	7.6	10.2
Heptasaccharide (DP7)		1.5	5.4	6.3	7.0
Octasaccharide (DP8)		1.4	4.0	4.4	4.1
Nonasaccharide (DP9)		1.4	3.2	3.5	3.1
Decasaccharide (DP10)		1.3	2.7	3.0	2.5
Higher saccarides (DP11+)	99.4	87.8	64.0	54.0	44.6
Theoretical molecular weight	**18,000**	**3600**	**1800**	**1200**	**1000**

Source: Reproduced by kind permission of Tate & Lyle Americas Inc.

agglomerated to form a porous granular product, which produces less dust when handled and is more easily dissolved in water. And very rarely, a high DE maltodextrin is sold as a very viscous syrup, with a low solids, which requires a preservative, such as sorbic acid, to be added.

Maltodextrins are produced in a process similar to glucose syrups, but there can be subtle variations in functional properties, depending upon the type of starch used in their production – maize, waxy maize, potato, rice, tapioca and the method of starch hydrolysis which has been used – acid, acid enzyme or enzyme enzyme. Solutions of maltodextrins made from waxy starches tend to be more stable than maltodextrins made from other starches because they do not retrograde. The reason for this is that waxy starches contain predominately amylopectin, which does not suffer from retrogradation. Solutions of maltodextrins made from starches which contain the amylose fraction tend to form cloudy solutions, because the amylose comes out of solution, that is retrogradation. Interestingly, maltodextrins made from tapioca tend to be more bland than those made from other starches.

Because maltodextrins are made from very low DE syrups, they have the same functional properties as a low DE glucose syrup, namely:

- Very low dextrose content
- Low browning
- Bland taste
- Lack of sweetness
- Very high viscosity
- Good film forming properties
- Low hygroscopicity
- Good moisture control
- A large molecular weight, generally in excess of 1000
- Made up of complex, yet very soluble and nutritious carbohydrates
- Being spray dried, they have a relatively large surface area

These properties make maltodextrins suitable for many different applications.

Brewing

Because maltodextrins contain very few fermentable sugars, they can be used in the brewing of low alcohol beers. They will produce a minimal amount of alcohol, but due to the high viscosity of maltodextrins, they will contribute body and mouthfeel to the beer, as well as improving the flavour.

Bulking agent

Being bland and soluble maltodextrins are ideal bulking agents in dry formulations. A typical application is in powdered domestic low-calorie sweeteners, where a high intensity artificial sweetener is blended with the maltodextrin. The maltodextrin acts as a bulking agent and effectively dilutes the concentrated sweetness from the high intensity sweetener to a more acceptable level.

Fat and calorie reduction

Because of the high viscosity of maltodextrins, they have a texture similar to a fat, which suggests that they could be used as a possible fat replacer, especially as carbohydrates are generally cheaper than fats. The problem in using maltodextrins as a total fat replacer is that they are bland and do not have the same flavour and sensory characteristics of a fat, because maltodextrins are water soluble, whilst fats are oil soluble, and some flavours are water soluble, whilst others are oil soluble. Therefore, only a partial fat replacement is possible, generally about 30–40%. At this level of inclusion, it is possible to produce a stable emulsion of fat and maltodextrin, which gives an acceptable balance of both water and oil soluble flavours together with an acceptable texture, with the higher sugars of the maltodextrin acting as an emulsifier and stabiliser.

Maltodextrins are carbohydrates, and therefore have a caloric value of 4.0 kcal/gram, compared to fats which have a caloric value of 9.0 kcal/gram. By replacing some of the fat with a maltodextrin, it should be possible to reduce the number of calories in a product, and at the same time reduce the fat content, which could result in a more healthy product.

Carriers

As maltodextrins are bland, soluble free flowing powders, with a relatively large surface area, they are ideal carriers for flavours. The large surface area means that they are able to carry a lot of flavour, and because maltodextrins are very soluble, the flavour is quickly released when it is added to a liquid. Additionally, because of their bland nature, maltodextrins would not contribute to any background flavour.

Flavours can be added either to the powdered maltodextrin or to the maltodextrin syrup prior to spray drying. If added to the syrup, there is the risk that some of the flavour could be lost during the drying process, but because of the high viscosity of a maltodextrin syrup, this will aid the encapsulation of the flavour during the drying process.

Cereal coatings

Being viscous and bland, the film forming properties of maltodextrins can be successfully used in cereal coatings, where the viscosity can be used to 'stick' other food ingredients to the base cereal. Additionally, since they have a low DE, they will produce less browning.

Confectionery

Maltodextrins have many potential applications in confectionery. The high viscosity of maltodextrins can be used to advantage to change the texture of confectionery during its manufacture. The high viscosity will act as a foam stabiliser, partially replacing gelatine or egg albumen. It can control crystallisation, and increase the chewiness of soft confectionery by increasing the viscosity of the confectionery in the mouth. They can be used in the making of hard gums. Maltodextrins because of their good film forming properties can be used as glazes in the production of panned goods. They can also be used as a binder in granola bars, and in tabletting. Maltodextrins can be used to make an instant fondant by blending it with icing sugar, avoiding the use of conventional fondant making equipment.

Dairy fillings

The combination of viscosity and blandness makes it ideal for use in 'cream fillings', acting as a carrier, and bulking agent for other dairy ingredients (reconstituted milk, milk fat, lactose, etc.).

Dry mixes

Maltodextrins being free-flowing bland powders, with low hygroscopicity and good solubility, can be easily incorporated in many instant food formulations where they can add viscosity and act as a carrier for flavours. Typical applications would be instant dry soups, sauces, cake mixes, instant desserts. Where a clear soup is required, then a low DE maltodextrin should be used. In some formulations, the maltodextrin is co-spray dried with other ingredients.

Because maltodextrins can add viscosity to a drink, low DE maltodextrins can be used in a powdered drink formulation to stabilise the foam of the final drink.

Encapsulation

Because of the high viscosity of maltodextrins, they are good film formers, which makes them very suitable for encapsulating liquids and flavours during spray drying. The film which they form dries, as opposed to crystallising, and hence the films are more stable.

Film forming

As mentioned previously, maltodextrins, because of their high viscosity, are excellent film formers. Since maltodextrins are composed mainly of higher sugars, the resulting film dries, as opposed to crystallising. As a coating, a maltodextrin film is the ideal barrier to moisture movement or oil migration.

Frozen foods

Due to their large molecular weight and high viscosity, maltodextrins can be used to control the formation of ice crystals, thereby improving the mouthfeel of frozen desserts.

Geriatric, invalid and baby foods

Because maltodextrins are composed of large soluble carbohydrate molecules, they provide body and improve the mouthfeel of geriatric, invalid and baby foods, making them an ideal source of nutritious solids. They can be easily consumed, and are then slowly broken down by the gastric juices, avoiding any overload to a delicate digestive system, thereby providing a slow but steady release of energy over a long period of time. Unlike dextrose, which has a molecular weight of 180, maltodextrins have a large molecular weight of over 1000, which means they have a minimal effect on the osmotic pressure, and therefore are less likely to cause diarrhoea. Typical inclusion rates in a geriatric food would be 20–40%, in a baby food about 2% and in a slimming food about 15%. Sometimes, maltodextrins will be incorporated in foods for babies who suffer from lactose intolerance.

Ice cream

Maltodextrins are very useful for the small ice cream manufacturer, who does not have a sufficiently large requirement to warrant a bulk tank installation. Being supplied in bags, they are easy to handle. Maltodextrins can be used in conjunction with dextrose to control the texture of an ice cream. Since ice creams are an aerated emulsion, the high viscosity of a maltodextrin will act as a stabiliser.

Being a powder, maltodextrins can be used in powdered ice cream mixes which just require the addition of water. In this type of product, viscosity of the maltodextrin helps to improve the emulsion stability.

Margarine

The minimum fat content of margarine is 80%, and additionally, it must not contain more than 16% water. This leaves a deficit of 4%, some of which could be made up with a maltodextrin, which would have the advantage of supplying nutritious solids, and because of its viscosity, will help to act as an emulsion stabiliser.

Meat analogues

Maltodextrins can be used as binders in meat analogues, and as well as holding the product together, maltodextrins can also be used to replace part of the fat.

Since maltodextrins are good film formers, they can be used in conjunction with dextrose as a surface coating for meats, so as to improve the browning during cooking.

Personal hygiene

Because maltodextrins are non-toxic, they are ideal for increasing the viscosity of liquid hand soaps, and since they have virtually no smell, they do not detract from the fragrance of the product.

Pharmaceuticals

As in confectionery, maltodextrins can be used as a binder in tabletting.

Snacks

Being bland, a good film former, and viscous, maltodextrins are ideal as a coating for snacks. Their adhesive film forming properties help to prevent oxidation of the oils in nut products, and acts as an adhesive layer and carrier for other ingredients, which can contribute flavour and texture to the coating.

Sports drinks

Being large soluble carbohydrates, maltodextrins can be used in sports drinks, where they can supply a slow but steady release of energy. Because of the large molecular weight of maltodextrins, it is possible to increase the carbohydrate availability, with only a very slight increase in the osmolality of the drink.

Spreads

The high viscosity of maltodextrins, together with their blandness, means that they can be used in spreads. The high viscosity acts as an emulsion stabiliser, whilst the blandness does not affect the flavour of the spread.

Chapter 7
Trehalose

7.1 Introduction

This is a relatively new sugar to be produced by the glucose industry, being commercially first produced in Japan in 1994.

Trehalose, sometimes called mycose, mushroom sugar or ergot sugar, is a naturally occurring non-reducing disaccharide despite being composed of two dextrose molecules. It is found mainly in fungi, particularly mushrooms. It is a white crystalline powder, readily soluble in water and is very stable over extremes of both temperature and pH.

7.2 Production

Trehalose is produced by a complex enzyme process from starch. Starch is basically a polymer of dextrose, and is composed of two fractions – amylose and amylopectin. Amylose is composed of a straight chain of dextrose molecules, joined at the 1–4 linkage. Amylopectin is similar, except that there are additional branches at the 1–6 sites. Trehalose is composed of two dextrose molecules, except that they are joined together at their respective 1–1 linkage (reducing site) so that trehalose becomes a non-reducing sugar. Therefore, the object of the enzyme process is to produce amylose which can then be further converted to trehalose.

Figure 7.1 is a possible process.

1. The first stages of the process up to the second enzyme tank are similar to a conventional PEE process for making a dextrose or maltose syrup.
2. At the second enzyme tank, because the enzyme conversion is dependent upon all three enzymes working together, all three enzymes are added concurrently.
   (i) The debranching enzyme isoamylase is added to the liquefied starch to hydrolyse the 1–6 linkages to produce linear chains of amylose.
   (ii) The enzyme MTSae – malto-oligosyl trehalose synthase – converts the α 1–4 links of the amylose into α 1–1 links.
   (iii) The enzyme MTHase – malto-oligosl trehalose trehalohydorolase – then liberates the trehalose.

As trehalose is a relatively new commercial sugar, the process for its production like the early days of HFGS is still evolving and it is interesting to speculate on possible future process changes.

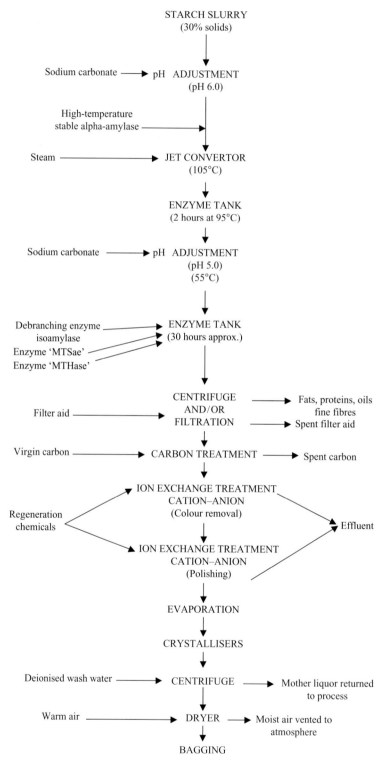

Figure 7.1 Schematic production of trehalose.

In some processes, so as to increase the trehalose yield, amyloglucodase is added to hydrolyse any remaining 1–6 linkages, thereby making more 1–4 links available for conversion. However, the amount of amyloglucosidase to add is critical – too little and it has no effect on the yield, too much and the yield is reduced because of the preferential formation of dextrose at the expense of the amylose units.

Since trehalose is made up of two dextrose units similar to maltose, and as the production of maltose is well established, there might be the possibility in the future that trehalose could be made by the enzyme conversion of maltose.

Also in the area of enzymes, there is the possibility of using immobilised enzymes as used in the production of HFGS to reduce production costs, and with the advances in DNA techniques, the possibility of using a recombinant enzyme should not be dismissed.

The production of trehalose involves amylose. Perhaps, the industry might become involved in a plant breeding programme to create starch plants which produce predominantly amylose rather like the industry did for the production of waxy starch. Previous attempts to produced high yielding amylose plants have resulted in plants with reduced yields per acre, making the starch more expensive.

Trehalose is still a new sweetener, but if it becomes a successful mainstream product such as HFGS, then perhaps new dedicated refineries will be built which will only use commercially priced amylose starch, instead of 'piggy backing' on a conventional syrup refinery using ordinary starch.

7.3   Properties

Being a disaccharide trehalose is frequently compared with other disaccharides particularly maltose and sucrose. Like maltose, trehalose is composed of two dextrose molecules (see figure 7.2), but whereas maltose is a reducing sugar (see figure 7.3), trehalose is a non-reducing sugar. Therefore, in that respect, it is more comparable to sucrose, which is also made up of two reducing sugars – namely dextrose and fructose (see figure 7.4).

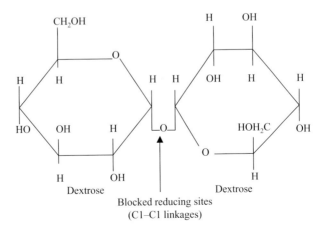

Figure 7.2   Trehalose molecule composed of two dextrose molecules, showing the blocked reducing site.

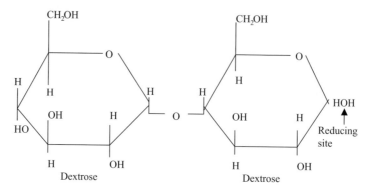

Figure 7.3 Maltose molecule composed of two dextrose molecules.

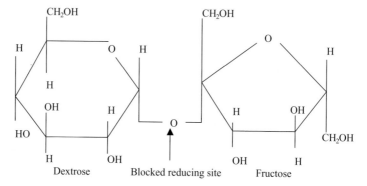

Figure 7.4 Sucrose molecule composed of one dextrose molecule and one fructose molecule.

Table 7.1 compares the differences between trehalose, maltose and sucrose, and hence the unique application potential of trehalose.

Table 7.1 Comparison of properties.

	Trehalose	Maltose	Sucrose
Molecular weight	342	342	342
Melting point (°C)	203	102	190
T_g value (°C)	79	43	52
Sweetness	50	50	100
Reducing sugar	No	Yes	No
pH stability	Very good	Good	Poor
Solubility (at 20°C, in g/100 ml)	68.9	44.0	66.7

Trehalose being composed of two dextrose molecules is metabolised in the same way as maltose, and therefore has the same caloric value as other carbohydrates namely 4 kcal/g.

Trehalose has the interesting ability to bind water in products and to help organisms to resist drought. Preliminary work would also suggest that trehalose has an 'affinity' for certain protein structures, possibly using a similar mechanism to that of organisms which can survive droughts. Another interesting property is the ability to slow down the oxidation of fatty acids.

Preliminary work suggests that trehalose can enhance flavours. This flavour enhancement could possibly be due to its reduced sweetness compared to sucrose, and therefore allows the flavour to come through, instead of the flavour being over-powered by the sweetness of the sucrose.

7.4 Applications

Because trehalose has only become available in commercial quantities since the mid-1990s, its full application potential is still being developed, and long-term applications will be very price dependent. Since trehalose is a non-reducing disaccharide like sucrose but with half its sweetness, the obvious initial application areas would be for replacing sucrose in products where a reduced sweetness would be beneficial. By replacing sucrose on a weight for weight basis, there would not be a yield loss. Because trehalose has other unique properties, there are several speculative potential areas in the food and pharmaceutical industries as well as possible medical applications.

7.4.1 Confectionery

With its high glass transition temperature (T_g), the addition of trehalose to a hard-boiled sweet will alter the thermoplastic properties of the sucrose/glucose mix, which could be useful where the confectionery mass has to be manipulated or where the cooked mass is difficult to process.

One of the problems encountered with confectionery products is moisture pick up, particularly in humid or tropical conditions. The main reason for this is that dextrose is more hygroscopic than either maltose or sucrose. Most sugar confectionery uses an acid-converted 42 DE syrup which contains about 19% dextrose. By using either a lower DE syrup such as a 35 DE or a maltose syrup with a low dextrose content, the problem of moisture pick up can be reduced but syrups lower than 35 DE are too viscous to process due to their higher sugars content. Increasing the sucrose content can result, if acid is present, in an increase in dextrose due to any process inversion.

Work with trehalose suggests that using a combination of a maltose syrup which has a low dextrose content, together with a partial or total replacement of sucrose by trehalose, could reduce the moisture pick up even further. Since trehalose is stable under low-pH conditions, it will not, unlike sucrose, be 'inverted' to produce additional dextrose. The result is a sweet less prone to moisture pick up, will have reduced graining, be less susceptible to cold flow and will have an improved product shelf life.

The fact that trehalose is less sweet than sucrose will help to enhance flavours, and might even allow for a cost reduction by reducing the flavour addition.

7.4.2 Dairy

Having the same molecular weight as sucrose, trehalose will have minimal effect on the freezing point depression of a frozen dessert or ice cream. Being less sweet than sucrose, the sweetness of the product will be reduced, but as mentioned previously, this reduction in sweetness could enhance the flavour of the product.

Conventional glucose syrups cannot be used in condensed milk because they contain reducing sugars which will react with the milk protein to produce a brown-coloured condensed milk. However, trehalose being a non-reducing sugar like sucrose could be used in the manufacture of sweetened condensed milk, although the sweetness of the end product would be slightly reduced depending upon the level of sucrose replacement. The typical sucrose content in a condensed milk is about 40%. There are two reasons for using sucrose. The first is to increase the solids and the second is that sucrose will increase the osmotic pressure and will therefore improve the keeping qualities of the product. Since the solubility of trehalose is similar to that of sucrose and its molecular weight is the same, trehalose meets both of these requirements. Additionally, trehalose could help to protect the milk proteins from heat damage.

The typical solids of evaporated (or unsweetened) milk are about 32%, compared to 73% for condensed milk.

7.4.3 Jams and fruit fillings

Trehalose is stable in low-pH conditions, which makes it suitable for use in jams and fruit fillings where it would reduced the risk of dextrose crystallisation resulting from process inversion of the sucrose. With its reduced sweetness, trehalose would enhance the fruit flavours.

7.4.4 Cosmetic and personal hygiene products

It has been suggested that trehalose can be used in skin cosmetics because it is claimed it could protect the skins fibroblasts from dehydration, by replacing water lost from the skin. Because trehalose can reduce the oxidation of fatty acids, it could be used in deodorants to suppress body odours.

7.4.5 Pharmaceuticals

As trehalose can protect proteins from heat damage, it could be used to stabilise proteins in the processing of antibiotics. Being non-hygroscopic, trehalose could be used in tabletting as an excipient.

7.4.6 Medical applications

Speculatively, because trehalose can stabilise proteins, there is the possibility that it could be used in the treatment of degenerative neurological diseases such as Huntington's disease. It might also be useful in suppressing the development of osteoporosis by reducing the loss of the protein matrix of the bone.

Chapter 8
Sugar alcohols: an overview

8.1 Introduction

Sugar alcohols also known as 'polyols' are polyhydric alcohols, and therefore are not a glucose syrup. Whilst sorbitol, maltitol, mannitol and erythritol are all naturally occurring polyols, commercially they are now produced using syrups from the glucose industry and are of interest for several reasons – their sweetness, their inability to be readily absorbed by the body, and for their technical and chemical properties.

Possibly, the most commonly known polyol is sorbitol, which occurs in the berries of the rowan or mountain ash tree, from which it was originally extracted in 1872. Mannitol is present in seaweed whilst maltitol is present in certain vegetables and erythritol is present in melons.

Other polyols, such as isomalt, lactitol and xylitol, will not be discussed because they are made from sucrose, lactose and xylose, respectively, and therefore do not use a glucose syrup in their production.

With the exception of erythritol, most polyols are produced by the hydrogenation of glucose syrups. Hydrogenation is the name given to the process of adding hydrogen gas, in the presence of a catalyst, to a substance at elevated temperatures and pressures. This technology is well established and is typically used to modify edible oils to make them suitable for use in margarine and shortening, etc.

During hydrogenation, a molecule of hydrogen is added to the aldehyde group of the sugar, which is reduced to a hydroxyl or alcohol group – hence the name 'sugar alcohol'.

When a glucose syrup is hydrogenated, each particular sugar in the glucose syrup will produce a different polyol, and each polyol will have different properties. When a glucose syrup which contains both dextrose and maltose is hydrogenated, the result will be a polyol containing both sorbitol and maltitol.

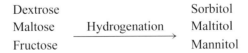

Figure 8.1 below illustrates the molecular structural change when dextrose is hydrogenated to make sorbitol.

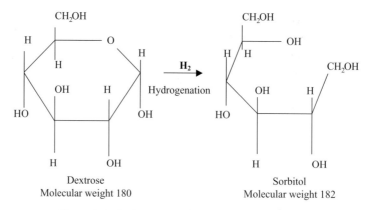

Figure 8.1 Hydrogenation of dextrose to produce sorbitol.

8.2 Production

To make a polyol, for example sorbitol from dextrose, the hydrogenation process would typically involve the following process stages.

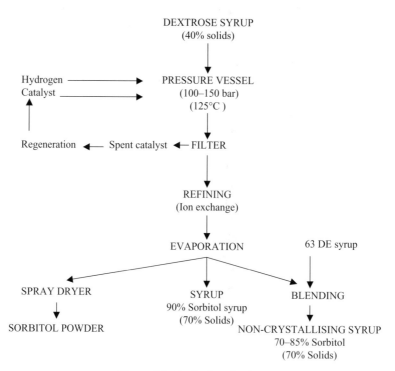

Figure 8.2 Production of sorbitol.

1. So as to ensure that only sorbitol is produced, the dextrose content of the syrup should be 99.9% dextrose. If there are any other sugars present, then the resulting sorbitol will be contaminated with other polyols. Ideally, the syrup should be made by dissolving crystalline dextrose in de-ionised water. The syrup is diluted to about 40–50% solids. So as not to poison the catalyst, it is very important that the dextrose syrup is of the highest

purity, having been both carbon and ion exchanged treated and has a conductivity of less than 10 micro-Siemens.
2. Most catalysts used in hydrogenation are expensive and therefore they are recovered and regenerated when spent. The catalysts are generally in the form of very fine porous metallic granules. The porous structure increases the surface area of the catalyst and improves the efficiency of the catalyst. A typical catalyst would be Raney nickel.
3. Like the syrup, the hydrogen must also be of the highest quality, so that it also does not poison the catalyst, therefore the hydrogen is produced on site by the electrolysis of de-ionised water. In theory, about 125 litres of hydrogen at atmospheric pressure is required to produce 1 kilogram of sorbitol.
4. The process of hydrogenation can be either a batch or a continuous process. In a batch process, hydrogenation is carried out in a stirred pressure vessel, with hydrogen gas being continually introduced. A typical reaction time would be about 2 hours. Since hydrogenation is an exothermic reaction, that is it produces heat, once the operating temperature has been reached, minimal additional heat is required. Where the hydrogenation is a continuous process, the catalyst is being continually recovered and replaced with fresh.
5. When the reaction is complete, the sorbitol syrup is filtered to recover the catalyst, which is then regenerated and returned to the process.
6. The sorbitol syrup is then refined using both activated carbon and ion exchange to improve the colour.
7. After refining, the sorbitol syrup is evaporated up to 70% solids and sold in one of three forms, namely:
    (i) As a syrup containing a minimum of 90% sorbitol at 70% solids, however at this sorbitol concentration, the syrup is liable to crystallise.
    (ii) As a non-crystallising syrup with a solids of 70%, which is made by the addition of 63 DE syrup to the sorbitol syrup. The ratio of sorbitol syrup to 63 DE syrup is usually about 70:30. This addition of 63 DE syrup will of course reduce the sorbitol content to 70–85%.
    (iii) As a spray-dried powder.

The process described above for making sorbitol is similar for the production of maltitol from maltose syrup and for mannitol from fructose syrup. In the case of mannitol, because it has a low solubility, it is recovered as a solid by crystallisation as opposed to a syrup.

It is interesting to note that by changing the catalyst and the reaction conditions, other products such as glycerol can be produced.

Whilst most polyols are produced using hydrogenation, there is however one exception, namely erythritol, which is unique in that it is produced by the fermentation of dextrose by the osmophilic yeast *Moniliella pollinis* or the fungi *Aureobasidum* sp. with a yield of about 50% (see Fig. 8.3). It is well known that the addition of corn steep liquor can improve the yields of certain fermentations, so the addition of corn steep liquor might also have a similar beneficial effect on improving the yields of erythritol.

1. The production of erythritol is interesting in that the osmophilic nature of the yeast allows the solids of the dextrose solution to be higher than those for normal fermentations. Approximately, 50% of the dextrose will be fermented.

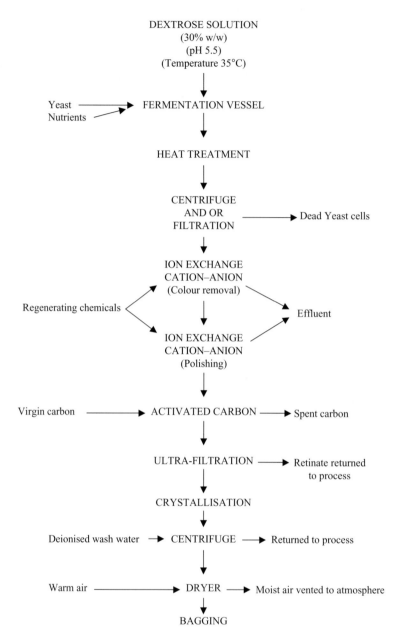

Figure 8.3 Schematic diagram for the production of erythritol.

2. There are two reasons for the heat treatment stage. Firstly, it kills the yeast and coagulates the proteins which are then removed by the centrifuge and filters, so that they cannot blind the ion exchange reins or carbon. Secondly, the heat treatment pasteurises the product producing a sterile product.
3. The ion exchange resins and activated carbon remove minerals, colour and any 'off flavours'.

4. The retinate (or rejected stream) from the ultra-filtration stage will contain any remaining particulate and colloidal solids, together with dead bacteria. The retinate will also contain unfermented dextrose which can be returned to the process. This stage therefore effectively purifies, separates and concentrates the erythritol solution prior to crystallisation.
5. With a solubility of only 37% w/w, erythritol is relatively easily crystallised with the crystals being recovered by centrifuging. At the end of the centrifuge cycle, the crystals are washed 'in situ' with de-ionised water, prior to drying using warm air. Basically, the recovery of the crystalline erythritol is very similar to that of the dextrose process.

Mannitol could also be produced by fermentation using an osmophilic yeast and the mould aspergillus, but it has yet to be commercialised, possibly due to low yields and the fact that the present process for producing mannitol by hydrogenation is satisfactory.

The advantages of using enzymes to produce polyols are that they are specific in their action and the process does not require the use of hydrogen and expensive pressure vessels. Therefore, in the future, enzyme production might become the preferred process for polyol production, providing high yields can be obtained.

Currently, the glucose industry does not make xylitol from dextrose – it is made by the hydrogenation of xylose, which is obtained from hemicellulose. According to Riviere (Industrial applications of microbiology, Surrey University Press, 1975), theoretically, xylitol could be made by a triple enzyme process from dextrose.

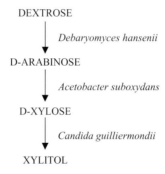

Figure 8.4 Theoretical enzymatic production of xylose from dextrose (according to Riviere).

8.3 Overview of polyol properties

Polyols are of interest not only to the food industry but also to the pharmaceutical, health care and chemical industries because of their interesting combination of properties.

They are sweet substances with a clean pleasant taste and are available either as a syrupy liquid or as a powder, and because of their chemical structure, they are neither readily metabolised by the body nor easily fermented by bacteria making them virtually non-calorific. The EC Directive 90/496/EEC gives the caloric value of polyols as 2.4 kcal/g (10 kJ/g). This compares to carbohydrates which have a caloric value of 4 kcal/g (17 kJ/g). Since different polyols are metabolised differently, the caloric value given by the EU is not always accepted by non-EU countries, and can therefore result in different caloric values being quoted in different countries. One exception to this EU directive is erythritol, which tentatively has been given a caloric value of 1.0 kcal/g (4.2 kJ/g). It is therefore important to always check the local legislation.

The combination of sweetness and not being readily metabolised by the body means they will contribute minimal calories to a food. This makes them an ideal ingredient for use in low calorie and sugar free foods where they can be used to replace calorie containing ingredients such as sucrose and glucose syrups. Table 8.1 shows the approximate sweetness of starch derived polyols compared to sucrose and polyols derived from other raw materials.

Table 8.1 Approximate sweetness values of polyols.

Starch-derived polyols		Polyols derived from other raw materials	
Erythritol	70	Isomalt	55
Maltitol	90	Lactitol	40
Mannitol	70	Xylitol	100
Sorbitol	50		
Sucrose	100		

Besides being slightly sweet, polyols also have a negative heat of solution, which produces a cooling effect in the mouth, with sorbitol exhibiting the greatest effect followed by xylitol, and then maltitol. The cooling effect of mannitol is less than the other polyols because of its low solubility (Table 8.2).

Table 8.2 Approximate solubilities (% w/w) of polyols at 25°C.

Starch-derived polyols		Polyols derived from other raw materials	
Erythritol	37%	Isomalt	25%
Maltitol	60%	Lactitol	57%
Mannitol	20%	Xylitol	64%
Sorbitol	70%		
Sucrose	67%		

Because polyols are not fermented by the bacteria present in the mouth, they are ideal ingredients for use in oral hygiene products.

Since polyols are not metabolised, they do not require insulin, and therefore can be used in diabetic foods.

As polyols are not metabolised, they do not affect the body's glycaemic index, and can be used to help control body weight.

Polyols can be used as a low calorie bulking agent or as a carrier for high intensity sweeteners.

Table 8.3 Molecular weights of polyols.

Starch-derived polyols		Polyols derived from other raw materials	
Erythritol	122	Isomalt	344
Maltitol	344	Lactitol	344
Mannitol	182	Xylitol	152
Sorbitol	182		
Sucrose	342		

Polyols, because of their molecular weights, can be used to alter the freezing point of a product (Table 8.3).

Unlike sucrose which is inverted by acid, starch-derived polyols are stable in the pH range 2–10.

Polyols have anti-oxidant properties.

Whilst polyols do not react with proteins, where any residual reducing sugars from the original glucose syrup are still present, then the brown Maillard colour will occur.

With the exception of erythritol, when formulating a product which will contain a polyol, always remember that polyols are only partially metabolised by the body with the unmetabolised polyol acting as roughage. Erythritol, on the other hand, is excreted by the body in the urine and is therefore considered not to be a laxative. Whilst polyols have the potential to be a diabetic safe sweetener, excessive consumption of a product containing a polyol will result in diarrhoea. The degree of the discomfort will vary from person to person and from polyol to polyol. It is for this reason that polyols are given an ADI (acceptable daily intake) rating which, like their caloric values, will vary from country to country, and in some countries will vary from polyol to polyol.

8.4 Applications overview

As a general statement, most polyols are used mainly for one of four possible reasons – a straight forward calorie reduction claim, a 'sugar' free claim, for diabetic reasons or because they are dental friendly, however some polyols are more suitable for a particular application than another. Several polyols are available as powders, which makes them ideal for use in chocolates with reduced calories or for use in dry mixes. The following is only a general overview.

8.4.1 Sorbitol

Sorbitol can be considered a 'first generation' polyol since it was the first polyol to be commercially developed and so its applications are well established.

Ascorbic acid (vitamin C)

Possibly, the largest application for sorbitol is in the production of ascorbic acid. Not only is ascorbic acid an essential vitamin, but it is also an important anti-oxidant.

Bakery products

Sorbitol is not fermented by yeast and does not react with proteins, so it will not react with flour proteins to give the Maillard reaction.

Sorbitol is an excellent humectant and is used to extend the shelf life of many baked goods and fillings. Typical use level is about 5%. Powdered sorbitol can also be used to replace sucrose in sugar reduced products such as biscuits.

Confectionery

Whilst sorbitol has in the past been used for many years in different 'sugar free' confectionery products, other polyols such as maltitol and mannitol are probably now more suitable. Sorbitol is however a very efficient plasticiser and humectant and can be used as an alternative to glycerine or invert. When used in a cream paste at about 0.5% addition, it will give the paste a softer texture. A 5–10% addition of sorbitol to fondants and fudge reduces drying out, controls water activity and can help control crystallisation.

Powdered sorbitol can be used in the production of sugar free chewing gum. Powdered sorbitol can also be used in tabletting and chocolates, but a high intensity sweetener has to be included to give the required sweetness. Because sorbitol is hygroscopic, ideally, the production area should be air conditioned.

Deposited hard boiled confectionery can be made with sorbitol replacing all the sucrose and glucose syrup. Because of the low viscosity of cooked sorbitol, it is difficult to work the cooked mass, which makes it unsuitable for making hard boiled sweets by 'roping'.

Unlike dextrose, powdered sorbitol is anhydrous, and can therefore be used in the production of chocolate, but the conching temperature must not exceed 42°C.

Sorbitol syrup and powdered sorbitol can be used in dragee pan work to produce a smooth coating.

Sorbitol can improve the freshness and keeping qualities of desiccated coconut, candied fruits and similar types of products by reducing the loss of moisture.

Counter-medicines

The traditional carrier for the active ingredients in a lot of medicines for young children has been a sucrose syrup (Syrup Simplex BP). The disadvantage of using a sucrose syrup is that it is not tooth friendly because it is fermented by the bacteria in the mouth producing acids, which lead to tooth decay. By using sorbitol as the medicinal carrier, as it is not fermented by oral bacteria, the possibility of dental decay is substantially reduced.

Diabetic foods

As sorbitol is only partially metabolised by the body, it can be used in foods for diabetics as part of a programme to help manage blood sugar levels.

Humectant

Sorbitol has been used as a humectant in many different applications from food to tobacco and adhesives.

Jams and preserves

When making diabetic and reduced calorie jams and preserves, sorbitol syrup can be used to replace sucrose and glucose syrups. Since the molecular weight of sorbitol is less than that of sucrose, the osmotic pressure is increased giving improved microbiological stability to the end product.

Industrial uses

Sorbitol is used in the manufacture of the Tween (polyoxythylene sorbitan monostearate) and Span (sorbitan tristearate) range of emulsifiers and detergents.

Ice cream

Sorbitol, because of its low molecular weight, can be used as an alternative to glycerine in the production of soft scoop ice creams. Typical addition rate would be about 5%.

Oral hygiene products

Because sorbitol is not fermented by the bacteria in the mouth, it is used in oral hygiene products such as toothpastes where it acts as a carrier for the insoluble ingredients and also as a plasticiser for the toothpaste. It is also used in mouthwashes.

Surimi

During the production of surimi, sorbitol can be used to protect the protein from being denatured during the freezing process. The advantage of using sorbitol over the traditional use of sucrose is that sucrose produces a sweet product, whilst sorbitol being less sweet makes a product which is less sweet and allows the subtle flavours of the fish to come through.

8.4.2 Maltitol

Maltitol is a disaccharide polyol composed of dextrose and sorbitol and is usually available in three grades containing varying amounts of maltitol, and either as a syrup or as a crystalline powder. Typical maltitol contents are 50–55%, 72–77% and 80–90%, depending upon the maltose content of the syrup from which the maltitol was made.

Bakery products

The recipes of most baked products contain minimal amounts of water, therefore crystalline maltitol can be used to replace part or all of the sucrose in cakes and biscuits. Because it is not fermented by yeast, it is not suitable for leavened goods.

A combination of a maltitol syrup and crystalline maltitol can be used to replace sucrose and glucose syrup in marzipan, almond and coconut pastes also in icing and fondants.

Confectionery

Maltitol is a confectionery friendly ingredient, which can be used in several types of confectionery, particularly diabetic confectionery, often with minimal alterations to the existing formulation or process. Typically, it can be used in the manufacture of chocolate using normal conching parameters.

Maltitol syrup containing 72% of maltitol or higher can be used to replace sucrose and glucose syrup in gums, jellies pastilles and hard boiled sweets.

Whilst maltitol contains a dextrose molecule, the reducing site of the dextrose is blocked by the sorbitol molecule which makes maltitol non-reducing. Therefore, when maltitol is used to make toffees, there is no Maillard reaction with the protein of the milk. This results in toffees which lack colour and flavour, which means that both colour and flavour have to be added. If a claim of 'sugar free' is to be made, then lactose-free milk should be used.

Maltitol can be used as a binder in cereal bars in the same way as a glucose syrup.

Crystalline maltitol can be used in tabletting, whilst powdered maltitol can be used in the production of sugar free chewing gum and in dragee pan work.

Ice cream

Maltitol syrup can be used to make a sugar-free ice cream, by replacing both the sucrose and glucose syrup, and using lactose-free milk powder. Since the molecular weight of maltitol is very similar to that of sucrose, there is little or no change to the texture of the end product. The sweetness could be adjusted by the addition of a high-intensity sweetener.

Since this type of 'ice cream' does not contain any sucrose, it would not meet the legal compositional requirements, and therefore could not be called an ice cream but a frozen dessert.

Jams and preserves

When making diabetic and reduced calorie jams and preserves, maltitol syrup can be used to replace both sucrose and glucose syrups. Because there is no process inversion, the osmotic pressure is lower than that in a conventional jam which could result in microbiological spoilage. Therefore, if the soluble solids are below 65%, then a preservative could be added.

8.4.3 Mannitol

Like other polyols, mannitol can be used in reduced calorie and diabetic foods and pharmaceutical applications. However, since it is has a low solubility, its application uses are limited.

Confectionery

Because mannitol has low moisture absorption properties, it is mainly used as a dusting agent during the manufacture of chewing gum, and for coatings to prevent products from sticking together. When making hard-boiled confectionery with maltitol, the end product can be sticky due to the hygroscopicity of the maltitol, but by the addition of about 7% mannitol to the formulation, this stickiness can be substantially reduced. Powdered mannitol could also be used in the manufacture of reduced calorie chocolate.

Pharmaceutical

Mannitol is used as a bulking agent in tablet formulations, where its low hygroscopicity helps to stabilise moisture sensitive ingredients. Additionally, having a low solubility, it can help to control the release of an active ingredient in chewable tablets.

Pyrogen-free mannitol is used as an intravenous diuretic in the treatment of cerebral oedemas.

8.4.4   Erythritol

Erythritol is a unique polyol. It is slightly less sweet than sucrose being about 70% that of sucrose, and whilst it is absorbed by the body, it is not metabolised, being totally excreted, unchanged, in the urine and contributes zero calories – a truly non caloric polyol. Because of its non caloric properties, the EU are considering giving it a caloric value of 1.0 kcal/g or less, as opposed to the blanket 2.4 kcal/g for all polyols.

As erythritol is excreted from the body in the urine, it does not have any laxative effects, which are usually associated with polyols.

Because erythritol is not metabolised, it does not affect the glycaemic or insulin requirements of the body, making it an ideal ingredient for diabetic products and as an aid in controlling body weight. Erythritol because of its antioxidant properties is a free radical scavenger.

Erythritol being non-cariogenic can be used to advantage in oral hygiene products such as in a toothpaste and a mouthwash where a tooth friendly ingredient is required.

The molecular weight of erythritol is approximately a third of the molecular weight of sucrose. This means that only 3.5 grams of erythritol will give the same depression of freezing as 10 grams of sucrose. Additionally, because of its low molecular weight, it has a low viscosity.

Erythritol is not as soluble as some other polyols, and therefore it is frequently used in combination with maltitol particularly in confectionery and bakery products. One area of interest could be in the manufacture of reduced calorie chocolate.

Chapter 9
Glucose syrups in baking and biscuit products

9.1 Introduction

Whilst sucrose is the main sugar in most bakery applications, a variety of starch derived sweeteners such as glucose syrups and dextrose can in some cases be effective as a partial replacement for sucrose where they can offer a cost saving. When used at low levels in pastry, glucose syrups can have a plasticising effect which can improve the machineability of the pastry.

Baked goods can be divided into two groups, namely fermented and non-fermented products. Typical fermented goods are bread and morning goods such as rolls. Non-fermented goods are sponges, cakes and pastry. Glucose syrups can be used in both groups of products, but for completely different reasons.

For yeast fermented goods to be successful, it is essential that there is an adequate source of 'sugar' for the yeast. During fermentation, yeast will convert sugars such as dextrose, fructose, sucrose and maltose into carbon dioxide and alcohol. The gluten in the flour forms a viscoelastic mass within the dough, and the carbon dioxide produced during the fermentation is trapped in the gluten to form bubbles of gas. It is these bubbles of gas which gives the bread its characteristic open texture – no sugars, no gas bubbles. The result is solid dough.

In non-fermented goods such as sponges, cakes and pastry, the sucrose is there mainly to provide a sweet taste to the end product. In the case of pastry, the low level of sucrose will aid the machineability. With biscuits, the sucrose is there not only to provide sweetness but also to give the biscuit a crisp bite. Sucrose is also used in both fillings and decorations for use with sponges and biscuits.

In most of these applications, starch derived sweeteners can be used to partially replace some of the sucrose, additionally both glucose syrups and dextrose will also act as a humectant and enhance the colour and improve the volume.

When considering the use of glucose syrups in any baking application, it must be borne in mind that most bakery recipes contain only a minimal amount of water. Glucose syrups, however, contain between 20% and 30% water. Therefore, the extra water from the syrup can often be the limiting factor into how much syrup that can be used in a recipe – too much syrup will mean excess water being added to a semi-dry recipe. Dextrose and maltodextrins, being dry products, will contribute minimal moisture.

Another possible limiting factor for using glucose syrups in baked goods could be that most bakeries are designed to handle dry products such as flour, crystalline sucrose and fat, but with the baking industry moving to larger and sometimes one product manufacturing units, the bulk handling of syrups is becoming less of a problem, especially with the use of automation.

One major difference between sucrose and starch derived sweeteners is that sucrose is a non-reducing sugar and therefore will not react with the protein in the flour to produce the well known Maillard browning reaction. Starch derived sweeteners, on the other hand, contain reducing sugars which will react with the protein in the flour to give the Maillard brown colour. This colour production can be both an advantage and a disadvantage. The only starch derived sweetener which is not a reducing sugar is trehalose. Trehalose, like sucrose, is a disaccharide sugar, with the same molecular weight as sucrose, but has only half the sweetness of sucrose.

Two syrups which are often used in the baking industry are invert and Golden Syrup. Where invert syrup is being used for its sweetness and humectant properties, it can be effectively replaced by high fructose glucose syrup (HFGS). To replace Golden Syrup is more difficult, because it is frequently used for its characteristic flavour as well as for its sweetness.

9.2 Fermented goods

There are many different ways of making yeast fermented baked goods such as bread, but they all contain three basic stages – mixing, fermentation and baking.

1. When the flour, water and yeast are mixed to make the dough, two reactions occur simultaneously. One concerns the amylase enzymes in the flour which starts to convert the starch of the flour into sugars, predominantly maltose. The other concerns the proteins gliadin and glutenin of the flour which are hydrated to form the viscoelastic mass gluten.
2. During the fermentation stage, the yeast converts the maltose first into dextrose and then into carbon dioxide and alcohol.

$$\text{STARCH} + \text{WATER} \xrightarrow{\text{Flour Amylases}} \text{MALTOSE}$$

$$\text{MALTOSE} + \text{WATER} \xrightarrow{\text{Yeast Enzymes}} \text{DEXTROSE}$$

$$\text{DEXTROSE} + \text{WATER} \xrightarrow{\text{Yeast Enzymes}} \text{CARBON DIOXIDE} + \text{ETHYL ALCOHOL}$$

The carbon dioxide produced during the fermentation is trapped in the gluten matrix forming bubbles, and because the gluten is elastic, the bubbles can expand. As the gluten is evenly spread throughout the dough, a series of interconnected gas bubbles are formed, which gives the product its characteristic open structure.

3. During the baking, several changes take place. As the dough is baked, the gas inside the bubbles expands, increasing the volume of the dough. The carbon dioxide together with the alcohol and water escape, whilst the starch is gelatinised and the gluten is denatured. With the gluten being denatured, it loses its elasticity becoming semi-rigid, and together with the gelled starch, it gives the baked dough its final aerated-like structure. Additionally, the sugars on the surface of the dough react with the protein of the flour to give the crust a nice rich brown colour so characteristic of the Maillard reaction.

In some situations, extra fermentable sugars are required, for example in the continuous dough making process which depends upon the addition of extra fermentable sugars. A typical sugar addition in the continuous dough making process would be about 6–12% based on the weight of flour. The choice of which sugar to use will depend upon their fermentability and relative costs, and with both dextrose and HFGS being highly fermentable and frequently cheaper than sucrose, they are often the preferred sugars.

9.3 Non-fermented goods

A variety of starch based sweeteners can be used in non-fermented goods such as cakes, where they can replace some of the sucrose or reduce the overall sweetness of the cake. They can also increase the shelf life of the product with their humectant properties. Being reducing sugars, starch derived sweeteners will enrich the crust colour. They will also increase the volume of the cake.

There are however some limiting factors in using glucose syrups in cakes. Because glucose syrups contain water, there can sometimes be an excess of water in the final mix. Another limiting problem is the 'brown base phenomenon'.

When a cake containing too much glucose syrup is baked and then cut vertically in half, there will often be a dense brown layer at the base of the cake – the brown base phenomenon. This brown-coloured layer is due to the Maillard reaction between the reducing sugars of the syrup and the protein of the flour. To understand why this occurs at the bottom of the cake, it should be appreciated that sucrose is a crystalline solid as opposed to a glucose syrup which is a viscous non-crystallising liquid. So when the cake is baked, the syrup becomes hot and its viscosity is reduced, changing the syrup from a viscous liquid to a free-flowing liquid. As a result of this change in viscosity, the syrup separates from the rest of the ingredients, and now being a free-flowing liquid, it percolates to the bottom of the cake mix. Sucrose, on the other hand, because it is a crystalline solid, tends not to migrate but remains relatively static within the cake mix. Hence the brown layer at the base of the cake.

To reduce the migration of the syrup, emulsifiers and pre-gelled starch or some other thickener can be added but these can have an adverse effect on the texture of the cake.

Since the 1950s, the baking industry has used a 63 DE syrup in cakes, especially in dark cakes, because it can offer sweetness with an acceptable colour formation. A 63 DE syrup has an approximate sweetness of 70 compared to a sucrose sweetness value of 100. The syrup solids can be as high as 82% which can minimise any excess water in a recipe.

With its balanced sugar spectrum, it can replace invert and can be an equally effective humectant. A typical sugar spectrum of a 63 DE syrup is as follows:

Dextrose	34%
Maltose	33%
Maltotriose	10%
Higher Sugars	23%

The amount of sucrose which can be replaced by a 63 DE syrup is limited by the amount of colour the syrup will give to the baked cake, therefore it is important to experiment to find the appropriate replacement levels. Typical sucrose replacement levels on a dry basis could be as follows:

High-ratio slab cake	10%
Sponge sandwich cake	10%
Rich fruit cake	10%
Madera slab cake	15%
Swiss roll	25%

If dextrose is used as an alternative to a 63 DE syrup, the possible sucrose replacement levels will be slightly less because the reducing power of dextrose is greater at 100 DE as opposed to 63 DE for the syrup.

Where minimal colour formation is more important than sweetness, then a maltose syrup should be considered. Maltose syrups have approximately half the sweetness of sucrose, but produces less Maillard colour compared to either a 63 or 42 DE syrup because it contains less dextrose than either of the other two syrups.

With the availability of HFGS, which have a similar sweetness to sucrose, it is now possible to replace higher levels of sucrose, especially in cakes which use a small amount of sucrose. Below is a starting recipe for the replacement of 75% of the sucrose in a high-ratio genoese-type cake.

High-ratio flour	224 grams
High-ratio shortening	170 grams
HFGS (42% fructose) (70.0% solids)	200 grams
63 DE Glucose Syrup (82% solids)	57 grams
Castor sugar	63 grams
Skimmed milk powder	17 grams
Pasteurised eggs	179 grams
Salt	4 grams
Baking powder	11 grams
Water	65 grams
Emulsifier	10 grams
Colour	As required
Flavour	As required

The use of sodium aluminium phosphate (SALP) in conjunction with the baking powder will reduce some of the colour formation, whilst the emulsifier will improve the volume and at the same time reducing the 'brown base phenomenon' by keeping the syrup in the body of the cake.

9.4 Biscuits

There are many different types of biscuits, however the main ingredients of most biscuits are flour, sugar, fat and water or sometimes milk. By varying the different amounts of these major ingredients, the texture of the biscuits can be changed. Further changes can be obtained by altering the type of flour, sugar or fat. Of these ingredients, the sugar content and importantly the type of sugar and its particle size have probably the most influence on the texture, sweetness, flavour and particularly the bite of the biscuit. Typical sucrose inclusion rates vary from less than 1% for wafers and water biscuits to 16% for digestives and up to 35% for ginger nuts. The lower sucrose inclusion rates, that is 16%, contribute sweetness, whilst the higher rates such as 35% in ginger nuts is responsible for the hard bite as well as the sweetness.

To totally replace a major crystalline ingredient such as sucrose with a liquid glucose syrup is difficult especially in biscuits such as ginger nuts with their high sucrose content and still achieve a hard bite. Another problem is that biscuit recipes contain little water, but syrups contain about 20–30% water. A further problem is to maintain the sweetness. These three factors suggest that successful total sucrose replacement should only be in biscuits with a low sucrose content, and only a partial sucrose replacement in biscuits with a high sucrose content. Because of the major part which sucrose has on the structure of a biscuit, it is important to monitor any change which can affect the weight or dimension of the biscuit as well as the sweetness and bite. A contributing factor to the dimension stability of a biscuit is the particle size of the sucrose – coarse crystal size produces less spread, which is another technical problem to overcome but not impossible.

Changes to the weight and dimension will have an affect on the packaging, such as the number of biscuits to a packet, and hence the weight of the packet, and also on the dimensions of the outer wrapping material. There is also the appearance of the biscuit, such as colour and checking. The reducing sugars in starch derived sweeteners can cause the biscuit to be darker than normal, but this darker than normal appearance should not be taken as an indication that the baking temperature is too high. If that approach is taken, and the temperature is reduced, there is the possibility that too much moisture will remain in the biscuit, which could result in checking. Checking normally occurs in the packet during storage due to a moisture gradient from the centre to the edge of the biscuit. Checking is most common in semi-sweet biscuits and some types of cream crackers.

A possible starting recipe for the total replacement of sucrose in a digestive biscuit using a blend of HFGS and a 63 DE Glucose syrup could be as follows, and if the biscuit is to be chocolate coated, any marginal reductions in sweetness will be masked.

Biscuit flour	2.750 Kg
Wholemeal flour	900 grams
Vegetable shortening	1.150 Kg
HFGS (42% fructose @ 70% solids)	876 grams
63 DE Glucose Syrup (@ 82% solids)	250 grams
Malt powder	50 grams
Whey powder	100 grams
Salt	50 grams
Sodium bicarbonate	40 grams
Ammonium bicarbonate	40 grams
Pre-gelled maize starch	50 grams

9.5 Biscuit fillings

Biscuit fillings are generally fat (shortening) based. If they were water based, the water would migrate into the biscuit to produce a biscuit which would be soft and lack the desirable sharp snap. An extra fine dextrose can be used to replace up to 50% of the icing sugar. If the icing sugar is to be totally replaced by dextrose, then 5% of a 15 DE maltodextrin should be added. The reason for the maltodextrin addition is that dextrose contains about 7–9.5% moisture which will separate from the shortening, but the higher sugars of the maltodextrin would increase the viscosity of the dextrose phase to help form a stable emulsion. The use of dextrose in a cream filling is beneficial where citrus or mint flavours are used. The slight reduction in sweetness allows the sharp acid bite of the citrus to come through, whilst the cooling effect of the dextrose enhances the mint flavour.

The particle size of either the icing sugar or dextrose should be 75 microns (200 mesh) or less so as to ensure that the cream has a smooth texture on the tongue, however the finer the sugar or dextrose, the more fat will be required to cover the increased surface area of the sugar and dextrose. Typical recipe could be as follows:

Shortening	400 grams
Icing sugar	300 grams
Extra fine dextrose	300 grams
Flavour and colour	As required

9.6 Wafer fillings

Most wafer bars consist of several layers of wafers which have been coated with a filling then laminated, followed by chocolate enrobing.

There are many different types of fillings which can be used, but possibly the most common are either caramel or chocolate. For a caramel, a soft confectionery caramel would be suitable but replacing the normal confectioners 42 DE syrup with a 63 DE syrup and using a fat with a slip point of 32°C. By using a 63 DE syrup, a softer less viscous caramel is obtained which is easier to spread on the thin wafer base. HFGS could also be used to replace any invert. The moisture content of the finished filling should be 10% or less, which will ensure minimal moisture migration into the wafer. Because jams can have a moisture content of 20% or higher, moisture migration becomes more of a problem and it is common for the wafer to be coated with a fine film of fat to prevent moisture migration.

For a chocolate filling, the above biscuit cream recipe could be a suitable initial recipe with the addition of about 5–10% chocolate powder or chocolate paste.

A typical recipe for a date or fig filling could be as follows, where the 63 DE syrup can act as a humectant and binder:

Whole date/fig	360 grams
Date/fig paste	180 grams
Brown sugar	200 grams
63 DE Glucose Syrup (80% solids)	220 grams
Salt	2 grams
Citric acid	1 gram
Water	4 grams

To give the filling more body, a 42 DE syrup can be used to replace the 63 DE syrup.

9.7 Bakery sundries

Glucose syrups are used in the production of bakery sundries such as fondant, hundreds and thousands, piping jellies, biscuit fillings, bakery and biscuit jams, bakery glazes, marzipan, 'fruit slices', etc, where they modify the sweetness, act as an humectant, and control sucrose crystallisation.

For details of bakery jams and fillings, see Chapter 14 (Glucose syrups in jam).

9.7.1 Fondant

There are several different recipes for a fondant. Some are based on (1) invert and a glucose syrup, (2) sucrose and glucose syrup and (3) icing sugar and maltodextrin. The two most important properties are that the particle size of the solids are in the range 12–17 microns to give a fine smooth mouthfeel. The second is that the solids must be sufficiently high to prevent microbiological spoilage. Both glucose syrup and maltodextrin are used in fondants because they contain higher sugars, which control the formation of sucrose crystals ensuring the formation of very fine sucrose crystals. A typical recipe could be as follows:

Sucrose	400 grams
63 DE Glucose Syrup (82% solids)	100 grams
Water	100 grams
Flavouring	As required

Method

1. Heat all the ingredients to 120°C with constant stirring.
2. Cool to 65°C with constant stirring to induce sucrose crystallisation.
3. Allow 24 hours before use.
4. If required, melt at 60°C and add extra 63 DE to make the fondant more easy to work.

Where a thin fondant or cream is required, the amount of glucose syrup in the original recipe can be increased or the fondant can be melted and then diluted with glucose syrup prior to use.

Fondants can also be made using dextrose to replace the sucrose and to ensure a smooth fine texture, when the mix starts to crystallise it is mechanically sheared. The final texture will depend upon the degree of shearing.

An addition of about 5% chocolate to a fondant will give the fondant both flavour and colour. An alternative to chocolate could be a chocolate colouring and flavouring.

9.7.2 Hundreds and thousands

These are small threads (about 0.5 cm long) of dried extruded sugar paste and are used for decorating the tops of cakes. The 63 DE syrup and HFGS act as binders for the icing sugar as well as a processing plasticiser. A possible recipe could be as follows:

Icing sugar	100 grams
63 DE Glucose Syrup (82% solids)	15 grams
HFGS (42% fructose)	10 grams
50% Gelatine solution (160 bloom)	4 grams
Colour	As required

Method

1. Mix the 63 DE syrup, HFGS and gelatine solution with the icing sugar to give a smooth paste which can be easily extruded through a die, with a diameter of about 1.0–2.0 mm. N.B. Because of the small diameter of the die, it is essential that the particle size of the icing sugar will allow the paste to pass through the die.
   It is possible to replace about 50% of the icing sugar with extra fine dextrose which has a similar particle size to icing sugar, that is 75 microns (200 mesh) or less.
2. Dry the threads with warm air, and dust with icing sugar if necessary.

9.7.3 Icings

The addition of 2–5% of a glucose syrup such as a 42 DE, 63 DE or HFGS to a conventional icing will give it body and retard drying out.

Whilst most icings are based on icing sugar, icings can also be made with blends of icing sugar and extra fine dextrose or by using 100% dextrose. To ensure a smooth texture, the particle size of the dextrose should be about 75 microns (200 mesh). Dextrose icings, being less sweet than an icing sugar one, are ideal where a citrus flavour is incorporated. A suitable recipe could be as follows:

Dextrose monohydrate (extra fine)	87 grams
Water	12 grams
Flavour	As required

A softer icing can be made by mixing one part of marshmallow with three parts of fondant. Alternatively, the icing can be thinned with a glucose syrup.

In a butter icing, 50% of the icing sugar can be replaced with extra fine dextrose, but if all the icing sugar is replaced with extra fine dextrose, then a 5% addition of a maltodextrin (15 DE or lower) should be added. The reason for this is that dextrose contains about 8–9% water which will separate out. By adding a maltodextrin, the higher sugars of the maltodextrin will increase the viscosity to form an emulsion with the butter.

To make a chocolate icing, add about 5% drinking chocolate powder.

9.7.4 Marshmallows

Marshmallows are an aerated product, made by heating glucose syrup and sucrose to about 80% solids, cooling then adding gelatine followed by whipping in air to produce a sponge-like product.

When considering which sweetener to use in a marshmallow, there are two considerations – viscosity and sweetness. Because marshmallows are an aerated product, the sweetener should be viscous to help produce a stable foam, which would suggest using a low DE, high viscosity syrup such as a 42 DE glucose syrup. The other quality of a marshmallow is sweetness, which would suggest using a 95 DE syrup, but a 95 DE syrup lacks viscosity and because of the high solids required to ensure good microbiological stability, the 95 DE syrup would crystallise to give the marshmallow a gritty texture, therefore a compromise syrup has to be used, namely a 63 DE.

By using a 63 DE syrup as opposed to a 42 DE syrup, a manageable viscosity at higher solids, ideally about 80–82%, can be obtained. Additionally, by using a syrup to increase the solids, as opposed to sucrose, the risk of sucrose crystallisation in the end product is reduced. The 63 DE syrup will also act as a humectant retarding the drying out of the mallow making it unnecessary to use invert or glycerine. The shelf life of a marshmallow can be further increased by using limed gelatine or alternatively by the addition of 0.1% sodium bicarbonate to raise the pH of the mix to about 7.2. Marshmallows produced by using acid produced gelatine without the addition of sodium bicarbonate are less stable. A possible starting recipe could be as follows:

Part 1	Granulated sugar	250 grams
	63 DE Glucose Syrup (81% solids)	550 grams
	Water	200 grams
Part 2	Limed gelatine (200 bloom)	20 grams
	Water	50 grams
Part 3	Sodium bicarbonate	1 gram

Method

1. Heat Part 1 to about 81% solids.
2. Pre-swell the gelatine in Part 2.
3. When Part 1 has cooled to 60°C, add the gelatine with constant stirring, whipping in as much air as possible.
4. Dissolve the sodium bicarbonate in a minimal amount of water and add.
5. When the consistency is correct, deposit onto the biscuit base. *N.B.* The consistency will depend upon how much air has been incorporated, that is the density or overrun of the mix. Typical density of a marshmallow is in the range of 0.20–0.45.

9.7.5 Marzipan

A possible recipe for marzipan could be as follows:

Sugar	510 grams
HFGS (42% fructose @ 70% solids)	220 grams
Ground almonds	255 grams
Glycerine	10 grams
Sorbic acid	1 gram
Flavour	As required

Method

Blend together all the ingredients, then pass through a triple roll mill several times until a smooth paste is obtained.

An economy marzipan can be made by replacing about 50% of the almonds with ground rice.

The HFGS acts as a binder and plasticiser in the above recipe. An alternative to HFGS could be either a 42 or 63 DE syrup depending upon the required plasticity of the mix.

9.7.6 Fruit flavoured pieces

These can be made from pectin. A possible recipe could be as follows:

Buffered rapid set pectin	3.5 grams
Sucrose	250.0 grams
63 DE Glucose Syrup (82% solids)	105.0 grams
Water	200.0 grams
Citric acid	0.05–0.1 grams
Colour and flavour	As required

Method

1. Dry blend the pectin with 50 grams of sucrose.
2. The remaining sucrose and water are mixed with the 63 DE syrup and heated to 60°C with constant stirring.
3. The dry blended sucrose and pectin are added with constant stirring, and the temperature is increased to 106°C.
4. The solids are checked with a refractometer and should be about 68%. The pH should be 3.0–3.2.
5. The citric acid is dissolved in 10 grams of water, with constant stirring and added. *N.B.* The lower the pH, the quicker the set.
6. The cooked mix is now poured into moulds or cast onto a cold surface, prior to being cut to shape, and then sanded.

Agar agar can also be used to make fruit slices. Where a harder and more firm fruit piece is required, then a 42 DE syrup can be used.

9.7.7 Piping jelly

These can also be made from pectin using a similar recipe to fruit solids by increasing the 63 DE content, and reducing the final solids.

9.7.8 Bakery glaze

A bakery glaze can be made using a blend of a maltodextrin and sucrose. The higher sugars present in the maltodextrin, being non-crystallising, prevent the sucrose from crystallising, and instead the mixture forms a glass rather like in a hard-boiled sweet. It is this glassy layer which gives the glaze its gloss.

15 DE Maltodextrin	200 grams
Sucrose	540 grams
Water	260 grams

Method

1. Dry blend the maltodextrin and sucrose.
2. Heat the water to about 45°C, and slowly add the dry powder mix with constant stirring, until all the solids have been dissolved. Increase the temperature if difficulty is experienced in obtaining total solution.
3. Aim for a refractometer solids of 72–75%.
4. Apply after baking.

Bakery glazes can also be made using pectin.

9.8 Reduced calorie products

Cakes and biscuits are mainly composed of starch, sucrose and fat, and all three have a high calorie content. Therefore, to reduce the calories in cakes and biscuits, they must be reformulated to contain less starch, sugar and fat. However, when replacing major ingredients such as fat and sucrose in a bakery product, expect to make other ingredient changes if an acceptable product is to be produced.

Fat can be replaced with a low DE maltodextrin. Fat has a caloric value of 9.0 calories but a maltodextrin being a carbohydrate has a caloric value of only 4.0. Because of the high viscosity of maltodextrins, they have a similar texture to that of fat. Unfortunately, maltodextrins are bland and because some flavours are oil soluble, maltodextrins lack the same sensory and flavour characteristics of a fat. Therefore, only a 40% fat replacement is considered as acceptable.

There are several possibilities for replacing the sucrose in a recipe however, care must be taken to ensure that the balance of sweetness with a reduction in calories is achieved, and without adversely affecting the yield. The use of a polyol, with or without a sweetener is possibly the easiest approach, but by using a polyol, there is the risk that the end product could be laxative. As a general rule, polyols, with the exception of erythritol, are only partially metabolised by the body and will therefore contribute 40% less calories than sucrose. Whilst polyols will reduce the calories, they would, with the exception of maltitol, contribute only half the sweetness of sucrose, and therefore when used, the recipe will require a sweetening agent. Because maltitol is approximately 10% less sweet than sucrose, it is possibly the preferred polyol to use if the calorie reduction is acceptable. If a further calorie reduction is required, then the maltitol could be replaced with erythritol. Erythritol is considered to be non-caloric, but it is about 30% less sweet than sucrose, therefore a heat stable high-intensity sweetener such as saccharin or sucralose would have to be used. Depending upon the calorie reduction required and the type of product to be made, other possibilities could be blends of polyols with fructose (Table 9.1).

Table 9.1 Relative caloric and sweetness values for selected sugars and polyols.

	Calories per gram	Sweetness
Sucrose	4.0	100
Fructose	4.0	150
Erythritol	0.2	65
Maltitol	2.4	90
Sorbitol	2.4	50

A possible reduced calorie biscuit can be made using powdered maltitol to replace sucrose. Since maltotol is slightly less sweet than sucrose, the need to add an artificial sweetener is greatly reduced. To reduce the calorie content further, the maltitol could be replaced with erythritol, but a heat-stable artificial sweetener would be required. A possible recipe could be as follows:

Biscuit flour	440 grams
Fat	270 grams
Powdered Maltitol	220 grams
Skimmed milk powder	20 grams
Water	50 grams

A reduced calorie fondant could be made by replacing icing sugar with either erythritol powder or mannitol. The calorie count could be further reduced by replacing the syrup phase with maltitol or some other sugar alcohol. The sweetness could be increased by the addition of an artificial sweetener. Such a fondant would, however, have laxative properties.

A reduced calorie marzipan could be made by replacing the sugar with erythritol powder, and the HFGS with maltitol syrup, whilst the sweetness could be increased by the addition of an artificial sweetener. Similarly, to reduce the calorie count of a marshmallow, maltitol could be used to replace the 63 DE.

Calorie reductions in a biscuit filling can be achieved by replacing the icing sugar and dextrose with erythritol.

Where a polyol has been used in any recipe, there is always the risk that the final product could be laxative.

9.9 Breakfast cereals

Approximately one third of breakfast cereals are pre-sweetened and most are made from maize (corn), wheat, oats or rice, and are either puffed or made into flakes with some being extruded.

Puffed products are made by heating the cereal with infrared heat, which turns the moisture inside the cereal into steam. The pressure of the steam then causes the cereal to expand and eventually ruptures to give the cereal its characteristic puffed appearance. In some processes, the cereal is pre-coated before being puffed.

Flaked products are made by passing steam heated cereals through a series of heated rollers, which results in thin flakes. The flakes are then cooked or toasted to give them a brown colour and flavour. The coating is then applied at a later stage.

There are several different ways to coat the cereals. If the cereal is roughly spherical like puffed products, then the coating can be applied using a dragee pan, or some similar rotating pan. As the cereal rotates, the molten coating is sprayed onto the cereal, with a stream of hot air being blown into the pan to remove the moisture. This process can also be used to agglomerate smaller pieces, such as broken nut and cereals pieces, so as to produce larger pieces for subsequent addition to either a breakfast cereal or to cocktail type nibbles.

Another method is to spray the coating onto the cereal as it passes down an inclined rotating kiln. Hot air is introduced in a counter current flow to the movement of the cereal so as to remove the moisture. Fluid bed dryers have also been used for drying the coating. It is important to keep the coated cereals moving until the coating has completely dried.

The coating process can be either a single or multiple application. When applying a secondary coating, it is important that the original coating is sufficiently receptive to retain the secondary coating. Icing sugar, extra fine dextrose or maltodextrins can be used as a dusting to protect the cereal pieces from sticking can together.

Typical coatings will be based on a molten sucrose–glucose mix, similar to that used for a hard-boiled sweet and cooked to about 95% solids. Either a 42 DE or a maltose syrup can be used. The reason for using a glucose syrup is the same as for when making a high-boiled sweet – to prevent the sucrose from crystallising. Therefore, by altering the ratio of sucrose and glucose syrup so that the appearance of the coating can be changed, a high sucrose content can result in an opaque coating, whilst a high-glucose syrup will result in a clear glass like coating. The advantage of using a maltose syrup containing 70% maltose is that it contains few higher sugars, and therefore has a lower viscosity than a 42 DE syrup, which allows the solids of the coating syrup to be higher than normal and makes the coating more suitable for spray application. Additionally, the malty flavour of the malt syrup blends very well with the flavour of the cereal. Other ingredients such as honey or brown sugar can be used to give flavour to the coating, or to make a marketing claim. It is important that there is minimal amount of water in the coating when it goes onto the cereal pieces, hence the 95% solids. If there is too much water in the coating, the water will be absorbed by the cereal, making it difficult to remove and results in a product which is limp and which lacks the characteristic crispness associated with a breakfast cereal.

A typical coating could have the following composition (UK Patent GB 2282049 B).

Sucrose	20%
Honey	5%
Maltose Syrup (70% maltose)	64%
HFGS (42% fructose)	11%

All percentages are on a dry weight basis.

Opaque coatings can be produced by dusting icing sugar over the coating, before it is completely dried. Another way of producing an opaque coating is to apply a glucose syrup which contains starch which has been modified so that it does not gel and hence results in an opaque appearance.

Chapter 10
Glucose syrups in brewing

10.1 Introduction

Traditionally, beer has been made from water, malt, hops and yeast, with the malt supplying the sugars for the yeast to convert into alcohol. However, due to scarcity, price or for quality considerations (e.g. management of nitrogen content), some of the malt might be replaced with other cereals, sugar or brewing syrups. These replacement materials are referred to as 'adjuncts'.

In most countries what can and cannot be used as an adjunct is strictly controlled by the revenue collection service. Permission is normally granted providing that the adjunct will not affect the health of the consumer and will not give a false ABV (alcohol by volume) reading of the beer. Since the collection of duty in the United Kingdom is based on the alcohol content of the beer as it leaves the brewery, this last requirement is exceptionally important. Fortunately, glucose syrups meet both of those requirements.

Plate 1 Inside a typical brewhouse.
Photograph courtesy of Briggs of Burton PLC and Carlton and United Breweries, Australia.

10.2 Brewing process

A simplified brewing process is as follows:

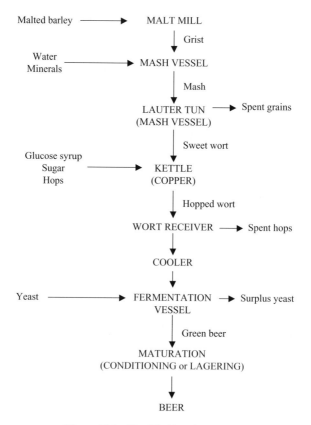

Figure 10.1 Simplified brewing process.

The object of the malt mill is to produce a grist which has the optimum particle size for mashing, conversion and wort extraction.

At the mashing stage, warm water is mixed with the grist which allows the enzymes present in the crushed malted barley to convert the starch into soluble sugars, which together with other soluble products will go to make up the wort. Sometimes at this stage, enzymes are added so as to improve the conversion of the starch, thereby increasing the fermentability of the wort.

The mash (wort and spent grains) now passes to the lauter tun where the spent grains and fibres are separated from the wort to produce 'sweet wort'.

The sweet wort passes to the kettle, where glucose syrups, sugar and hops are added. The hops and hop-based products are added at this stage to impart flavour. Glucose syrup and sugar are added to supplement the soluble sugars from the grist. During the boiling, flavours are released from the hops; the boiling sterilises the sweet wort and coagulates any soluble proteins. It is therefore important that the design of the kettle allows for adequate heating and agitation to ensure an even distribution of flavour and sugars and maximum protein coagulation. Since proteins contain nitrogen, it is important that they are removed.

Any excess of nitrogen will have a detrimental effect on the stability of the final beer. The wort at this stage is now referred to as 'hopped wort'.

The hopped wort passes to the wort receiver where the hops and coagulated proteins are removed prior to fermentation.

Typical temperatures for fermentation are between 10°C and 17°C, depending upon whether the hopped wort is for lagers or ales. Therefore, the hopped wort from the kettle passes through a cooler to bring it to the optimum temperature.

During fermentation, the yeast converts the fermentable sugars into alcohol.

Trisaccharides: Slowly fermentable sugars, e.g. maltotriose.

$$C_{18}H_{32}O_{16} + 2H_2O \longrightarrow 3C_6H_{12}O_6$$
$$100 \text{ grams} \qquad \qquad 107 \text{ grams}$$

Disaccharides: Readily fermentable sugars, e.g. maltose and sucrose.

$$C_{12}H_{22}O_{11} + H_2O \longrightarrow 2C_6H_{12}O_6$$
$$100 \text{ grams} \qquad \qquad 105 \text{ grams}$$

Monosaccharides: Readily fermentable sugars, e.g. dextrose.

$$C_6H_{12}O_6 \longrightarrow 2C_2H_5OH + 2CO_2$$
$$\text{Dextrose} \qquad \text{Ethanol} + \text{Carbon dioxide}$$

It is interesting to note that whilst both sucrose and maltose are disaccharide sugars, they are hydrolysed differently. Sucrose is hydrolysed outside the yeast cell into dextrose and fructose by the enzyme invertase which the yeast cell secretes, but maltose is hydrolysed inside the yeast cell. However, fermentation always takes place inside the yeast cell, therefore the products of sucrose hydrolysis have to pass through the cell wall before they can be fermented.

During the fermentation stage, there is a chemical gain due to the inclusion of water into the dextrose molecule. This gain is known as the hydrolysis gain, and will affect the gravity of the wort. Since the amount of alcohol produced is monitored by determining the change in the specific gravity of the hopped wort, this gain in weight must be taken into account.

Having reached the required alcohol content, the yeast is separated from the fermented wort to give 'green beer' which now undergoes maturation, that is conditioning or lagering. The object of maturation is to produce a clear product with a stable flavour, and involves storing the green beer at a low temperature followed by filtration prior to packaging into a suitable container.

10.3 Historical use of glucose syrups

Whilst glucose syrups have been used on and off in the UK beer since 1836, in the early days, their use was either for adding colour to the beer or blended with invert, and sometimes made into solid blocks. These blocks would have been put into the kettle (also called a copper, hence the name 'copper sugars'). It must be realised that both the fermentability of these early glucose syrups and their colour were a bit of an unknown, and probably very variable. Since those early days, both the colour and quality of invert containing brewing syrups did improve to become a useful adjunct.

It is a little more difficult to establish just when glucose syrups were first used in the United States for brewing. During the years of Prohibition, which existed from 1920 to 1933, the maximum alcohol content allowed in a beer or any other drink was 0.5%. To get around this law, drinking clubs known as 'speakeasies' were established, where stronger liquor was illegally served, and it is known that fermentable products from the glucose industry were used in the making of these stronger drinks. These fermentable products were not obtained directly from the glucose industry, but from a third or fourth party, with the preferred product being solid blocks of glucose, because they could be easily transported. These solid blocks of syrup are sometimes referred to as 'chip sugar'.

With the repeal of Prohibition in 1933, glucose syrups continued to be used by the brewers, although it was not until the 1950s, with the production of new types of syrup, that they became an acceptable adjunct. One reason for this acceptance was that the malt produced from American-grown barley contains more nitrogen than malt produced from European barley. This higher nitrogen content gave rise to unstable beers, but by adding glucose syrup to the sweet wort at the kettle, the high nitrogen level could be diluted to a more acceptable level, which resulted in the stability of the beer being greatly improved. With the stability of the beer being improved, brewers were then able to increase their sales from a local to national distribution.

The notable exception to the acceptance of adjuncts in beer is in Germany, where the German Purity Law of 1516 – the Reinheitsgebot – forbids the use of adjuncts, except in beers for export.

Within Europe, the acceptance of glucose syrups as a brewing adjunct was slower than in either the United Kingdom or America. A possible reason for this could be that continental breweries were often small family run breweries with their brewers being trained in the German or Czech way of brewing, where glucose syrups were not allowed. This state of affairs was to change after the Second World War with the formation of national and international brewing groups.

In the late 1930's and early 1940's, American glucose producers had started to use a combination of both acid and enzymes to produce glucose syrups, but due to the Second World War, this technology did not become available in the United Kingdom until the early 1950's at a time when, as a result of the war, sugar was scarce. One of the glucose syrups produced using this new technology was a high DE syrup, for example 90 DE, which was used to replace invert in brewing. However, the big breakthrough in the use of glucose syrups in brewing, particularly in the United Kingdom, came in 1958 when this acid enzyme technology was used to produce a 63 DE syrup. This particular syrup had a fermentability very similar to that of a malt wort, making it ideal for brewing where it could replace some of the more expensive malt and scarce sugar.

During this period, many of the small local breweries were being taken over by larger regional brewers who were building large modern breweries. With these larger breweries came the need to improve the storage and handling of their raw materials, and the new glucose syrups were ideal – they could be delivered in road tankers, stored in bulk tanks and when required pumped directly to the kettle. This amalgamation and expansion of breweries was not limited to the United Kingdom. It was a worldwide activity, including Europe, and has resulted in the formation of several large international brewing groups and the introduction of franchise brewing.

After the war, many Britons started taking their holidays abroad, where they were introduced to lagers and other light coloured beers. This introduction to new types of beer opened up new marketing opportunities for the brewers, with the UK public moving away from the traditional dark ales to the lighter coloured lagers. With this move away from traditional UK beers, the brewers started the franchise brewing of beers from other countries. Frequently, the sugar spectrum required for these franchised beers was often different to normal UK beers, and so the UK brewers started to ask for glucose syrups with different fermentabilities.

10.4 The role of glucose syrups

So what can glucose syrups offer the brewer? Ideally, the brewer is looking for a fermentable sugar spectrum which can match any particular wort and composed of the following:

- Readily fermentable sugars, such as dextrose and maltose, to produce alcohol.
- Slowly fermentable sugars such as maltotriose, which can also produce alcohol, but at a slower rate than the readily fermentable sugars.
- Non-fermentable sugars, which will provide body and mouthfeel to a beer, together with calories.

All of these sugar requirements can be supplied by a glucose syrup in different proportions, allowing the brewer to modify the fermentation characteristics of a wort so that a greater variety of beers can be produced and to manage any day to day recipe changes.

- Glucose syrups are an economical and consistent source of fermentable extract, which is immediately available upon delivery. Malt, on the other hand, once in the brewer's silos ties up capital and subsequently picks up extra processing costs with milling, mashing, filtration, etc. Similarly, granulated sugar has to be dissolved before use.
- Because syrups can be added directly to the kettle, it is possible to extend the brewhouse capacity without additional major capital expansion. By adding syrup directly to the kettle, it reduces any extract loss during mash filtration and allows for faster lautering times.
- By adding the syrup directly to the kettle, the gravity of the wort can be increased, enabling the wort to be used in high-gravity brewing.
- Because of the consistent quality of glucose syrups, they produce a more uniform wort and therefore a beer with a smooth, palatable and consistent taste.
- Since glucose syrups are low in nitrogen, their use will dilute both the nitrogen and non-starch contents of the wort, improve the protein break, and give improved beer stability.
- Glucose syrups have a low colour, so they will dilute any excess colour from the malts, which is important when making lagers or other light-coloured beers.

- Because of the refining process used in the manufacture of glucose syrups, they are very clean and not contaminated by mycotoxins, pesticides, nitrosamines or any undesirable metallic ions.
- Modern glucose refining ensures that no flavour is added to the beer from the syrup.
- Due to a combination of high sugar solids and sugar spectra, glucose syrups have a high osmotic pressure, which reduces microbiological spoilage of the syrup during storage.
- Controls the level of esters produced during fermentation, particularly in stronger beers which are not high-gravity brewed.

Table 10.1 illustrates the sugar analysis and fermentability of some typical glucose brewing syrups, however because of the way in which glucose syrups are produced, different syrups can be easily blended so as to match the fermentation characteristics of most worts regardless of the grist used to make that particular wort.

Table 10.1 Sugar analysis and fermentability of typical glucose brewing syrups.

	Malt Wort	62 DE	Maltose	High Maltose	95 DE
Sugars					
Percentage of dextrose	13	34	5	2	93
Percentage of maltose	52	33	50	70	4
Percentage of maltotriose	12	10	20	19	1
Percentage of higher sugars	23	23	25	9	2
Fermentable sugars					
Percentage of readily fermentable (dextrose and maltose)	65	67	55	72	97
Percentage of slowly fermentable (maltotriose)	12	10	20	19	1
Percentage of non-fermentable (higher sugars)	23	23	25	9	2
Minimum fermentability	77	77	75	92	98

Typical glucose syrup inclusion rates would be 10–35% on an extract basis, but some inclusion rates could be as high as 50%. However, there is always the risk that too high an inclusion rate could dilute the nitrogen level leaving the yeast insufficient nutrients to grow, which could result in a hung fermentation, and possibly off flavours being produced.

10.5 Low alcohol and low calorie beer

Low DE glucose syrups and maltodextrins, because they have very few fermentable sugars, can be used in the production of low-alcohol beers, and since they contain predominately higher sugars, which being non-fermentable will only add body and mouthfeel to the final beer but little or no alcohol, however the unfermentable higher sugars will contribute calories.

Since the calorie content of a beer is derived from both the alcohol and any unfermented sugars, one possible way to produce a low calorie, low alcohol beer might be a two stage approach. First, a totally fermentable syrup would have to be used such as a 95 DE syrup or dextrose in the fermentation. The resulting beer would contain a lot of alcohol, but with very little residual carbohydrate. With minimal calories coming from the carbohydrates, the alcohol could then be removed by evaporation, dialysis or reverse osmosis, however care would have to be taken to recover and return to the beer any flavours.

10.6 De-ionised glucose syrups

Why use a glucose syrup which has had all the minerals removed, when minerals are often added to the brewing water, i.e. Burtonising the water? The short answer is that it is not necessary to use de-ionised glucose syrup to produce a good beer, but there are certain advantages.

To understand why the UK brewers became interested in de-ionised syrups, one has to go back to the end of the Second World War. Because of the war effort, there had been little or no investment in new brewing equipment. Additionally, small breweries were being taken over or amalgamated, and there was a need to rebuild and expand production. These expansion programmes included the installation of bulk storage tanks for their raw materials, particularly for the delivery of glucose syrups by road tankers, thereby replacing the use of drums.

Traditionally, breweries have used copper for a lot of pipework, mash tuns, wort kettles, etc, but due to the Cold War, governments were stock piling strategic metals, such as copper, which now became very scarce. The breweries, therefore, looked to use stainless steel for their pipes and tanks.

At this time, the glucose industry was still making syrups using hydrochloric acid to hydrolyse the starch, with the reaction being stopped by neutralising the acid with sodium carbonate to form sodium chloride. This resulted in the glucose syrups having a chloride content of about 2000 ppm.

Unfortunately, the stainless steel used for the new storage tanks and pipework was the 314 grade, which was not resistant to chloride attack. It was therefore only a matter of time before the chloride from the syrup started to attack the welds of the stainless steel joints, particularly around stressed flanged areas, resulting in tanks and pipework collapsing. The solution therefore was to either have the expense of replacing all the 314 stainless steel with the chloride-resistant 316 stainless steel, or to use de-ionised glucose syrups, which would have a chloride content of less than 30 ppm.

Prior to 1975, de-ionised syrups were only available in very small quantities, and at a premium, but after 1975, they became freely available at an affordable price. This overcame the problem of the chloride attacking the stainless steel, and as this happened at a time when more and more breweries were using glucose syrups, the brewers changed their glucose specifications to include de-ionised syrup.

The downside of using de-ionised syrups is that chlorides and other ions have a beneficial effect on the flavour of the beer. But, by using de-ionised syrups, it was now easier to calculate the amount of minerals to add when 'Burtonising' the water. A further advantage is that de-ionised syrups have improved colour and colour stability, and because they do not contain any trace minerals necessary for yeast and bacteria to live, the syrups are virtually self-sterilising, thereby reducing the risk of a possible source of 'foreign yeast' infecting the breweries own yeast.

10.7 High gravity brewing

High gravity brewing was originally a technique used by brewers to increase the brewhouse capacity, without having to invest in new gristmills, mash vessels or lauter tuns, but over time, high-gravity brewing has proved so successful that it is now regarded as an economic necessity. The only new pieces of equipment required are extra storage tanks and a plant to produce sterile, deaerated and carbonated water.

Basically, the soluble solids of the wort are increased, and after fermentation, the beer is then diluted with sterile, deareated and carbonated water prior to bottling or canning. The skill in using high-gravity brewing is to be able to produce a wort which when fermented and diluted still tastes like beer and not like beer which has been diluted with water!

It is very difficult to easily increase the solids of a wort using traditional brewing techniques. One approach is to recycle low gravity wort, but a far easier way to increase the solids is by the addition of glucose syrup to the wort in the kettle.

Typical percentage solids of a traditional 1048 wort (12 degrees Plato) would be 12%, but glucose syrups, on the other hand, have a solids content of up to 82%, therefore, by adding a glucose syrup at the kettle, the solids of the wort can be significantly increased. Additionally, the fermentability of the glucose syrup can be tailored to exactly match the fermentability of the wort. For example, both wort and a 63 DE syrup have the same minimum fermentability. Alternatively, the addition of a glucose syrup can be used as a way of changing the fermentation characteristics of the wort.

Whilst increasing brewery capacity is often the main reason for high-gravity brewing, there are several other advantages. As well as allowing a higher proportion of unmalted carbohydrate adjuncts, for example glucose syrup, to be used, some brewers also consider high-gravity brewing to be more environmentally acceptable.

High-gravity brewing however requires careful yeast management to prevent the formation of esters, particularly ethyl acetate and isoamyl acetate. When a high-maltose syrup is used to make the wort, then the formation of esters is greatly reduced.

By using high-gravity brewing, a typical increase in capacity could be in the order of 10–30%.

10.7.1 High gravity brewing calculations

For the non-brewer, the following is an illustration of how to calculate the amount of syrup required to increase the solids of a wort.

A brewery produces 1500 HL of wort with a specific gravity of 1048 (12 degrees Plato) and wishes to increase the gravity to 1072 (18 degrees Plato) by the addition of some 63 DE glucose syrup. The solids of the glucose syrup are 82% and the specific gravity is 1420. How much syrup should be added? This can be calculated by using Pearson's Square.

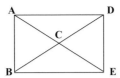

A = Original specific gravity = 1048
B = Specific gravity of 63 DE syrup = 1420
C = Required final specific gravity = 1072
D = Difference between **B** and **C** = 348
E = Difference between **C** and **A** = 24

$$\text{Amount of syrup to be added} = \frac{\text{Original volume} \times E}{D}$$

$$= \frac{1500 \times 24}{348}$$

$$= 103.45 \text{ HL}$$

$$\text{New volume of wort} = 1500 + 103.45$$

$$= 1603.45 \text{ HL at 1072 specific gravity}$$

This gives an increase in wort production of 6.9%, which when diluted back to 1048 gravity, the increase would be about 60%.

10.8 Brewer's extract – cost calculations

During the mashing process, the crushed malt is mixed with warm water, and the enzymes activated during the malting process degrade the starch and protein into soluble products. The solution of these soluble products is called **wort**.

The quantity and quality of the solubles which make up the wort is known as the **extract**, or **brewer's extract**, and is composed mainly of sugars (glucose, maltose, etc.) and soluble proteins.

The amount of extract available from either malt or glucose syrup is measured in **litre degrees per kilo**. In the case of glucose syrup, the brewer's extract can be calculated from the solids of the syrup. Extract is solids dependent – the higher the solids, the higher the extract. Solids can be measured using a refractometer calibrated to measure sucrose solids.

$$\text{Extract} = \frac{\text{Syrup solids}}{0.2549} \text{ litre degrees per kilo}$$

Basically, the sugar composition of wort is very similar to an unrefined glucose syrup.

Generally, the amount of extract in malt is about 80–82%, but not all of this will be extracted during the mashing process, possibly only 95% will be recovered. As it is the sugars which the yeast will convert into alcohol, it becomes commercially very important that the yield is as high as possible. In the case of a glucose syrup or sugar, all of the sugars present are totally soluble and available, and so the extract yield is 100%.

Different types of malt will have different extract values.

When a brewer talks about a malt with 80% extract, it refers to the soluble solids it contains, and they should be considered in the same way as glucose syrup solids. Therefore, a malt with 80% extract is the same as an 80 solids glucose syrup, and the brewer's extract can be calculated in exactly the same way.

$$\text{Extract} = \frac{80}{0.2549}$$
$$= 314 \text{ litre degrees per kilo}$$

The problem is knowing how efficient is the mashing process, and hence the yield. Table 10.2 shows the relationship between the % malt extract and yields, and the effect on the brewer's extract.

Table 10.2 The relationship between % malt extract and yield.

Percentage of malt extract	Yield			
	100%	95%	92%	90%
80%	314	298	289	283
81%	318	302	293	286
82%	322	306	296	290

Some brewers will add enzymes to the mashing, so as to increase the yields, but enzymes cost money!

The brewer knowing the cost of his malt and the amount of extract which can be obtained will be able to calculate the price of the extract.

$$\frac{\text{Market price of malt}}{\text{Brewer's Extract}} = \text{Price of extract litre degrees per kilo}$$

The brewer will then compare the cost of the extract with the price of a glucose syrup using the same calculation.

N.B. In the following examples, the prices used are by way of illustrating the principles involved in costing an extract.

Example 1: Glucose syrup

(a) Glucose syrup cost $355.
(b) Glucose syrup solids = 75% (all solids are 100% available).
(c) Extract of glucose syrup = $\frac{75}{0.2549}$
= 294.2 litre degrees per kilo

(d) Cost of extract $= \dfrac{355}{294.2}$

$= \$1.207$ per litre degrees per kilo of solids

Example 2: Malt

Assume a lager malt, with an extract of 302 litre degrees per kilo.

(a) Cost of malt $311

(b) Cost of extract $= \dfrac{311}{302}$

$= \$1.03$

Example 3: Malt

If the brewer had paid $311 for malt with 81% extract, and the mashing yield is only 95%, then from Table 10.2 the extract would be 302 litre degrees per kilo. The price however would **not** be $311, but 5% higher, that is $327.

$$\text{Cost of extract} = \dfrac{327}{302}$$

$= \$1.08$ per litre degrees per kilo of solids

Example 4

Some brewers will be using many different ingredients in their mashing and they will calculate the total cost of the extract as follows:

Malt   – 80% @ $1.08 per litre degrees per kilo of solid = $0.864

Sugar   – 10% @ $1.18 per litre degrees per kilo of solid = $0.118

Glucose – 10% @ $1.21 per litre degrees per kilo of solid = $0.121

Total cost of extract = $1.103

Therefore, total cost of extract would be $1.103 per litre degrees per kilo of solids.

10.8.1 Typical extract values (hot water)

Source: Adapted from Brewing, by Lewis & Young, published by Chapman & Hall.

Sugars

Invert sugar	320 litre degrees per kilo
'Solid sugar' from maize	310
Sugar (dry)	384
Malt extract	300
Liquid sugar (66% w/w solids)	260

Glucose syrups

The brewer's extract of a glucose syrup is dependent upon the solids:

$$\text{Extract} = \frac{\text{Solids}}{0.2549}$$

Table 10.3 shows the brewer's extract of some typical glucose syrups used in brewing.

Table 10.3 Brewer's extract for some typical glucose syrups.

Syrup type	Minimum fermentability	True solids	Brewer's extract (litre degrees per kilo)
95 DE	98	67.7–69.1	266–271
95 DE	98	73.4–75.4	288–295
63 DE	77	78.7–79.9	309–313
50% Maltose	75	80.1–81.1	314–318
70% Maltose	92	68.6–69.6	269–274
70% Maltose	92	78.9–79.9	309–313

Cereal

Maize grits	340 litre degrees per kilo, dry basis
Maize flakes	340
Maize starch	352
Rice flake	360
Wheat flour	340
Wheat torrified	300
Barley flakes	280
Barley torrified	270

Flakes: These are produced by steam heating a cereal grain which is then passed through heated rollers (85°C). The resulting thin flakes and the destruction of the cellular structure allows for easy penetration by the enzymes of the malt.

Torrified: This is a process in which cereals grains are subjected to infrared heat. The heat turns the moisture inside the cereal grain into steam causing the cereal grain to expand and eventually rupture. The internal structure of the cereal grain is disrupted and the starch is partly gelatinised, opening up the grain for subsequent attack by the enzymes of the malt during the mashing process.

Special malts and roasted barleys

Measurements in litre degrees per kilo, on dry basis.

Crystal malt	268	Colour = 200 EBC
Carapils	260	Colour = 30 EBC
Amber	275	Colour = 50 EBC

Chocolate	270	Colour = 1000 EBC
Roasted	265	Colour = 1200 EBC
Roasted barley	270	Colour = 1350 EBC
Pale or lager malt	302	Colour = 3–8 EBC

Roasting: This is a process whereby the properties of a barley are changed by heating the barley in a slowly rotating drum. The heating causes both a Maillard reaction between the proteins and the sugars of the barley together with the caramelisation of the sugars to give the malted and roasted barley their characteristic colour and flavour. By altering the heating regime, both the colour and flavour can be modified to give a product suitable for use in light-coloured beers such as lagers or a product for use in very dark beers such as pale ales.

10.8.2 Brewing syrup addition calculations

Typical analysis of a maltose syrup

Total solids = 78.8%
Density at 20°C = 1.394 (This will vary with temperature.)

HPLC sugar analysis:
Dextrose = 18.1%
Maltose = 55.0%
Maltotriose = 8.8%
Higher sugars = 18.1%

1. 'Total solids' means that one tonne will contain 788 kilos of syrup.
2. The HPLC sugar analysis shows the amount of sugars present in the syrup, expressed as a percentage on a dry basis.
   (a) 18.1% Dextrose means that of the 788 kilos of solids
       18.1% is dextrose = 143 kilos.
   (b) 55.0% Maltose means that of the 788 kilos of solids
       55.0% is maltose = 433 kilos.
   (c) 8.8% Maltotriose means that of the 788 kilos of solids
       8.8% is maltotriose = 69 kilos.
   (d) 18.1% Higher sugars means that of the 788 kilos of solids
       18.1% are higher sugars = 143 kilos.
       Total = 788 kilos

A. **Determining syrup addition on a weight basis**

1. If you require 500 kilos of maltose solids.
   From the HPLC results, there are 433 kilos of maltose present in 1000 kilos of syrup. Therefore, 500 kilos of maltose will come from

   $$\frac{1000 \times 500}{433} = 1155 \text{ kilos} = 1.155 \text{ tonnes of syrup @ 78.8\% solids.}$$

2. If you require 500 kilos of fermentable solids.
   The fermentable solids of the maltose syrup are made up of the following:

   $$\begin{aligned} \text{Dextrose} &= 143 \text{ kilos} \\ \text{Maltose} &= 433 \text{ kilos} \\ \text{Maltotriose} &= 69 \text{ kilos} \end{aligned}$$

   This gives a total of 645 kilos of fermentable solids in one tonne of syrup. Therefore, 500 kilos of fermentable solids will come from

   $$\frac{1000 \times 500}{645} = 775 \text{ kilos} = 0.775 \text{ tonnes of syrup @ 78.8 solids.}$$

B. **Determining syrup addition on a volume basis**

The density of the syrup is 1.394. This means that 1000 litres of syrup will weigh 1394 kilos. Therefore, 1394 kilos of syrup at 78.8% solids will contain 1098 kilos of solids, and from the HPLC analysis, we have the following:

(a) 18.1% dextrose will be present in 1098 kilos of solids $= 199$ kilos
(b) 55.0% maltose will be present in 1098 kilos of solids $= 604$ kilos
(c) 8.8% maltotriose will be present in 1098 kilos of solids $= 96$ kilos
(d) 18.1% higher sugars will be present in 1098 kilos of solids $= 199$ kilos
$\text{Total} = 1098$ kilos

1. If you require 500 kilos of maltose solids.
   Hundred litres will contain 604 kilos of maltose. Therefore, 500 kilos will come from

   $$\frac{1000 \times 500}{604} = 828 \text{ litres of syrup with a density of 1.394.}$$

2. If you require 500 kilos of fermentable solids.
   The fermentable solids of maltose syrup are made up of the following:

   $$\begin{aligned} \text{Dextrose} &= 199 \text{ kilos} \\ \text{Maltose} &= 604 \text{ kilos} \\ \text{Maltotriose} &= 96 \text{ kilos} \end{aligned}$$

   This gives a total of 899 kilos of fermentable solids in 1000 litres of syrup. Therefore, 500 kilos of fermentable solids will come from

   $$\frac{1000 \times 500}{899} = 556 \text{ litres of syrup with a density of 1.394}$$

10.9 Chip sugar

Before glucose syrups were readily available, the brewers would use as an adjunct a product called 'chip sugar'. This was a blend of invert and 42 DE glucose syrup, which solidified into a block, weighing about 56 lb (25 Kg). When required, the blocks were either put into the kettle with the hope that the turbulence of the boiling wort would dissolve the chip sugar or alternatively the block of sugar would be placed in a wire cage, which would be lowered into the boiling wort. The reason for using chip sugar in those days was exactly the same as when using any brewing adjunct today – namely scarcity of ingredients, price or quality considerations. With the development of larger breweries, it became more convenient to have the glucose syrup delivered in bulk tankers, and since glucose syrups could be pumped directly into the copper, there was a cost saving in not having to handle the blocks of chip sugar. With these new developments, there was no longer a demand for chip sugar. Additionally, the old fashioned chip sugar could not offer the total versatility that a modern syrup can offer.

With the advent of micro-brewing, the possibility of using chip sugar could once again become of interest to the small brewer. Whilst micro-brewers tend to be craft brewers with an empathy for the German Reinheitsgebot purity approach, there could be some commercial and operational advantages for using chip sugar especially if they are small, with limited space and a small syrup requirement which could not justify a bulk tank installation. By using chip sugar, there would be no major capital investment required, and it would be easier to produce different types of beers.

10.9.1 How to make chip sugar

Instead of using invert to make chip sugar, a 95 DE syrup can be used, however it is important that the dextrose content of the syrup is 95% minimum. If the dextrose content is less than 95%, the chip sugar might not form a hard solid block. Chip sugar can be made as follows:

1. Make a syrup blend containing 70% 95 DE syrup and 30% 42 DE on a dry basis.
   95 DE Syrup (75% solids) = 933 grams
   42 DE Syrup (82% solids) = 366 grams
2. The total solids of the above blend will be about 77%, therefore the blend will have to be evaporated up to 85% solids.
3. After evaporating up to 85% solids, the blend is cooled to about 60°C. Stir in 50 grams of dextrose (5% dextrose based on solids) to act as a seed for the 95 DE to crystallise. If the temperature is too high, the dextrose will dissolve, and will no longer be able to act as a seed.
4. Pour the mix into suitable moulds. Solidification should be complete within 48 hours. Store in a cool dry area.
   A suitable mould can be made using a plastic bag as a liner, which is placed inside a cardboard container. A suitable size container could measure about 50 cm × 30 cm × 30 cm. It is important that the solid block can be lifted without causing injury to the operator. A typical solid block would weigh about 25 kilos.
5. To use the block of chip sugar, remove it from its container and plastic liner, and place into the kettle.
   In the above recipe, a 42 DE syrup was used but other glucose syrups can be used to change the fermentable characteristics of the chip sugar. Table 10.4 illustrates the

possible sugar spectra and fermentabilities which can be obtained by using maltose and 63 DE syrups, at a 30% dry basis, to replace the 42 DE syrup. The four different maltose syrups contained 28%, 41%, 50% and 70% maltose, respectively.

Table 10.4 Sugar spectra and fermentabilities for chip sugar, using different syrup blends.

	42 DE Glucose Syrup	63 DE Glucose Syrup	Maltose Syrup (28% maltose)	Maltose Syrup (41% maltose)	Maltose Syrup (50% maltose)	Maltose Syrup (70% maltose)
Percentage of sugars						
Dextrose	72.0	77.2	67.3	68.2	67.3	66.4
Maltose	5.0	12.7	11.2	15.1	17.8	23.8
Maltotriose	40	2.5	8.2	5.2	6.7	6.4
Higher sugars	19.0	7.6	13.3	11.5	8.2	3.4
Percentage of fermentable sugars						
Readily	77.0	89.9	78.5	83.3	85.1	90.2
Slowly	4.0	2.5	8.2	5.2	6.7	6.4
Non-fermentable	19.0	7.6	13.3	11.5	8.2	3.4
Minimum fermentability						
	81.0	92.4	86.7	88.5	91.8	96.6

Chapter 11
Glucose syrups in confectionery

11.1 Introduction

The confectionery industry can be divided into two categories – chocolate or sugar confectionery.

The chocolate companies are not big users of glucose syrups compared to sugar confectionery companies, because the water present in glucose syrups is incompatible with the chocolate process – even the 10% moisture in dextrose will cause problems if used to replace sucrose in chocolate crum, but spray-dried glucose syrups, which have a moisture content of less than 1%, have been successfully used as a partial replacement for sucrose. As a general rule, the moisture content of ingredients used in chocolate work has to be less than 1%, however, where chocolate companies use glucose syrups, it is mainly for the centres of chocolates such as chocolate-covered soft caramels, pectin jellies, etc.

The sugar confectionery companies, that is those companies which make hard boiled sweets, toffees, caramels, fudge, fruit gums, etc, are possibly one of the largest users of glucose syrup. In Europe, confectionery companies mainly use acid converted 42 DE (dextrose equivalent) glucose syrup, often referred to as 'Regular' or 'Confectioner's Glucose', with lesser amounts of other syrups. In America, however, both 42–49 DE and 63–70 DE syrups are extensively used. The other main ingredient used by sugar confectioners is sucrose. So why are glucose syrups used in sugar confectionery when they have half the sweetness of sucrose?

11.2 What can glucose syrups offer the confectioner?

Glucose syrups can offer the following functional properties:

- Control sucrose crystallisation and graining
- Reduce moisture pickup
- Reduce cold flow
- Improve processing
- Modify the sweetness
- Modify the texture

11.2.1 Control of sucrose crystallisation and graining

The crystallisation of sucrose in confectionery is essential in developing the texture of many types of confectionery such as fondants, fudge, grained mints and short nougat. The use of a glucose syrup helps in the development of the very fine sucrose crystals in a controlled manner, as well as adding viscosity and soluble saccharide solids to the product.

On the other hand, sucrose crystals are undesirable in other types of confectionery where they can produce a gritty texture, or in the case of a high boiling mask the clarity of the product.

It is for these reasons that sucrose crystallisation should wherever possible be carefully controlled.

A traditional high boiled confection will have a moisture content of less than 2% and will typically contain about 60% sucrose and 40% glucose on a dry basis, although this ratio varies. The maximum solubility of sucrose at 80°C is only 79%. Above 79% the sucrose, solution becomes supersaturated, so if only sucrose was used to make the high-boiled sweet, it would crystallise out of solution before the final moisture content of 2% could be reached. But by adding a glucose syrup, sucrose crystallisation is prevented in two ways. Firstly, the higher sugars of the syrup increase the viscosity of the mix and this increase in viscosity prevents the sucrose molecules coming together to form crystals. Secondly, the presence of the other sugars in the glucose syrup increases the solubility of the sucrose, thereby reducing the risk of sucrose crystallisation. However, if a syrup with a high dextrose content were to be used, because the solubility of dextrose is less than that of sucrose, then the additional increase in dextrose would cause problems of dextrose crystallisation.

Originally, sucrose crystallisation was controlled by inverting some of the sucrose by using either an acid such as tartaric acid or an acidic salt such as cream of tartar (potassium hydrogen tartrate) to give an equal amount of dextrose and fructose. See figure 11.1.

$$\text{Sucrose + Water} \xrightarrow{\text{Heat + Acid}} \text{Dextrose + Fructose}$$

Figure 11.1 Inversion of sucrose.

As mentioned previously, in a high boiled sweet, because of the solubility of sucrose, there is the risk of it crystallising, but by inverting some of the sucrose, the concentration of sucrose is effectively reduced, thereby reducing the risk of sucrose crystallisation. If, however, too much dextrose and fructose are formed, then there is the risk that dextrose crystallisation might become the problem.

Tartaric acid was the preferred acid because the inversion process was more controllable than if citric or some other acid had been used. However, this method of controlling crystallisation was unreliable because the inversion process could not be easily controlled, so the addition of an acid converted 42 DE glucose syrup became the accepted practice.

Another reason for using a glucose syrup in preference to using invert is that invert does not contain any higher sugars and lacks viscosity compared to a glucose syrup. (Invert or invert syrup is a mixture of dextrose and fructose or a mixture of sucrose, dextrose and fructose, depending upon the proportion of sucrose which has been inverted). The reason why an acid-converted syrup was originally used as opposed to an enzyme converted syrup was because the production of enzyme converted syrups was not sufficiently advanced until the 1950s.

The products originally used for inverting the sucrose, that is cream of tartare etc, were referred to as 'doctors', because they controlled sucrose crystallisation, and as a glucose syrup also controls crystallisation, they too are sometimes referred to as a 'doctor'.

Undesirable graining in confectionery can be of two types. One is when the confectionery piece has a gritty texture due to undissolved sucrose crystals and the other which is common with high boiled sweets is when the sweet becomes cloudy and sticky. Both of these types of graining are due to an imbalance of sugars in the sweet. This imbalance can be due to several reasons – the incorrect choice of glucose syrup, an incorrect sugar/syrup ratio or too high a process inversion from the cooking process. Whatever the cause, the end result is that the reducing sugar content is too high, and this sets off a chain reaction.

- Too high a reducing sugar content makes the sweet hygroscopic.
- Hygroscopic sweets pick up moisture from their surroundings.
- This increase in moisture content forms a dilute sugar solution on the surface of the sweet making the surface sticky. This dilute sugar solution also reduces the surface viscosity, which allows the sucrose molecules to migrate and nucleate, and then crystallise.
- This surface crystallisation gradually, over a period of time, advances further into the centre of the sweet, giving the sweet the characteristic cloudy appearance of graining.
- In some instances, if there is a predominance of dextrose in a recipe, the dextrose will crystallise.

11.2.2 Reduce moisture pick up

Whilst moisture resistant packaging will go some way to reduce moisture pick up, the part played by the sugar spectrum of the sweet is equally important as discussed previously. Therefore, the correct choice of glucose syrup is important together with the correct manufacturing parameters, particularly the pH so as to reduce process inversion – if the pH is too low there will be excessive process inversion. Ideally, the glucose syrup should contain minimal amounts of high DE sugars, that is dextrose and fructose. Maltose syrups, on the other hand, with their low dextrose content, pick up moisture at a slower rate than a conventional acid converted glucose syrup, which makes them the ideal syrup to use in confectionery production in countries which have a high humidity.

In some cases, where a dilute syrup has been formed due to moisture pick up, there is the possibility that moulds will grow.

11.2.3 Reduce cold flow

Cold flow occurs when a product changes its shape over a period of time, for example during storage, and is often temperature and moisture related. A good example of cold flow can be seen in confectionery products which contain a fat such as chocolate. At low temperatures, the fat is solid, but if the temperature is increased, the fat changes from a solid to a liquid. If the fat was acting as a binder when cold giving rigidity to the product, when it turns into a liquid, it can no longer give rigidity to the product or have any binding capacity, and so the product is no longer held together and the product becomes deformed.

With a glucose syrup, it is the large molecules of the higher sugars which give rigidity to the sweet, and prevents cold flow. Also, glucose syrups which have a low DE, that is syrups which contain lots of higher sugars, tend to produce a sweet with a tougher bite.

As mentioned previously, temperature also affects cold flow and this is connected to the glass transition temperature (T_g) values of the various sugars. The glass transition value is the temperature at which an amorphous glass softens, and as a typical high boiled sweet is a non-crystalline glassy product, it too will soften with increasing temperature. However, the higher the T_g value of the sugars used to make the sweet, the greater will be the heat stability of the sweet to thermal deformation. Typical T_g values for different sugars are as follows:

Sugar	T_g
Sucrose	52°C
Dextrose	31°C
Maltose	43°C
Trehalose	79°C

The cold flow can however be changed by using sugars with different T_g values. For example, by adding trehalose to sucrose, the thermal stability of the blend would be increased.

Cold flow will also occur when the moisture content of the sweet is too high. This can happen when either the solids are too low due to insufficient cooking which is frequently caused by a faulty thermometer, or if the sweet contains too high a proportion of high DE sugars, such as dextrose and fructose, which make the sweet hygroscopic. This increase in moisture allows the sugars to form a syrup which allows the sweet to become deformed, and this deformation is greatly increased if the temperature is increased. The reason for this is because the viscosity of the syrup is reduced, which allows it to flow at a quicker rate.

11.2.4 Improve processing

The viscosity and hence the ease of processing a high boiled sugar mass depends upon its temperature, solids and the sugars which it contains. Therefore, the correct choice of which glucose syrup to use is important.

High boiled sweets are produced by either depositing the cooked mass into non-stick metal moulds (Plate 2) or by producing a 'rope' from which individual sweets are made.

Where a high boiled sweet is made by depositing, the hot mass is deposited into non-stick moulds which are carried on a chain conveyor (Plate 3). As the moulds pass under the hopper containing the hot liquid mass, a predetermined quantity of the cooked mass is deposited into the non-stick moulds (Plate 4). It is important that there is no tailing from the hopper after each mould is filled. The moulds then pass through a cooling tunnel (Plate 5), where the mass solidifies, and is then ejected from the mould by a small plastic rod, which is situated at the bottom of the mould. For there to be no tailing from the hopper, there must be minimal higher sugars and hence minimal viscosity in the cooked mass, and this can only be achieved by choosing the most suitable glucose syrup for the equipment. The use of a very high maltose syrup which contains minimal higher sugars will reduce the tailing problem. The correct moisture content of the cooked mix is also important to reduce tailing.

Glucose syrups in confectionery 153

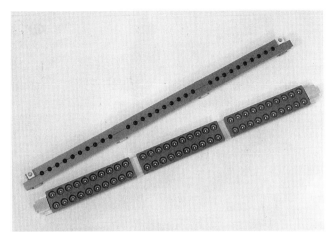

Plate 2 Individual non stick moulds. The white spots at the centre of the moulds are the plastic rods which remove the sweets from the moulds. Photograph courtesy of Baker Perkins Ltd.

Plate 3 Non stick moulds on the chain conveyor. Photograph courtesy of Baker Perkins Ltd.

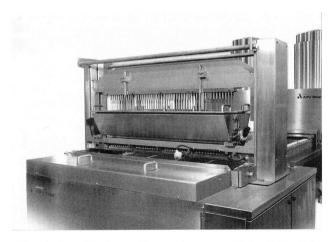

Plate 4 Depositing hopper. Photograph courtesy of Baker Perkins Ltd.

Plate 5 Depositing hopper with cooling tunnel in the background. Photograph courtesy of Baker Perkins Ltd.

The reason why it is important to prevent tailing is because tailing produces deformed sweets which cannot be sold.

This type of process can be used to make nearly all deposited confectionery (Plate 6), except for jelly confectionery which is deposited into starch. However, for confectionery made using starch moulds, the requirement of minimal tailing is still applicable. Starch moulds are used instead of metal moulds where a skin is required to form on the outside of the confectionery, and this is achieved by the starch absorbing moisture from the surface of the confectionery.

Plate 6 A selection of deposited confectionery. Photograph courtesy of Baker Perkins Ltd.

When sweets are made using a roping process, the semi-solid cooked mass is placed in batch rollers, which consists of three or four conical shaped oscillating rollers. These rollers manipulate the hot mass in such a way so as to produce a 'sausage' which is tapered at one end. The tapered end then passes through several rollers gradually further reducing the diameter of the 'sausage' to a rope of the required size. The rope now passes through a roller which forms individual sweets by pressing and cutting it into individual pieces. In this process, tailing is not important. What is important in this process, however, is that that the semi-solid cooked mass is sufficiently plastic for the conical rollers to manipulate it to produce a rope – if the viscosity is too low, the conical rollers would be unable to manipulate the cooked mass to produce a rope. So as to produce a cooked mass with sufficient viscosity, it is best to use a glucose syrup containing a high proportion of higher sugars, such as a 42 or 35 DE syrup.

This roping process is frequently used to make confectionery with soft centres such as toffee caramels. The soft centre is injected into the 'sausage' so that it is co-extruded prior to it being made into a rope. This process is also used for making striped confectionery such as black and white striped humbugs, where two or more semi-solid cooked masses of different colours are made up into layers prior to being placed into the conical batch rollers.

11.2.5 Modify the sweetness

As glucose syrups, with the exception of fructose and HFGS, are less sweet than sucrose, the addition of a glucose syrup to a recipe will generally reduce the apparent sweetness of the confectionery. However, this reduction in sweetness will often enhance and allow other subtle flavours to come through, which would otherwise be overwhelmed by the sweetness of the sucrose.

11.2.6 Modifying texture

By using syrups with different properties, it is possible to alter the texture of a piece of confectionery. For example, a low DE syrup with lots of higher sugars will make a product more chewy. In the case of a fondant, it is the viscosity of the 42 DE syrup which helps to give the fondant its smooth texture, and if a 63 DE syrup was used to make a pectin jelly, it will have a softer bite than if it had been made with a 42 DE syrup.

11.3 Which glucose syrup to use?

Because different glucose syrups have different sugar compositions, the correct choice of glucose syrup is critical and will affect both the processing and shelf life of the product.

When considering which glucose syrup to use in a confectionery product, there are several different factors which should be taken into consideration. Ask the question, 'What is the objective?' As a general rule, consider the following:

- Low DE syrups contain lots of higher sugars, and higher sugars mean high viscosity, and high viscosity means a chewy product. Low DE syrups also tend to pick up less moisture.

- High DE syrups have less higher sugars and therefore have lower viscosity, and will produce a softer bite, but the higher the dextrose content of the syrup, the more hygroscopic will be the syrup together with the risk of dextrose crystallisation.

Table 11.1 Summary of the syrups most commonly used in sugar confectionery.

	Dextrose	Maltose	Maltotriose	Higher sugars	Fructose
35 DE	15%	12%	11%	62%	–
42 DE acid	19%	14%	11%	56%	–
42 DE enzyme	6%	44%	13%	37%	–
50 DE	30%	18%	13%	39%	–
63 DE	34%	33%	10%	23%	–
Maltose 70	2%	70%	20%	8%	–
HFGS 42	52%	4%	1%	1%	42%
Spray dried 95 DE syrup	93%	5%	1%	1%	–

35 DE Syrup

This has slightly more higher sugars than an acid produced 42 DE which makes it more viscous, but because it contains less dextrose, it is less hygroscopic than an acid produced 42 DE syrup. This makes the syrup ideal for confectionery which is either produced or sold in countries which have a high relative humidity. A 35 DE syrup is also equally good as an acid produced 42 DE at reducing sucrose crystallisation.

An increase in the hot viscosity of a freshly cooked high boiling made with a 35 DE syrup may be apparent. If so, this can be compensated by a slight increase in the working temperature through the forming equipment.

Acid produced 42 DE Syrup

This syrup is the most frequently used syrup in sugar confectionery, because of its ability to reduce sucrose crystallisation and have a manageable processing viscosity, which makes it ideal for use in most types of confectionery production. However, one exception could be in the manufacture of liquid-deposited jellies and marshmallows, where the lower viscosity of a 63 DE syrup may be an advantage.

Enzyme produced 42 DE Syrup

This syrup is strictly a maltose syrup, and contains less dextrose than a conventional acid produced 42 DE syrup. This reduction in dextrose gives this syrup several interesting properties:

- It is less hygroscopic making it very suitable for use in confectionery which is to be made and sold in countries with a high relative humidity.
- It is slightly less sweet.

- Because it contains less dextrose, it produces less Maillard Browning, resulting in lighter coloured confectionery compared to those made with a conventional acid 42 DE. To produce a darker sweet using a batch process, leave one-third of the previous batch in the cooker, which can then be included into the next batch, so that it is cooked twice, thereby enabling more colour and flavour to be developed. This is a particularly useful approach where the addition of caramel or other colours is to be avoided.

Additionally, because this syrup contains less higher sugars, it is less viscous than a conventional acid converted 42 DE syrup, and therefore is often used to advantage when depositing high boilings.

N.B. The totally different sugar composition of these two 42 DE syrups illustrates how important it is to compare the sugar composition of a syrup, rather than rely purely on the DE.

50 DE Syrup

This has extra sweetness compared to a 42 DE syrup, and being less viscous, it can be used in soft confectionery centres.

63 DE Syrup

This syrup is more sweet and less viscous than a 50 DE syrup. When used in confectionery, a 63 DE syrup will give the product a softer, less chewy texture for two reasons. First, a 63 DE syrup contains less higher sugars than an acid converted 42 DE syrup. The second reason is that a 63 DE syrup will act as a humectant and by picking up moisture from the air, the increase in moisture content of the product could result in a sticky product. This syrup can be used in the confectionery soft centres and fillings, also in jellies, and marshmallows.

70% Maltose Syrup

Similar to a 42 DE maltose syrup, but as it has minimal higher sugars, it has less viscosity and therefore produces very little tailing when used in the production of deposited confectionery. Because the syrup contains less dextrose, there is less Maillard Browning resulting in lighter coloured products. A darker coloured sweet can be produced by following the suggestion made for the enzyme produced 42 DE syrup.

HFGS 42

Being made up of dextrose and fructose, this syrup can be used to replace invert, acting as an excellent humectant. However, because of the high dextrose content, HFGS will crystallise unless stored at a temperature of 27–32°C.

Spray dried 95 DE Syrup

Because spray dried 95 DE syrup has a moisture content of less than 1%, it can be used in chocolate confectionery, where it can replace about 10% of the sucrose.

Spray dried 95 DE syrup can also be used to replace up to 50% of the icing sugar in a French paste, where moisture content in the final product is critical to ensure good storage properties.

Maltodextrins

These are products with a DE of less than 20, and are generally available as spray dried powders and have a low sweetness. Because they have a low DE, solutions of maltodextrins have a high viscosity. By choosing the correct maltodextrin, their high viscosity can be used to reduce sucrose crystallisation, act as a stabiliser for aerated products such as chews, and improve the body and chewy texture of confectionery products. Because of their low DE and hence their low hygroscopicity, solutions of maltodextrins can be used as a protective glaze on dragee-produced products.

11.4 Typical glucose syrup inclusion rates

When confectioners discuss sucrose to glucose syrup ratios in a high boiling, sucrose is always the first mentioned ingredient and expressed on a dry basis, with the glucose syrup always being second and expressed on an 'as is' basis. With other types of confectionery, the ingredients are usually declared in order of magnitude. High-pressure liquid chromatography (HPLC) results, however, express both the sucrose and glucose syrup on a dry basis. So beware – always compare like with like!

If a confectioner making high boilings is using a 50/50 blend of sucrose and glucose syrup and the glucose syrup solids are 80%, it will mean that the confectioner is using 50 parts of dry sucrose and 40 parts of syrup solids (i.e. 80% of 50), which means that there are only 90 parts of solids in the mix and not 100. By expressing the sucrose and glucose solids as a percentage of the total solids, the confectioner's 50/50 blend is a 55.5% of sucrose and 44.5% of glucose solids, and that is how the analysis would be reported by HPLC.

The amount of glucose syrup or sucrose used in a piece of confectionery will depend upon several different factors. Whilst technically the amount of glucose syrup used in some products can vary from 30% to 70%, the amount actually used might be decided not by technical considerations, but by economics – the price of either the syrup or the sucrose.

In some confectionery, the ratio of sucrose to glucose cannot be changed due to technical reasons. For example, in grained products such as fudge, sucrose has to be the major ingredient, because glucose syrup will not crystallise to form a grainy texture. In other grained products such as a fondant, glucose syrup is important because it controls the size of the sucrose crystals, and hence the texture.

The following simple summary chart shows how by starting with sucrose and glucose syrup, and by the addition of other ingredients, or by changing the processing conditions, different types of confectionery can be made. Confectionery production is still an art, with each manufacturer having their own particular recipes ranging from expensive quality to budget products, which means there is no one recipe for a given product but several, therefore the glucose syrup inclusion rates (given in brackets) should only be taken as a guide. The examples given in this simple chart are an illustration of some of the many uses of glucose syrup in confectionery products.

Typical glucose inclusion rates as a percentage of the total sweetener could be as follows (Table 11.2):

Table 11.2 Typical glucose syrup inclusion rates.

Confection	Ingredients	Process	Result
High boilings	Sucrose + 42 DE glucose syrup (30–70%)	Boil to 99% solids, and deposited or made into a rope	Clear hard boiled sweet
Pulled sugar confection, for example rock	Sucrose + 42 DE/35 DE glucose syrup (30–60%)	Boil to 99% solids. Then pull to induce sucrose crystallisation	An opaque, hard boiled sweet
Fondant	Sucrose + 42 DE glucose syrup (10–30%)	Boiled to 88% solids, then cooled with agitation to induce sucrose crystallisation	An opaque paste, containing sucrose crystals, suspended in glucose syrup
Toffee and caramels	Sucrose + 42 DE glucose syrup (30–70%) + milk solids + fat	Heat to 118°C to obtain a soft caramel, or to 131°C for a hard toffee	A brown paste or solid, with a characteristic flavour, due to the reaction of milk proteins and reducing sugars
Fudge	Basically a blend of a toffee with fondant (42 DE glucose syrup 20–40%)	The mixture is allowed to cool, so that the sucrose crystallises to give fudge its texture	A toffee, with a crystalline texture
Gums and jellies (including Turkish Delight) • Agar • Gelatine • Pectin • Starch	Sucrose + 42 DE/63 DE glucose syrup (30–70%), thickened with the appropriate gelling agent	The mixture is boiled to about 80% solids, thickened, then deposited into starch	Depending upon which gelling agent is used, and the cooking process, a range of sweets with different textures can be made

(Continued)

Table 11.2 (*Continued*)

Confection	Ingredients	Process	Result
Chews	Sucrose + 42 DE glucose syrup (60–65%) + gelatine + fat + whipping agent (also maltodextrin)	The mix is heated to about 125°C, and then whipped to incorporate as much air as possible. Moisture content 6–8%	An opaque chewy sweet
Marshmallows	Sucrose + 63 DE glucose syrup (50–55%) + gelatine	The mix is heated to about 107°C, then whipped to incorporate as much air as possible. Final moisture is 12–18%	A soft white mass which is frequently put onto a biscuit base, then enrobed with chocolate
Muesli bars	A mixture of vine fruits, nuts and cereals, bound together with sucrose + 42 DE glucose (also fat) (1 part of sugar solids, and/or fat, to 4 parts of muesli)	The sucrose and syrup are heated to about 135°C, which is then mixed into the other ingredients, and formed into bars	There are many variations on the ingredients added, also on the type of syrup used. If required, the bars can then be covered with chocolate

From these very basic products, there is then the possibility of making more products by either changing some of the ingredients, or combining two or more different products, such as hard boiled sweet with a soft caramel centre.

Whilst it might be desirable to have a laboratory pilot plant for making confectionery, this is not essential. A lot of basic initial laboratory work can be carried out using very simple equipment such as a stainless steel domestic milk saucepan with a pouring lip, hot plate, wooden spatula, digital thermometer and weighing scales, etc. There are two problems when using a pilot plant – firstly, they are expensive on ingredients because large quantities of ingredients are required, and secondly, there is the problem of disposing of failed experiments – and in application development work, there can be many failed experiments.

When making any confectionery, it is important to check the moisture content of the end product – incorrect moisture content will affect the storage life. The moisture content can be checked by cooking to a certain temperature and then checking the solids using a refractometer. Another approach is to weigh the saucepan and ingredients and then calculate the amount of water which has to be boiled off. Next continually weigh and re-weigh the saucepan during the cooking process until the correct weight has been reached.

When the moisture content of confectionery is less than 2.5 % as in a high boiling, then the moisture content is best determined using a Karl Fischer apparatus.

11.5 Some basic confectionery recipes

As glucose syrups are the versatile 'backbone' of sugar confectionery, the following examples are by way of illustration and in no way limiting and should only be considered as a starting point for future work. The rest is up to the skill and imagination of the technologist.

11.5.1 High boilings

Sucrose	400 grams
42 DE Glucose Syrup (82% solids)	400 grams
Water	200 grams
Citric acid	As required
Flavour	As required
Colour	As required

Method

1. Weigh sucrose, glucose syrup and water into a suitable saucepan.
2. Heat to dissolve the sucrose with continuous stirring, then raise the temperature ideally to about 155°C. (If the temperature is any higher when using an open pan, discolouration will become a problem.)
3. Check solids on a refractometer. Aim for a solids content of 98–99% refractive solids.
4. Mix into the cooked mix, as quickly as possible, the citric acid, flavour and colour.
5. Deposited into moulds.

Notes

1. It is necessary to add water to ensure that all the sucrose is dissolved.
2. Citric acid is usually added to fruit flavoured high boilings. Typical addition rate would be 0.5–1.2% based on the total sweetener solids. As citric acid will invert some of the sucrose, it should be added just prior to the mix being deposited into the moulds so as to reduce inverting the sucrose.
3. A more accurate way to determine the final solids is to calculate the amount of water which has to be evaporated in order to reach a solids of 98–99%. To reach these solids is very difficult when using an open pan. It is easier to reach these solids if the cooking is carried out under vacuum.

11.5.2 Pulled sugar confectionery

The recipe is basically the same as for a high boiling, but once the required solids have been reached, the cooked mass is pulled on a rotary pulling machine (or by hand) where the cooked mass (now taking on a plastic texture) is continuously being stretched, then folded over, then stretched, then folded, and so on, for several minutes until the mass takes on an opaque appearance. The opaque appearance is due to the shearing action during the pulling which makes the sucrose crystallise, together with the folding action incorporating lots of fine bubbles of air. Once the desired texture has been obtained, the mass can then be used in the making of rock; as part of another sweet, for example a casing or as a sweet in its own right. A 35 DE syrup can also be used in this type of product, where the slightly higher proportion of higher sugars increase the plasticity of the mass.

11.5.3 Fondant

This type of product is an example of the positive side of sucrose crystallisation and plays an important part in the textural characteristics of several types of confectionery. Here a glucose syrup, normally a 42 DE, is a controlling factor in the production of the highly desirable fine crystals.

In making grained confectionery, the coarse sugar crystals are first completely dissolved, and then forced to recrystallise using a combination of controlled heating and cooling, together with mechanical stirring. This results in the formation of very fine crystals of sucrose suspended in the glucose syrup. So fine are these crystals that they are imperceptible to feel in the mouth.

Depending upon the formulation and processing techniques, the resulting consistency can vary from a short texture to a plastic like dough to a soft semi-flowing crème.

The development of fine sucrose crystals is influenced by a combination of the following factors:

- Percentage of sucrose
- Percentage of glucose syrup and type of syrup
- Moisture content
- Processing temperature regime
- Presence of seed crystals
- Mechanical processing
- Presence of any colloidal stabilisers
- Time and storage temperature

Sucrose	400 grams
42 DE Glucose Syrup (82% solids)	100 grams
Water	120 grams

Method

1. All of the ingredients are weighed out and mixed in a suitable saucepan and heated to 120°C (87% refractive solids).
2. The cooked mass is then beaten using a mechanical beater. If a mechanical beater is not available, then the mass can be poured onto a cool surface, and using a metal spatula continually turn over the hot mass to induce sucrose crystallisation, which will become apparent when the mass takes on an opaque appearance.

3. Store the fondant for about 24 hours before use. This is known as 'ripening'.
4. For use, heat the fondant to 60°C, before adding to other ingredients. If it is heated above 60°C, the sucrose crystals will be melted.

Notes

1. When a 63 DE is used, the fondant has a softer texture.
2. Some fondants might contain a fat or some type of hydrocolloid.
3. It is possible to make a fondant using dextrose to replace the sucrose, however, the crystalline fondant will require additional beating to produce an acceptable smooth product.
4. It is important that the moisture content does not exceed 12% to ensure good microbiological stability.
5. 'Instant Fondant' can be made by dry blending 90% icing sugar or superfine dextrose with 10% 15 DE maltodextrin. To use, just add water and mix.
6. In some applications, the fondant is thinned by blending in a proportion of 63 DE syrup.

11.5.4 Toffee and caramel

The main difference between toffees and caramels is that caramel contains a higher amount of fat, milk products and moisture than in a toffee. These differences can result in toffee having less flavour and a harder chew.

The brown colour and flavour of these products are mainly due to the Maillard Browning reaction between the reducing sugars of the glucose syrup and the milk proteins. Further changes to the flavour can be achieved by using different grades of sugar, for example brown sugar, or by adding flavours, for example mint.

42 DE Glucose Syrup (82% solids)	337 grams
Sweetened condensed milk	277 grams
Sucrose	228 grams
HPKO (melting point 32–35°C)	89 grams
Butter fat	59 grams
Emulsifier (GMS)	59 grams
Salt	6 grams
Flavour	4 grams

Method

1. All the ingredients, except the flavour, are weighed into a suitable saucepan and warmed with constant stirring to 60°C to form an emulsion.
2. Continue to heat to 124°C (89% refractive solids).
3. Stir in the flavour.
4. Deposited into non-stick moulds or onto an oiled surface, prior to cutting.

Variations

To improve the stability of toffee and caramel for use in hot or humid conditions, the melting point of the fat can be increased to 40°C. The moisture content of a soft caramel is about 9–10%, for a medium caramel 7–8%, and for a hard caramel 5–6% moisture. Further improvements can be achieved by replacing the 42 DE syrup with 35 DE syrup.

Since there is sucrose in sweetened condensed milk, it is possible to make a toffee without using any additional sucrose, by replacing the 42 DE syrup with a glucose fructose syrup blend (UK Patent 2,223156 B).

High Maltose Syrup – 70% maltose (80% solids)	270 grams
HFGS – 42% fructose (69% solids)	55 grams
Sweetened condensed milk	140 grams
HPKO (melting point 32–35°C)	45 grams
Butter fat	30 grams
Salt	3 grams
Emulsifier (GMS)	1 gram
Flavour and colour	As required

Method

As for a conventional toffee or caramel.

11.5.5 Fudge

Fudge is caramel to which fondant has been added so as to produce a grained texture. As with toffee and caramel, there are many variations. A basic fudge could be as follows:

Part 1

42 DE Glucose Syrup (82% solids)	180 grams
Sucrose	330 grams
Sweetened condensed milk	180 grams
HPKO (melting point 32–35°C)	40 grams
Emulsifier (GMS)	2 grams
Salt	8 grams

Part 2

Fondant	180 grams
Flavour	As required

Method

1. The ingredients in Part 1 are weighed into a suitable saucepan and heated to 60°C to form an emulsion.
2. The temperature is increased to 115°C (85% refractive solids) with constant stirring to develop both colour and flavour.

3. The mix is cooled with constant stirring to 85°C, and the fondant is added with constant stirring. The final moisture content should be about 7%.
4. The mix is then deposited into moulds and allowed to solidify.

Notes

1. To reduce drying out of the fudge, 1–3% of sorbitol can be added as a humectant.
2. Other additions can be desiccated coconut.

11.5.6 Gums and Jellies

These types of confectionery range from hard gums to soft jellies depending upon the type of gelling agent used and the grade of glucose syrup. Typical gelling agents which are used are gelatin, pectin and starch. Others are agar and gum arabic.

The texture of a gelatine sweet will be influenced by the grade of gelatine used. The different grades are identified by their Bloom strength – the higher the number, the stronger will be the gel. The starch most frequently used is a thin boiling starch (sometimes referred to as an acid-thinned starch), which has a 60 fluidity ('Fluidity' is a viscosity measurement of the cooked starch).

Both 42 DE and 63 DE syrups are used in gums and jellies. A 42 DE syrup gives the product a more chewy bite, whilst a 63 DE has a softer bite. The other consideration is the final moisture content which can vary from 15% to 24%.

All these variables will have an influence on the final texture. Generally, all these products are deposited into starch moulds for 24–48 hours. After being removed from the moulds, some might be further processed such as being 'sanded', that is covered with sugar, stoved or coated with a very thin film of mineral oil.

The following three recipes will show how different syrups and gelling agents can be used together.

(a) Gelatine Gum using 42 DE syrup

Part 1

Gelatin 250 Bloom	24 grams
Water	48 grams

Part 2

42 DE Glucose Syrup (% solids)	180 grams
Sucrose	148 grams
Water	56 grams
Citric acid, colour, flavour	As required

Method

1. The gelatine is hydrated in the water at 60°C for 2 hours.
2. The ingredients in Part 2 are weighed into a suitable saucepan, and heated to 116°C with constant stirring.
3. The mix is cooled to about 80°C and the gelatine solution is added with constant stirring.

4. Citric acid, colours and flavour are added. The citric acid is usually added as a 50% solution.
5. The refractive solids should be adjusted to 76–78%, and the mix is then deposited into suitable moulds.
6. It would be possible to stove these products which would reduce the moisture content to a level of 10–11%. Stoving is a process of storing the product in a fan oven at 40–50°C until the required moisture content has been reached, usually after 12–24 hours.

(b) Pectin Jelly using 63 DE syrup

Part 1

63 DE Glucose Syrup (82% Solids)	180 grams
Sucrose	120 grams
Water	144 grams

Part 2

High methoxy pectin	6 grams
Sucrose	20 grams
Sodium citrate	1.6 grams

Part 3

Citric acid, colour, flavour	As required

Method

1. Weigh ingredients in Part 2 into a suitable saucepan, and heat to 110°C with constant stirring.
2. Dry blend the pectin and sucrose in Part 1, and add to the mix with constant stirring so as to avoid lump formation.
3. Adjust solids to 76–78% refractive solids.
4. Cool to 95°C, add citric acid, colour and flavour, and deposit into suitable moulds, and after being removed, they are generally sanded.

(c) Starch Jelly using 63 DE syrup

Part 1

63 DE Glucose Syrup (82% solids)	240 grams
Sucrose	175 grams
Water	160 grams

Part 2

Water	175 grams
60 Fluidity, thin boiling starch	52 grams

Part 3

Colour and flavour	As required

Method

1. The ingredients for Part 1 are weighed into a suitable saucepan and heated to about 90°C with constant stirring.

2. The thin boiling starch in Part 2 is slurried in the water, and the starch slurry is then poured into Part 1, with constant stirring and keeping the temperature constant at about 90°C.
3. The mix is kept boiling to a refractive solids of 76–78%.
4. Add colour, and flavour then deposit into moulds.
5. After removal from the moulds, the refractive solids should be 80%.

The sucrose in all of these products can be replaced on a solid for solid basis using the same maltose HFGS blend that was used in the toffee and caramel examples.

11.5.7 Chews

There are several different recipes for aerated chews. This recipe contains gelatine, fat and a whipping agent.

Part 1
	Water	100 grams
	Sucrose	208 grams
	42 DE Glucose Syrup (82% solids)	277 grams
	HPKO (melting point 30–32°C)	10 grams

Part 2
	Water	15 grams
	Gelatine (200 Bloom)	7.5 grams

Part 3
	Water	15 grams
	Whipping agent	3.7 grams

Part 4
	Colour	As required
	Flavour	As required

Method

1. Hydrate Parts 2 and 3 in water at 60°C for 2 hours.
2. Weigh out ingredients in Part 1 into a suitable saucepan, and heat to 125°C with continuous stirring. Add Part 2.
3. Add Part 3 and Part 4, with continuous stirring.
4. The semi-plastic mass is now pulled to incorporate fine air bubbles and then formed into pieces by pressing or extruding.
5. Final moisture content 6–8%.

As with the previous recipes, the sucrose can be replaced with a maltose HFGS syrup blend. All replacements should be on a solid for solid basis. Some recipes will contain maltodextrin, which effectively increases the viscosity of the mix, improving air entrapment and product stability.

11.5.8 Marshmallows

These are essentially an aerated confectionery.

Part 1

Water	38 grams
Limed gelatine (200 Bloom)	7.7 grams
Sodium bicarbonate	0.4 grams

Part 2

63 DE Glucose Syrup (82% solids)	357 grams

Part 3

Flavour and colour	As required

Method

1. In Part 1 the gelatine is hydrated in water at 60°C for 2 hours. The sodium bicarbonate is added. The mix is then well stirred.
2. The syrup is heated to 107°C, and then the gelatine solution is whipped in, together with the flavour and colour.
3. The whipping should be continued at high speed, until the density of the mass is between 0.30 and 0.50, with a moisture content varying from 12% to 18%. Ideally, this stage should be carried out using a special machine such as an Oakes, which will force air under pressure into the product.
4. The aerated mass is either deposited into moulds or extruded.

For a more rigid product, a 42 DE syrup could be used.

11.5.9 Turkish Delight

In the United Kingdom, Turkish Delight is essentially an opaque starch/pectin jelly. The opaqueness is achieved by the addition of a thin boiling starch. The ratio of pectin to starch is 30:70, however Turkish Delight can be made using 100% thin boiling starch. Either a 42 DE or a 63 DE syrup can be used depending upon the required final texture. It is also possible to replace the sucrose in Part 3 (a) below, on a weight for weight basis, with a maltose HFGS syrup blend.

(a) Turkish Delight using starch and pectin with a 42 DE syrup

Part 1

Thin boiling starch (60 fluidity)	25 grams
Water	136 grams

Part 2

Low methoxy pectin	10 grams
Sucrose	45 grams
Water	143 grams

Part 3
        Sucrose   233 grams
        42 DE Glucose Syrup (82% solids)   197 grams

Part 4
        Colour   As required
        Flavour   As required

Method

1. The starch in Part 1 is made into a slurry with the water.
2. The pectin and sucrose in Part 2 are dry blended to prevent the pectin forming lumps when added to the water. The water is heated to about 60°C, and the pectin is slowly added with constant stirring. The mix is now boiled to give a clear solution.
3. The sucrose and syrup in Part 3 are now heated to boiling, and both Parts 1 and 2 are now added with constant stirring, making sure that the mix continues to boil.
4. Cook to a refractive solids of 77%.
5. The cooked mass is now deposited into suitable moulds. If a starch mould is used, then the solids will increase, as the starch absorbs some of the moisture from the cooked mass.
6. After 24–48 hours, remove the product from the moulds and dust the surface with a 50/50 mix of icing sugar and dry maize starch, to which 1% of a free flow agent (e.g. calcium phosphate) has been added.
7. The product can be cut and enrobed in chocolate.

N.B. The normal flavour is rose for rose coloured Turkish Delight, but other flavours can be used. In white coloured Turkish Delight, typical flavours could be lemon, or mint.

(b) Turkish Delight using a 63 DE syrup with a thin boiling starch

Part 1
        Water   92 grams
        63 DE Glucose Syrup (82% solids)   100 grams
        Dextrose monohydrate   20 grams
        Sucrose   180 grams

Part 2
        Water   20 grams
        Thin boiling starch (60 fluidity)   40 grams

Part 3
        Colour and flavour   As required

Method

1. Weigh ingredients of Part 1 into a suitable saucepan and heat to dissolve the solids. The heat is increased so that the mix gently boils.
2. Slurry the starch in the water in Part 2, and pour into Part 1 with constant stirring, allowing the mix to continue to boil.

3. Boil until the refractive solids are 78%.
4. Add colour and flavour. Stir, then deposit into suitable moulds.
5. After 24–48 hours, remove from moulds, cut and dust.

11.5.10 Muesli bars

There are many variations of these types of products, but basically they consist of two parts – a muesli and a binder.

In this recipe, the glucose syrup is acting as a binder together with a small proportion of fat. It is important that the fat sets at room temperature but melts at body temperature. By using both a syrup and a fat, it reduces the stickiness of the product compared to when sucrose is used on its own. This reduction in stickiness is useful during processing. To alter the flavour, brown sugar can be used instead of sucrose. Other additions could be molasses, honey or invert. The sucrose can be totally replaced with a maltose HFGS blend.

Muesli

Dried vine fruits	40 grams
Assorted chopped nuts	160 grams
Jumbo oats	200 grams
Crisped rice	100 grams

Binder

HPKO (melting point 35°C)	80 grams
Sucrose	160 grams
42 DE Glucose Syrup (82% solids)	260 grams

Method

1. The sucrose and syrup are heated to 135°C, and the fat is added and mixed into the hot syrup.
2. The muesli is then added to the hot syrup with constant but gentle stirring, so that all of the ingredients are evenly coated without being crushed.
3. The warm mix is placed into a mould, and light pressure is applied so that the ingredients are compacted, but not crushed.
4. When set, the bars are wrapped. Some bars might be enrobed with chocolate.

Notes

1. By changing the syrup, the texture of the bar could change – it could become sticky, harder or not bind the ingredients.
2. The cooking temperature is only a rough guide. A lower temperature might result in water being left behind in the ingredients and could result in a soft or sticky bar.

   The surface of the bar could be coated with other ingredients. The bar could be made up of different layers with each layer containing different ingredients – there are many possibilities.

3. Whilst vine fruits were used in this recipe, dates or chopped dry fruits such as apricots could be used.
4. One potential problem in using nuts is that nuts can become rancid during the storage life of the product.

11.5.11 Confectionery centres

There are many and varied centres used in confectionery. Here are two examples, one is a soft toffee or caramel and the other is a 'jam base'. Confectionery like chocolate éclairs which have a soft chocolate centre are usually made by the roping process. The centre is pumped into the centre of the cooked mass, just before the mass leaves the oscillating conical batch rollers, where it forms a central core with the outer portion of the cooked mass becoming the outer casing of the sweet.

(a) Soft toffee or caramel centre

42 DE Glucose Syrup (82% solids)	390 grams
Brown sugar	300 grams
Sweetened condensed milk	210 grams
HPKO (melting point 30–32°C)	15 grams
Emusifier	0.1 gram

Method

1. Weigh out the ingredients into a suitable saucepan and cook to 110°C (refractive solids 80%) with constant stirring.
2. Fill centres by pumping into the cooked outer shell.

(b) 'Jam base' centre

Sucrose solution (60% solids)	300 grams
42 DE Glucose Syrup (80% solids)	300 grams
Apple pulp	300 grams
Colour and flavour	As required

Method

1. Weigh out the ingredients into a suitable saucepan, and heat with continuous stirring to 120°C (87% refractive solids).
2. Add colour and flavour to match, for example yellow colour with an apricot flavour.
3. Fill centres by pumping into the cooked outer shell.

11.6 Calorie reduced products

Since confectionery products are often considered to be an indulgence product, the concept of a calorie reduced product could be considered a contradiction especially as the two main ingredients in sugar confectionery are the carbohydrates sucrose and glucose syrups which are full of calories.

To produce calorie-reduced confectionery, there are several technical problems to be overcome. Both sucrose and glucose syrups contribute not only sweetness but also bulk and structure to confectionery. They are also easily digested by the body, but as a general rule, bulk ingredients which are low in calories are not absorbed by the body and therefore tend to be laxative.

One advantage of sugar free confectionery is that products are considered to be 'dental friendly'.

One of the easiest ways to reduce calories is to replace the carbohydrate with a polyol such as sorbitol, maltitol or mannitol. The caloric value of a polyol in the EU is taken to be 2.4 cal/g, compared to carbohydrates which have a value of 4.0 cal/g, therefore by replacing the sucrose and glycose syrup with a polyol, the calories can be reduced by about 40%. Since polyols are less sweet than sucrose, this would have to be taken into account in any formulation (Table 11.3).

Table 11.3 Approximate calorific and sweetness values.

	Calories per gram	Approximate sweetness
Sucrose	4.0	100
Fructose	4.0	150
Glucose syrups	4.0	50–80
Sorbitol	2.4	50
Maltitol	2.4	90
Erythritol	0.2	65

The polyol erythritol is unique among polyols in that it is not metabolised by the body, being totally excreted unchanged in the urine and therefore contributes zero calories, and also has no laxative effects. It has approximately 70% of the sweetness of sucrose, but with a solubility of 37%, its possible applications in confectionery are limited. It might be of interest in calorie reduced chocolate and chewing gum, and possibly when used with other polyols in crystalline confectionery such as fondant and fudge.

Sorbitol can be used to totally replace both a glucose syrup and sucrose, on a one for one replacement in high boilings. Because of its low cooked viscosity, it is totally unsuitable for high boilings made by the rope process and therefore is only suitable for deposited high boilings. With the reduced sweetness of sorbitol, a heat stable artificial sweetener would have to be incorporated in the formulation.

Sorbitol syrup can be used for dragee pan coatings.

Sorbitol can also be produced as an anhydrous powder which means that it can be used in chocolate confectionery, although a lower conching temperature is required. Powdered sorbitol can also be used in the production of sugar free chewing gum and in tabletting.

Maltitol has several advantages over sorbitol in that it is 10% less sweet than sucrose and can be used to replace both glucose and sucrose in gums, jellies and high boilings. When used to make toffees, because it is a non-reducing sugar, there is no Maillard Browning, therefore the toffees lack both colour and flavour, which means that both colour and flavour would have to be added. Maltitol can also be used as a binder in cereal bars.

Like sorbitol, crystalline maltitol could be used to make sugar free chewing gum and in chocolate confectionery.

Mannitol has limited use in confectionery due to its relative low sweetness and solubility, but can be used in combination with other polyols to modify textures. It could be used in chewing gums, and as a blend with maltitol in high boilings.

So as to obtain a feel for the possibilities of making calorie reduced confectionery, a good starting point could be to replace the sucrose with a polyol in the above recipes.

Chapter 12
Glucose syrups in fermentations: an overview

12.1 Introduction

Fermentation processes use living organisms such as moulds (fungus), yeast and bacteria to produce a large range of chemicals. Because these organisms lack chlorophyll, they are unable to synthesise their own carbohydrates from carbon dioxide and water to satisfy their energy requirements. Therefore, they require a ready source of energy and glucose syrups being composed of soluble carbohydrates are their ideal energy source.

In the process of metabolising the carbohydrate, various by-products are produced, and when changes are made to the organism's environment, such as changes to the pH, temperature, aeration, concentrations, nutrients, etc, the organism will often react by producing different by-products, or where two or more by-products are being produced, then the ratio of these products can change. For example, yeasts are able to ferment sugar solutions in the presence or absence of oxygen, however appreciable growth of the yeast will only occur when oxygen is present, but at the expense of less alcohol being produced.

There are often many strains of a particular micro-organism, so it is important that all the strains are screened, so as to select the strain which produces the highest yields. And in the future, with a better understanding of the different metabolic pathways involved together with genetic modification of organisms, and improved microbiological, analytical and production techniques, it might be possible, given the correct conditions, for these types of organisms (and possibly others) to economically produce a whole range of different chemicals.

Whilst the economics of using fermentations to produce different chemicals is very dependent upon world economic conditions, there are some products which can only be produced by fermentation notably some antibiotics. In times of war when supplies of traditional raw materials are scarce, alternative methods of production such as fermentation have been used as a necessity.

In peacetime, however, the production of chemicals by fermentation can be uneconomic mainly due to the fermentations being slow and having relatively low yields. This has resulted in the decline of fermentations for the production of some chemicals in favour of using petrochemicals. With oil and natural gas becoming more expensive, it is possible that in the future, fermentations might once again come into their own.

This chapter therefore is an overview of some of the major products which can be produced by fermentation using glucose syrup as the substrate. It is not intended to mention every product – there are too many, it is just an illustration of the possibilities, and is sub-divided into four areas – pharmaceutical, enzymes, food grade chemicals and industrial products.

12.2 Choice of substrate

The choice of which carbohydrate to use for a particular fermentation will depend upon the following factors:

- What the organism can use
- Availability of carbohydrate
- Cost per unit of fermentable carbohydrate
- Yield
- Ease of recovery and post fermentation clean up of the end product
- Ease of effluent disposal

Commercially, there are two main carbohydrates used for fermentations – molasses and glucose syrups.

Molasses is the syrup produced from the crystallisation of sugar, by both cane and beet sugar producers, but the composition of the molasses from either source is different, and as a by-product, their composition will also vary, but in general, molasses is a mixture of sucrose, invert (dextrose and fructose), minerals and organic matter. It also contains small quantities of vitamins particularly biotin which are essential for the growth of yeast and several species of bacteria. Whilst molasses still contains some sugar, it has in the past been considered uneconomic to recover, so molasses was regarded as a cheap fermentation substrate but this situation has now changed for two reasons.

Table 12.1 Typical analysis of cane and beet molasses.

	Cane molasses	Beet molasses
Sucrose	30–40%	46–52%
Reducing sugars	10–20%	0.2–1.2%
Other sugars	1–3%	0.5–2.5%
Ash	8–12%	8–11%

First, with an increase in the world demand for sugar, there has been a stimulus to look at ways to recover as much sugar as possible from the molasses, and because of the improvements in recovering the sugar, the sugar content of the molasses has been reduced resulting in a pro rata increase in both minerals and organic matter, which has resulted in a decrease in fermentability. This increase in minerals means that effluent disposal becomes more expensive. Additionally, the colour has increased which has also made post fermentation clean up more expensive.

The second reason is that with the increasing price of oil, some sugar producing countries, particularly Brazil, are using molasses for fermentation to make bio-ethanol for use in motor fuel.

The result of these two changes is that potentially less molasses will be available, and therefore it will no longer be a cheap and plentiful fermentation substrate unless there is a substantial over supply due to an increase in sugar cane and beet planting.

Glucose syrups, because of the way they are produced, have consistent fermentability, and being de-ionised, they have a low mineral content with very good colour, making post fermentation clean up and disposal of the effluent both easier and cheaper. The preferred glucose syrup for fermentations is a 95 DE syrup, but syrups with a lower DE in the range 80–94 are sometimes used. Generally, the lower the DE, the cheaper the syrup because it contains less fermentable sugars and hence lower yields. These lower DE syrups originate from the dextrose crystallisation process and are sometimes referred to as greens or hydrol. Because the syrup originate, like molasses, from a crystallisation process, this type of syrup is sometimes considered to be the glucose industry's equivalent of molasses.

Some micro-organisms grow naturally on substrates low in readily fermentable sugars, so when these micro-organisms are used commercially, it is necessary to supply a substrate of a similar low fermentability, therefore the preferred glucose syrup might be a 42 or lower DE syrup.

In this chapter when discussing the different syrups used in fermentation, the syrups used for brewing and the production of alcohol will be omitted as they are covered in Chapter 10.

Table 12.2 The sugar composition and fermentability of different syrups commonly used in fermentations.

	Cane molasses (1)	Beet molasses (1)	42 DE	80 DE	95 DE
Sucrose	63%	75.0%	–	–	–
Fructose	11%	1.7%	–	–	–
Dextrose	9%	1.3%	19.0%	66.0%	94.0%
Maltose	–	–	14.0%	15.0%	4.0%
Maltotriose	–	–	11.0%	4.0%	1.0%
Higher sugars	–	–	56.0%	15.0%	1.0%
Other sugars (2)	17.0%	22.0%	–	–	–
Fermentability	83%	78%	44%	85%	100%

Notes

1. Because the composition of molasses is variable, depending upon origin and factory practice, these figures should only be taken as a guide.
2. Typical 'other sugars' could be xylose, arabinose, mannose, raffinose and galactose.

12.3 Basic fermentation process

A typical fermentation process will have the following stages:

- The inert organism will be propagated in a laboratory flask from a stock culture to produce larger samples.
- The larger samples are then transferred into culture tanks, and when a sufficient quantity of the organism has been grown, it is used to inoculate the substrate in the production tanks.

- Depending upon the particular organism, the fermentation process can take from 2 to 10 days or longer.
- At the end of the fermentation, the product is recovered. Typical recovery procedures could be distillation if it is a solvent such as ethyl alcohol. If the product is an acid such as citric acid, this can be recovered by forming an insoluble salt – calcium citrate – which is recovered and then acidified with sulphuric acid to form the insoluble calcium sulphate and citric acid which is then recovered.

    If it is a mould, then this can be recovered by filtration. In the case of penicillin production, the active ingredient is obtained by solvent extraction of the mould followed by crystallisation.
- To prevent contamination, it is essential that sterile conditions are maintained throughout the process.

A lot of fermentation products are produced by submerged fermentation which, it is claimed, gives a higher yield and is easier to control.

A typical fermentation vessel will be a cylindrical vertical tank made of stainless steel, with a working capacity of about 1000 hectolitres (22,000 UK gallons). The tank will be temperature controlled, and have a means of gentle agitation.

The substrate is run into the fermentation vessel, together with any nutrients. The pH and temperature are adjusted and the organism is added to the substrate and the fermentation is started. Frequently, sterile air will be used to aerate the substrate.

As fermentations use living organisms, the concentration of soluble sugars in the substrate is critical for a successful fermentation, and the reason is because the soluble sugars contribute to the osmotic pressure of the substrate. If the osmotic pressure of the substrate is greater than that of the organism, the osmotic pull from the cell to the substrate will dehydrate the organism which will then die. As a general rule, the uptake of sugars by organisms is dependent upon the size of the molecule, concentration and enzymes present in the organism for its metabolism. In the case of dextrose, the molecule is small and can passively pass across the cell wall of most organisms, which makes a high DE glucose syrup ideal for fermentations.

12.4   Products of fermentation

Fermentation products are many and varied, but most have one thing in common – they can all use products of the glucose industry as their energy source. The following four sections illustrate some of the typical applications of glucose syrups in fermentations.

12.4.1   Pharmaceutical

The main pharmaceutical fermentation products are 'antibiotics'. These can be defined as a chemical substance produced by a living organism, which has either an inhibitory (i.e. bactericidal) or a germicidal (i.e. bacteriostatic) effect towards other micro-organisms. To date, there are over 30 compounds produced by micro-organisms which are or could be of interest in both medicine and animal welfare. Penicillin, cephalosporin, streptomycin, tetracycline, griseofulvin and their derivates are some of the major antibiotics produced by fermentation.

The production of penicillin is a good example of a typical antibiotic production process, but it should be realised that different organisms will often require different conditions.

1. The inert spores of the penicillin fungus are cultured initially in a 100-ml flask, then into a 500-ml flask. After about 4 days, the inoculum is transferred into a 500 litres culture vessel and finally it is added to the substrate in the fermentation vessel, where it will remain for about 5–6 days at a temperature of 23–28°C and a pH of 4.0–5.0 with gentle agitation and aeration.
2. At the end of the fermentation, the broth will contain about 30 g/L of penicillin. The fungus is filtered off and the penicillin is solvent extracted using either amyl or butyl acetate. The penicillin is then recovered from the solvent, which is then re-used.
3. Since penicillin is an acid, it is treated with potassium hydroxide, to a pH of 5.0–7.0, followed by carbon treatment and crystallisation.

The carbohydrate solids in the substrate would be from a de-ionised 95 DE glucose syrup at an addition rate of about 3%. In addition to the carbohydrate, there will be a source of nitrogen – often corn steep liquor, together with amino acids and minerals.

12.4.2 Enzymes

Enzymes are biological catalysts derived from living organisms and are used by many industries.

As enzymes are derived from living organisms, their method of production and recovery is similar to that of antibiotics. How the enzyme is recovered will depend upon whether the enzyme is an intracellular enzyme, that is present inside the cell, or extracellular enzyme, that is present on the outside of the cell.

The process of recovering an intracellular enzyme would as follows:

1. Separate the micro-organism from the fermentation broth.
2. Place the cells in a sterile liquid and disintegrate the cells, usually using ultrasonics so as to release the enzyme.
3. Remove the cell debris, but retaining the liquid which contains the enzyme.
4. Purify and concentrate the enzyme.

For recovering an extracellular enzyme, the process would be as follows:

1. Separate the liquid from the micro-organism and retaining the liquid. Washing the micro-organism with dilute sodium chloride increases the recovery.
2. Purify and concentrate the liquid.

The modern glucose industry uses enzymes extensively in the production of its many syrups, and these enzymes would have been produced from organisms which have been grown on a glucose syrup substrate.

The following are typical enzymes used by the starch industry which could be produced from either bacteria or fungi.

Bacterial amylase

   α-amylase   – *Bacillus subtilis*
            Bacillus licheniformis
            Bacillus amyloliquifaccciens

   β-amylase   – *Bacillus megaterium*

Fungal amylase

   α-amylase   – *Aspergillus oryzae*
            Aspergillus niger
            Rhizopus oryzae

Other enzymes

   Amyloglucosidase   – *Aspergillus niger*
              Aspergillus awamori
              Rhizopus niveus
              Rhizopus delema

   Pullulanse   – *Bacillus acidopollulyticus*
            Klebsiella aerogens

   Isomerase   – *Streptomyces murinus*
            Streptomyces olivaceus

In addition to the enzymes which the starch industry uses, there are many more enzymes which are used by other industries and can be grown on a glucose syrup substrate.

12.4.3 Food grade products

The food industry uses many ingredients which have been made by fermentation such as bakers yeast, acidulants-like citric and lactic acids, gums and amino acids to mention just a few. Whilst the following illustrates the versatility of micro-organisms which can use glucose syrups as their substrate, each producer will have their own preferred strain of micro-organism.

Yeast

The seed culture for baker's yeast is *Saccharomyces cerevisiae*. Whilst molasses is the normal carbohydrate for yeast production, a high DE glucose syrup can be used equally well. For maximum yeast production as opposed to alcohol production, the carbohydrate concentration should be in the range 0.5–1.5%, with a pH 4.5–5.5 and a temperature of 30°C. Diammonium phosphate and biotin are usually added, together with other nutrients. It is important that the fermentation is well aerated. The fermentation is completed in about

12–24 hours, in which time the seed culture would have increased fivefold. Generally, the lower the carbohydrate concentration, the higher the yield of yeast. The reason for this is a survival mechanism on the part of the yeast. The yeast thinks that there is a shortage of 'food', and therefore starts to produce more cells so as to ensure that there is another generation to follow on.

Citric acid

Citric acid and its salts are used extensively in the manufacture of soft drinks and by the confectionery industry. Citric acid can be prepared by either surface or submerged culture, using strains of *Aspergillus niger*. The preferred carbohydrate is a high DE glucose syrup at a concentration in the substrate of about 15–20%, at pH 2.0, and a fermentation temperature of 26–30°C. The aerobic fermentation takes about 8 days.

$$2C_6H_{12}O_6 + 3O_2 \longrightarrow C_6H_8O_7 + 4H_2O$$
$$\text{Dextrose} \qquad\qquad \text{Citric acid}$$
$$360 \text{ grams} \qquad \longrightarrow 192 \text{ grams}$$

Fumaric acid

This is an acidulant used in many food products and can be produced by the fermentation of a 15% dextrose solution by *Rhizopus*.

Lactic acid

This can be produced by a large number of bacteria and moulds, but possibly *Lactobacillus deibrueckii* is the most commonly used. The preferred carbohydrate is a high DE glucose syrup at a concentration of 15%, and a pH of 5.8–6.0 at a temperature of 48°C, with a fermentation time of about 4–6 days.

Tartaric acid

This can be produced using *Acetobacter suboxdans*. The preferred carbohydrate substrate is a 10% high DE syrup, with a pH of 5.5–6.0 and a temperature of 28–30°C. The fermentation time can vary from 8 to 48 hours.

Mycoprotein

This is basically an edible fungus produced by growing *Fusarium graminearium* on a 95 DE glucose substrate. The protein-rich filamentous material is recovered by filtration, heat treated to destroy the nucleic acid, dried then made into a meat substitute by 'texturising' the fungal filaments and adding flavourings.

Monosodium glutamate

This is the sodium salt of glutamic acid, and despite being totally bland, it is a flavour enhancing amino acid.

The acid is obtained by fermenting a high DE syrup with *Micrococcus glutarnicus*. Since glutamic acid is an amino acid, it is recovered by adjusting the pH of the fermentation to 3.2, which is the isoelectric point for glutamic acid, which is precipitated out of solution and then crystallised. The glutamic acid is then neutralised with sodium hydroxide to give the sodium salt.

Lysine

This amino acid is used as a feed supplement and can be produced by a two-stage fermentation of a high DE syrup by a strain of *Escherichia coli* or *Gliocladium* followed by fermentation with *Aerobacter aerogenes* to give lysine.

Vinegar

Whilst technically, vinegar can be made by fermenting a dextrose syrup with *Acetopbacter aceti*, the end product cannot be legally called vinegar, because vinegar has to be made from fruits, malt or wine. (The word vinegar is derived from the latin vinum 'wine' + acer 'sour'.) Incidentally, acetic acid is now mainly produced by the carbonylation process, which uses methanol and carbon monoxide as its raw materials.

Vitamin B12

This vitamin is essential in the prevention of pernicious anaemia, and can be obtained from *Streptomyces olivaceus*, *Streptomyces griseus* and *Lactobacillus lactis*. Typical fermentation conditions are a substrate containing 1% dextrose, at a pH of 7.0–7.5 and a temperature of 25°C. The fermentation time is about 4 days.

Vitamin C (ascorbic acid)

Most vitamin C is produced using the classical Reichstein process developed in 1933 by Dr Reichstein at The Swiss Institute of Technology. This process starts with the hydrogenation of dextrose to produce sorbitol (see Chapter 8, for details on the production of sorbitol).

The sorbitol is then fermented with *Acetocbacter suboxdans* at 30–35°C to produce l-sorbose. This is then oxidised to produce 2-keto-L-gulonic acid, which is treated with acid to produce L-ascorbic acid. The overall yield from dextrose is about 60%.

Recently, the possibility of using fermentation processes for producing vitamin C have been investigated.

Work at The Shanghai Research Centre of Biotechnology have developed a two-stage fermentation process involving the direct fermentation of dextrose using a mutant strain of *Erwina* sp. *125*, to produce 2,5-diketo-gulonic acid (2,5-DKG). The 2,5-DKG is then deoxidised by another mutant strain of *Corynebacterum* sp. *SCB 30 58*.

The American company Genetech has been developing a one-stage fermentation process since 1989 using a recumbent cloned combination of *Corynebacterum* and *Erwina herbicola*.

There are also two research institutes looking at a one-stage fermentation process which uses yeast.

It is interesting to note that all these proposed fermentation processes to produce vitamin C use dextrose as their starting material.

Erythritol

The polyol erythritol is unique among polyols in that it is produced by a fermentation process. Details of its production are given in Chapter 8.

Pigments

The edible pigments β-carotene and xanthophylls can be produced by micro-organism.

12.4.4 Industrial products

Petrochemicals are the present source of many solvents, plastics and organic chemicals, but in the past, some of these would have been made using a fermentation process, and with possible shortages of petrochemicals, it might be worth revisiting these older processes and also see how fermentation processes could be used to make some of today's raw materials. (Sources: *Starch and Its Derivatives*, by J.A. Radley, 1954, pp. 83–106, and *Chemistry and Industry of Starch*, by R.W. Kerr, 1950, pp. 515–534).

Acetone and butanol can be produced together when *Clostridium acetobutylicum* ferments a 5% high DE syrup, at a pH of 5.7–6.2 and a temperature of 35°C. After about 2–3 days, the broth is filtered and the solvents are distilled off. A yield of about 25% or more is possible.

Butanol and isopropanol can be produced together when *Clostrium butylicum* or *B. technicus* ferments a high DE syrup. The carbohydrate concentration is between 7% and 10% depending upon the organism used.

Acetone and ethanol can also be produced by *B. acetoethylicus*.

During World War II, there was a shortage of rubber. This shortage was met by the production of synthetic rubber which was made using 2,3-butylenes glycol, obtained by fermentation using *Aerobacter aerogens* or *Aerobacillus polymyxa*. Yields of 40% were claimed on a 12% solids substrate. The original work used a starch mash but with the advances in producing high-grade low DE substrates, it is possible that the yields could be greatly improved. The 2,3-butylenes glycol can then be converted to butadiene.

A mention has already been made of producing lactic acid by fermentation for use in foods. Lactic acid and aliphatic lactates can also be used as intermediates to produce acrylic plastics, for example Perspex.

Another plastic which could be made indirectly by fermentation is nylon 6,6. Nylon is made from adipic acid which can be produced by fermentation of dextrose with a genetically modified strain of *E. coli* to produce cis,cis-muconic acid, which is then hydrogenated into adipic acid. (Source: article by John Frost, The Chemical Engineer, 16 May 1996).

Yoshiharu Doi in 'Microbial Polyesters' (VCH 1990) discusses the production of I.C.I's 'Biopol' plastic and other thermoplastic and biodegradable polyesters, using a two-stage fermentation of dextrose with *Alcaligenes eutrophus*. The fermentation conditions would be 2 days at 30°C and pH 7.0.

Itaconic acid is a product used in the manufacture of resins for use in the paint and polymer industries and can be made by the fermentation of a 15% dextrose solution with *Aspergillus terreus*.

Whilst enzymes produced on carbohydrate substrates are often used in washing powders, chemicals such as gluconates because of their low corrosive properties and chelating properties in alkaline solution are used in industrial cleaning solutions. They are particularly used in the dairy and beverage industries for the removal of calcium deposits. Gluconic acid is produced by the submerged fermentation of a highly aerated 30% dextrose solution by *Aspergillus niger*. The resulting acid is neutralised to give either the calcium or sodium gluconate. Both of these salts are also used in the baking industry in baking powders.

One interesting industrial product produced by fermentation is xanthan gum. This is used in food applications as a thickener and industrially in oil drilling mud. When used in a drilling mud, it produces a very stable viscous mud with good heat resistance and pH stability. Xanthan gum is produced by the fermentation of a glucose syrup using *Xanthomonas campestris*. This micro-organism can utilise a 42 DE syrup or a high DE syrup – the choice of substrate is a balance of raw material cost against yield.

Advances in fermentation technology would suggest that industrial fermentations using products of the glucose industry could still offer a viable future.

Chapter 13
Glucose syrups in ice creams and similar products

13.1 Introduction

Ice creams are essentially a stable emulsion composed of water, fat, milk solids and sugars, which has been aerated and then frozen, and glucose syrups can be used extensively to increase their acceptability, by contributing to sweetness and enhancing both texture and flavour. Glucose syrups are also a convenient source of nutritive solids, without introducing excessive sweetness, the risk of crystallisation or a gritty texture. By using different grades of glucose syrup, it is possible to alter the characteristics of most types of ice cream and frozen desserts, opening the way to many product variations and line extenders. Additionally, glucose syrups can frequently show a considerable cost saving when used in frozen products.

Because the solids of an ice cream are relatively low, typically between 30% and 40%, and with the ingredients being very susceptible to microbiological spoilage at these low solids, the manufacture of ice creams is strictly regulated. In the United Kingdom, the Food Labelling Regulations, 1996, Part 1, Schedule 8, states what can be used in an ice cream and the final composition, as well as what it can be called. The UK regulations also cover how ice creams are made and stored, with particular reference to the heat treatment, which will ensure a sterile product. It has been said that there are more regulations governing the manufacture of ice creams, than for any other manufactured food. Therefore, check the legislation, before carrying out any development work.

13.2 Ingredients and process

The main ingredients of an ice cream are fat, milk solids and sugar, and the following (see Fig. 13.1) is an outline of how these ingredients are put together to make an ice cream.

13.2.1 Fats

Fats can be either milk fats such as cream or butter, or vegetable fats such as hydrogenated coconut oil or margarine. In the United Kingdom, ice creams made with milk fats can be called 'dairy' ice cream, however most UK ice creams are made with vegetable fats for price reasons – vegetable fats being cheaper than milk fats, with the end product being

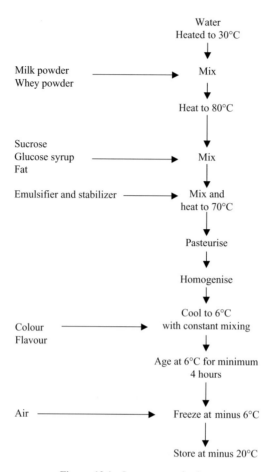

Figure 13.1 Ice cream production.

referred to as 'ice cream'. The origins for this differentiation can be traced back to the Second World War. Before the war, all ice creams were made using milk fat but during the war, milk fat was not allowed to be used in ice cream, so vegetable oils were used (i.e. margarine). When milk fat once again became available, the UK's dairy industry objected to ice creams made with vegetable oils being called 'ice cream'. Legislation was therefore introduced stating that only ice creams made with milk fat could be called 'dairy ice cream', and by implication that ice creams made with vegetable oils could be called 'non-dairy' ice cream. In other countries, ice creams made with vegetable fats are usually called 'mellorine' ice cream. As milk fats are generally more expensive than vegetable oils, dairy ice creams are considered a premium product, and are priced accordingly.

Fats contribute to the mouthfeel of an ice cream, and slight changes to the texture of an ice cream can be made by using fats with a higher or lower melting point. The normal melting point for an ice cream fat would be 34–36°C. Fats also act as carriers for flavours and help to stabilise the air cells. Typical fat content will be 6–12% w/w.

A low-fat ice cream, containing 2% fat (and only 78 calories), cannot be legally called an ice cream, because it does not comply with the legal fat content requirement.

13.2.2 Milk solids

The milk solids of an ice cream are usually referred to as 'milk solids, not fat' – usually abbreviated to MSNF. Sometimes, milk solids are also called 'serum solids'. The milk solids can come from skimmed milk, or whey, either as a liquid, or more usually as a powder. Both skimmed milk powder and whey powder are normally cheaper than either liquid or powder milk. The functional part of the MSNF is the protein content, which contributes to the foam and to the stabilisation of the emulsion. It also binds water and improves the body of the ice cream. MSNF also contains a small amount of the milk sugar lactose. Whilst in the United Kingdom, 'milk' is understood to mean cows milk, ice creams can be made using other types of milk such as goats or buffalo milk. Typical MSNF will be in the range 7.5–11.5% w/w.

During the ice cream making process, by heating the milk ingredients to 80°C, it ensures that the milk proteins are fully hydrated, which will in turn improve the stability of the ice cream.

13.2.3 Sugars

Sugars, depending on the type and quantity used, have the greatest influence on the texture of an ice cream. They can change the freezing point, as well as the sweetness. Typical sugar content, excluding the lactose from the milk, is 13–18% w/w. Artificial sweeteners are not permitted in any ice cream.

13.2.4 Emulsifiers and stabilisers

In addition to the above ingredients, there will be emulsifiers to ensure a good emulsion, and stabilisers which increase the viscosity and stabilise the emulsion. Normally, a combined emulsifier and stabiliser is used. The amounts of emulsifier and stabiliser used in an ice cream are critical since they will affect both the viscosity and texture – too little and the ice cream is too thin and unstable, too much and the ice cream becomes thick and heavy like a pudding. The inclusion rates would be 0.3–1.0% w/w, but typically 0.5% w/w. Typical emulsifiers are based on mono- and di-glycerides of fatty acids. Typical stabilisers are locust bean gum, guar gum, carrageenan and sodium alginate or blends of emulsifier and stabiliser.

13.2.5 Solids

The total solids of an ice cream are normally between 35% and 40% w/w. The amount of water present in an ice cream has to be sufficient to allow the sugars to be completely dissolved. If the sugars are not totally dissolved, the resultant ice cream will have a gritty texture. It is important that when the water freezes that it forms fine smooth ice crystals. At 6.0°C, about 50% of the water will be frozen.

13.2.6 Pasteurisation

The object of pasteurisation is to ensure that the ice cream is sterile and this is achieved by heating the ice cream mix to a sufficiently high temperature to kill any pathogenic micro-organisms. Typical pasteurisation temperature would be as follows:

- 65.0°C for 30 minutes
- 71.1°C for 10 minutes
- 79.8°C for 15 seconds

13.2.7 Homogenisation

The object of homogenisation is to produce very small fat globules of a uniform size, and this is achieved by forcing the ice cream mix under pressure through a homogenising valve or a very small orifice. A typical homogenisation pressure would be 2000 lb/sq. in. and the particle size of a fat globule can be reduced from 3 to 0.4 microns. By producing lots of very fine fat globules, the surface area of the fat is increased with the milk protein being absorbed onto the surface of the fat globules. This fat absorption increases the viscosity, which in turn stabilises the mix, and gives the ice cream a richer texture. Additionally, homogenisation will destroy the cell structure of micro-organisms, thereby contributing to the sterility of the mix.

13.2.8 Cooling, ageing and freezing

By cooling the ice cream to 6°C followed by ageing, the fat globules crystallise and agglomerate around the air cells contributing to the final texture of the ice cream.

During the freezing stage, air is incorporated using a scraped surface heat exchanger which lowers the temperature of the mix to minus 6°C, and as the blades of the scraper pass over the refrigerated surface, air is whipped into the mix. The amount of air which is added can be controlled and is responsible for the overrun in the ice cream – the greater the amount of air incorporated into the mix, the greater will be the overrun.

13.3 Glucose syrups – freezing point and relative sweetness values

The freezing point of an ice cream is controlled by the amount of soluble solids in the recipe – the greater the dissolved solids, the lower will be the freezing point (one of the colligative properties). Fortunately, both glucose syrups and sucrose are very soluble, and so by altering the amount of these ingredients in a recipe, the freezing point can be changed, enabling the freezing point to be tailored to meet any requirement. However, because glucose syrups and sucrose have different molecular weights, they will have different effects on the lowering of the freezing point. The lower the molecular weight of the sweetener, the greater the effect it will have on lowering the freezing point. In other words, dextrose having half the molecular weight of sucrose will be twice as effective at lowering the freezing point, than an equivalent weight of sucrose. Other glucose syrups, such as a 42 DE glucose syrup, which have a higher molecular weight than sucrose, will have a minimal effect on lowering the freezing point. Instead, they will impart body and mouthfeel to an ice cream. This is because they contain lots of higher sugars, which will increase the viscosity. This increase in viscosity will also help to reduce the risk of the lactose from the milk crystallising, which would result in an ice cream with a gritty texture. Table 13.1 illustrates the sugar spectra for some of the many glucose syrups used in ice creams.

Table 13.1 Sugar spectra of some of the glucose syrups used in ice creams.

	15 DE Maltodextrin	28 DE Syrup	35 DE Syrup	42 DE Syrup	Dextrose	HFGS (42% fructose)
Dextrose	1.3%	2%	15%	19%	99.5%	52 5
Maltose	4.1%	10%	12%	14%	0.25%	4%.
Maltotriose	6.0%	16%	11%	13%	0.25%	1%.
Higher sugars	88.6%	72%	62%	55%	–	1%
Fructose	–	–	–	–	–	42%

Both glucose syrups and sucrose will contribute to the sweetness of an ice cream, as well as affecting the freezing point and texture. The relationship between molecular weight, freezing point depression (FPD) and relative sweetness are illustrated in Table 13.2.

Table 13.2 Sweetener properties relating to ice cream and frozen desserts.

Sweetener	Average molecular weight	FPD	Relative sweetness
Sucrose	342	1.0	1.0
Glucose Syrup 42 DE	429	0.8	0.5
Glucose Syrup 35 DE	514	0.7	0.4
Glucose Syrup 28 DE	643	0.5	0.4
Maltodextrin 18 DE	1000	0.34	0.21
Maltodextrin 15 DE	1200	0.29	0.17
Maltodextrin 10 DE	1800	0.19	0.11
Maltodextrin 5 DE	3600	0.10	0.06
Maltose Syrup – 70% maltose	380	0.9	0.5
Maltose Syrup – 50% maltose	448	0.8	0.5
Maltose Syrup – 41% maltose	470	0.7	0.5
Maltose Syrup – 25% maltose	511	0.7	0.5
Glucose Syrup 63 DE	286	1.2	0.7
Dextrose	180	1.9	0.8
Fructose	180	1.9	1.7
HFGS (42% fructose)	190	1.8	1.0
Trehalose	342	1.0	0.5
Glycerol	92	3.7	0.8
Sorbitol	182	1.9	0.5
Maltitol	344	1.0	0.9
Alcohol (ethyl alcohol)	46	7.4	–

As there are many different glucose syrups and maltodextrins available and each syrup manufacturer will have their own suggestion about which syrup should be used. The following list is just an illustration, and is in no way limiting.

13.3.1 How to reformulate using glucose syrups

The basic rules for using glucose syrups in an ice cream are as follows. First, decide which type of ice cream you want to make – a hard brick or a soft scoop. Second, decide which sweetener you would like to use and why. *N.B.* All replacements must be calculated on a dry for dry basis, and if a syrup is used, then an adjustment must be made to the amount of water in the original recipe, so as to compensate for the amount of water present in the syrup.

- For body and mouthfeel, use a low DE syrup. The lower the DE, the greater will be the increase in body and mouthfeel.
- For a soft scoop ice cream, use a syrup with a high DE such as dextrose.
- To replace sucrose, use a fructose syrup, and a low DE syrup. The fructose syrup will not only supply the sweetness, but will also produce a soft ice cream. Therefore, the addition of a low DE syrup or maltodextrin will provide the body and mouthfeel.
- When replacing a major ingredient such as a sweetener, there might be a subtle change in the texture of the ice cream compared to the original. These changes can often be overcome by adjusting the amount of emulsifier and stabiliser used in the recipe.
- Also remember to consider the likely annual usage, together with the ease of handling and storage of the sweetener – syrup or powder. If you decide on using a syrup, is it available in drums or only in bulk tankers? For small manufacturers, a powder might be more convenient.
- Realistically, an ice cream manufacturer will want to use only one syrup in as many products as possible. The manufacturer will not like the idea of using lots of different syrups – just the one.
- Finally, check any likely cost changes to the recipe.

13.3.2 How sweeteners can be re-balanced

Calculate the FPD and relative sweetness for the existing sweeteners (include glycerol and sorbitol, etc., as sweeteners) from the values given in the above table. These two figures should now be the 'target' which the new sweeteners will have to match, and in some cases in order to match that 'target', it might be necessary to either increase or decrease the total amount of sweetener solids being used. Where an exact match for the 'target' is not possible, preference should be to match the sweetness. These calculations are not an 'exact science', due to the effects of the other ingredients and processing equipment – they are just an indication. Further small changes to the FPD, meltdown and texture can also be made by adjusting the amounts of stabiliser and emulsifier being used and by using a fat with a higher or lower melting point. A FPD value of about 14.0 will produce a hard ice cream, and a value of 20.0–25.0 will result in a soft ice cream.

Example 1

A soft ice cream contains 12.5% sucrose and 2.5% glycerol, and we would like to replace the glycerol, and if possible, some of the sucrose.

Original Freezing Point Depression

% of Sweetener × Freezing Point Depression of that sweetener.

$$\text{Sucrose} = 12.5 \times 1.0 = 12.5$$
$$\text{Glycerol} = 2.5 \times 3.7 = \underline{9.3}$$
$$\text{Total Freezing Point Depression} = \underline{21.8}$$

Original Relative Sweetness

% of Sweetener × Relative Sweetness of that sweetener.

$$\text{Sucrose} = 12.5 \times 1.0 = 12.5$$
$$\text{Glycerol} = 2.5 \times 0.8 = \underline{2.0}$$
$$\text{Total Relative Sweetness} = \underline{14.5}$$

These calculated FPD and relative sweetness figures now become the target figures which any alternative sweetener system must match.

These 'target figures' could be matched by using a combination of dextrose and fructose to replace the glycerol and at the same time reducing the amount of sucrose by 40%. Both dextrose and fructose are dry powders, but if the usage is sufficiently large, then it would be possible for the glucose syrup supplier to produce a specific syrup blend.

New Freezing Point Depression

$$\text{Sucrose} = 7.5 \times 1.0 = 7.5$$
$$\text{Dextrose} = 6.5 \times 1.9 = 12.35$$
$$\text{Fructose} = 1.0 \times 1.9 = \underline{1.9}$$
$$\text{Total Freezing Point Depression} = \underline{21.75}$$

(The original target was 21.8)

New Relative Sweetness.

$$\text{Sucrose} = 7.5 \times 1.0 = 7.5$$
$$\text{Dextrose} = 6.5 \times 0.8 = 5.2$$
$$\text{Fructose} = 1.0 \times 1.7 = \underline{1.7}$$
$$\text{Total Relative Sweetness} = \underline{14.4}$$

(The original target was 14.5)

Example 2

A hard ice cream contains 13.3% sucrose, and we would like to replace as much sucrose as possible.

Original Freezing Point Depression

% of sweetener × Freezing Point Depression of that sweetener.

$$= 13.3 \times 1.0 = 13.3$$

Original Relative Sweetness.

% of sweetener × Relative Sweetness of that sweetener.

$$= 13.3 \times 1.0 = 13.3$$

These are therefore the new target figures which any alternative sweeteners must match.

The sweetness will have to come from HFGS, but because the molecular weight of a 42% HFGS is less than that of sucrose, the 42% HFGS will depress the freezing point even lower, resulting not in a hard ice cream, but a soft ice cream. This difference in molecular weights would have to be balanced with some higher molecular weight sugars from a syrup containing lots of higher sugars, such as a 42 DE syrup. Unfortunately, syrups which contain lots of higher sugars are not sweet. This scenario would suggest that it might not always be possible to replace all the sucrose in an ice cream, even if the total solids of the ice cream are increased.

New Freezing Point Depression

$$
\begin{aligned}
\text{Sucrose} &= 4.4 \times 1.0 = 4.40 \\
\text{HFGS (42\% fructose)} &= 4.4 \times 1.8 = 7.92 \\
\text{42 D.E. Glucose Syrup} &= 4.4 \times 0.8 = \underline{3.52} \\
\text{Total Freezing Point Depression} &= \underline{15.8}
\end{aligned}
$$

(The original target was 13.3).

New Relative Sweetness

$$
\begin{aligned}
\text{Sucrose} &= 4.4 \times 1.00 = 4.40 \\
\text{HFGS (42\% fructose)} &= 4.4 \times 1.00 = 4.40 \\
\text{42 D.E. Glucose Syrup} &= 4.4 \times 0.50 = \underline{2.20} \\
\text{Total Relative Sweetness} &= \underline{11.00}
\end{aligned}
$$

(The original target was 13.3).

Example 2 illustrates that it is not always possible to obtain an identical match. By using HFGS to replace the sucrose, the freezing point has been depressed, changing the texture from a hard to a soft ice cream. Increasing the 42 DE has partially restored the hard texture, but the sweetness has be reduced. If less sucrose is replaced, then an ice cream nearer to the original would be obtained.

Now consider the possible cost savings, which might be achieved. As a general rule, it is easier to replace sucrose in an ice cream which has a high sucrose content and hence a greater potential cost saving, as opposed to replacing sucrose in a 'lean' ice cream to give minimal cost saving.

Since these types of calculation can be time consuming, it is useful to put the calculations into a computer spreadsheet, using the information in Table 13.2 on freezing point depression and relative sweetness and if the cost of the ingredients is also included in the spreadsheet, it would be possible to see the financial implications when an ingredient change is made – 'the what happens if' approach. This computer spreadsheet could also be used for other frozen products.

N.B. Since freezing point depression is dependent upon the molecular weight, an alternative approach to using FPD figures for the different sugars could be to substitute the molecular weight instead, and just match the molecular weight of each mix.

Table 13.3 below is a possible computer spread sheet format for calculating recipe changes. For simplicity in this example, only the sweeteners have been considered.

Table 13.3 Computer spread sheet format for calculating recipe changes.

Basic ice cream recipe					
Ingredients	% Solids	----Control---		-----Option 1---	
		As is %	DB %	As is %	DB %
Vegetable fat					
Skimmed milk powder					
Whey powder					
Emulsifier/Stabiliser					
Colour					
Flavour					
Sucrose	100%		12.5		7.5
Glucose Syrup 42 DE	80%				6.5
Glucose Syrup 35 DE	80%				
Glucose Syrup 28 DE	75%				
Maltodextrin 15 DE	95%				
Maltose Syrup – 70% Maltose	80%				
–50% Maltose	80%				
– 41% Maltose	80%				
– 25% Maltose	80%				
Glucose Syrup 63 DE	80%				
Dextrose Monohydrate	91.5%				
Crystalline Fructose	100%				1.0
HFGS (42% Fructose)	71%				
Glycerol			2.5		
Sorbitol					
Water					

% Total Solids			15.00		15.00
Freezing Point Depression			21.75		21.75
Relative Sweetness			14.50		14.00
% Sweetener Solids			15.00		15.00
Cost of mix per Kg					

13.4 Quick process checks

As mentioned previously, glucose syrups can change the texture of an ice cream. It might also affect how the mix will behave on the production line. Therefore, before starting to process the new mix, it is useful to carry out some simple basic quality checks.

13.4.1 Viscosity

Check the viscosity of the mix using a Ford Cup or something similar. A Ford Cup is a vessel containing a hole at the bottom, through which a known volume of liquid can flow. A finger is placed over the hole, and the cup is filled with the mix. The finger is removed, and the time it takes for the cup to empty is noted. This time is the viscosity of the mix in seconds / minutes. Experience will tell you if the mix is too thin or too thick for going through the process.

13.4.2 Overrun

$$\% \text{ Overrun} = \frac{\text{Volume of ice cream} - \text{volume of mix}}{\text{Volume of mix}} \times 100$$

13.4.3 Solids

The solids in an ice cream must meet the legal requirement and a quick way to check the solids of the mix is to use a refractometer, graduated to read % solids.

13.4.4 Fats

There is also a legal requirement for the amount of fat in an ice cream, and this can be quickly determined using a butyrometer tube.

13.5 Soft serve ice creams

In some small-scale situations, it is often more convenient to use a prepared ice cream mix. These prepared mixes contain all the ingredients which are present in a soft serve ice cream and are either in a liquid or a powder form.

Liquid mixes consist of an ice cream mix which has been pasteurised, homogenised and then aseptically packed and only requires passing through an in line freezer before being consumed. These types of products are typically used in mobile situations, for example ice cream vans.

The powdered mixes are produced either by dry bending or by co-spray drying of the ingredients. Dry mixes are normally produced for the domestic market, where they can be reconstituted by adding water, mixing in a domestic blender, followed by freezing in a domestic freezer. In powdered mixes dextrose, crystalline fructose, maltodextrins and low DE glucose syrup solids, for example 20 and 35 DE, can be used.

13.6 Other types of frozen dessert

Mellorine – A classification, used to describe an ice cream, in which the butterfat has been replaced with vegetable fat, such as HPKO (hardened palm kernel oil). This distinction is not made within the United Kingdom.

Mousse – A whipped cream made with sucrose and a thickener such as gelatine, which has been frozen but without additional agitation.

Parevine – A non-dairy frozen dessert, suitable for people who do not want to mix meat and dairy products for religious reasons.

Sherbet – A similar product in composition to an ice cream, except that it contains fruit (fruit puree) and a reduced amount of milk fat, but has nearly twice the amount of sweetener than that found in a typical ice cream.

Sorbet – A sweet fruit liquid, which has been thickened with gelatine, then whipped and frozen.

Water Ice – Similar to a sherbet, but does not contain any milk.

Ice lollies – Similar to a dilute fruit juice, with an added stabiliser such as pectin, and then frozen. Frequently, it will incorporate either a wooden or plastic stick for ease of eating. Some ice lollies are frozen in a plastic sachet which is peeled back before eating.

Ice Milk – Similar to an ice cream, but contains only 2–5% milk solids.

Fruit lollies are frequently sweetened with sucrose, but being acidic, and subjected to heat pasteurisation, the sucrose undergoes process inversion and changes into dextrose and fructose, both of which have a lower freezing point depression than sucrose. It is therefore useful to carry out a high-pressure liquid chromatographic (HPLC) sugar analysis on the end product before and after any sweetener replacement trials.

All of these products contain sucrose. If the sucrose is replaced, then both the sweetness and freezing characteristics will be affected, unless care is taken to re-balance the sweeteners.

13.7 Yoghurts

The only sugar used in the production of yoghurts is the sugar lactose present in the milk. It is this sugar which is fermented to produce the lactic acid, which gives yoghurt its characteristic flavour. Traditionally, an equal blend of *Lactobacillus bulgaris* and *Streptococcus thermophilus* is used for the fermentation stage. Other sugars are usually only added after the fermentation is complete, so as to increase the sweetness of the yoghurt. This extra sugar can also come from the addition of sweetened fruit. In some exceptional circumstances, extra sugar might be added if there is a deficiency of lactose. When fruit is added, the sugar and fruit are normally blended together, often with a modified starch, before being added to the yoghurt. These types of sweetened fruit blends are often referred to as 'fruit preps'. Fructose is often used to sweeten yoghurts. The reason for this is that fructose is 50% more sweet than an equivalent weight of sucrose, which means that less sweetener is required, resulting in a reduction in calories. Another advantage for using fructose is that it enhances the flavour of the fruit.

A typical fruit prep is similar in composition to a fruit pie filling, with a solids of about 35%. The stabiliser is frequently a blend of pectin, carob bean gum, and guar gum or just a freeze thaw acid stable modified starch. Citric acid and a flavouring are also added.

13.8 Sorbet

Water (1)	765 grams
Sucrose	100 grams
HFGS (42% fructose at 70% solids)	100 grams
28 DE Glucose Syrup (78% solids)	77 grams
50% Citric acid solution	9 grams
Stabiliser (2)	2 grams
Flavour	As required
Colour (3)	As required

Solids = 23.5%.
pH = 3.0.

Notes

1. Where a fruit puree has been used, correction must be made to the water addition, because of water from the fruit or fruit puree.
2. A typical stabiliser could be gelatine.
3. Where a fruit puree has been used, it might be possible to reduce or even omit the colour addition.

Method

Blend ingredients then freeze.

13.9 Mousse

Part 1

Gelatine 260 Bloom	6 grams
Water	10 grams

Part 2

Milk	662 grams
Sucrose	25 grams
Skimmed milk powder	40 grams
Fat-reduced cocoa powder	60 grams
HFGS (42% fructose, at 70% solids)	140 grams
42 DE Glucose Syrup (82% solids)	50 grams
Flavour	As required

Solids = 26.4%.

Method

1. Pre-soak the gelatine in warm water.
2. Warm the milk to 60°C.

3. Dissolve the sucrose in the warm milk, then add the cocoa powder and milk powder. When fully dispersed, add the HFGS, glucose syrup, with constant stirring. Heat to 85°C for three minutes to pasteurise.
4. Add the fully hydrated gelatine from Part 1, together with flavour, then beat air into the mix as it thickens. Aim for an overrun of about 100–120%.
5. Transfer aerated mix to containers, and store at 4°C.

13.10 Ice lollies

These are basically a soft drink which has been frozen, and a suitable sweetener would be a HFGS. However, because they are cold when consumed, the perceived sweetness might appear to be reduced, therefore this perceived loss in sweetness can be compensated for by increasing the sweetner solids, or by the addition of a high intensity sweetener.

So as to improve the mouthfeel, and shelf life stability, stabilisers such as pectin can be added, typically at about 0.5%. Sometimes the pectin is blended with carrageenan, or guar gum, locus bean gum, etc. These stabilisers by increasing the viscosity of the mix also help in the formation of fine ice crystals. Other ingredients would include fruit flavouring, citric acid and colouring. Because ice lollies have a low solids, typically about 10–12%, the mix is usually pasteurised. Where pasteurisation is not possible, then a preservative such as potassium sorbate can be added. A possible starting recipe could be as follows:

HFGS (70% solids), to give solids of	10–12% w/w
Citric acid	0.05–0.3%
Flavouring (depending whether they are natural or artificial)	0.1–0.5%
Stabiliser, e.g. pectin	0.5%
Preservative, e.g. potassium sorbate	1000 ppm
Colour	As required
Water to make	1000 ml

13.11 Fruit lollies

These are the same as an ice lolly, except that they contain fruit juice. In some cases, the fruit juice will be there to give flavour, for example orange juice, whilst other fruit juices such as apple, pear or grape are there for their sweetness. The amount of fruit juice which is added will vary from fruit to fruit, together with the required flavour.

13.12 Ripple Syrups

The variegated colouring found in ice creams is due to the addition of a ripple syrup. Ripple syrups are usually composed of a syrup, containing colouring, flavour and a thickener. When formulating a ripple syrup, there are several points to bear in mind:

1. The syrup must not affect the texture of the ice cream, and must not crystallise. This rules out high-DE syrups.

2. The ripple syrup must be sufficiently viscous that it does not become part of the ice cream mix.
3. The syrup must be pumpable.

A possible starting recipe for a ripple syrup could be as follows:

Part 1

Pectin (low methoxy)	0.4 Kg
Sucrose	0.9 Kg
Calcium phosphate di-basic	0.015 Kg
Water	18.0 Kg

Part 2

28 DE Glucose Syrup (75% solids)	10.5 Kg
HFGS (42% fructose at 70% solids)	11.3 Kg

Part 3

Colour	As required
Flavour	As required
Citric acid	As required

Method

1. Dry blend the pectin, sugar, and calcium phosphate in Part 1.
2. Slowly add the dry blended ingredients to water at room temperature, with constant stirring.
3. When well dispersed, bring to the boil, and add the 28 DE syrup followed by the HFGS, with constant stirring.
4. Continue boiling until the solids are about 42%, as measured on a refractometer. Add colour and flavour. Continue boiling until the solids are 45%. Add citric acid, if required.
5. Cool and fill into containers.

The use of a 28 DE syrup will give body and viscosity to the syrup, allowing the ripple syrup to become an obvious part of the ice cream, without it mixing with the ice cream mix. Additionally, 28 DE syrup will not crystallise. The use of HFGS will not only supply sweetness, but it will also have a greater effect on depressing the freezing point than a 28 DE syrup. A 28 DE syrup on the other hand would raise the freezing point. Therefore, by using a blend of the two, their respective effects on the freezing point will cancel out.

An alternative to using 28 DE and HFGS in Part 2 would be to use 9.8 Kg of 42 DE syrup (80% solids), and 7.9 Kg of sucrose, and to increase the water to 22.0 Kg.

Since the molecular weight of the 28 DE syrup and HFGS blend is very similar to that of the 42 DE syrup and sucrose blend, there would be minimal difference on how the two syrup blends would behave in use.

13.13 Topping or dessert syrup

When formulating these types of products, there are no hard and fast rules. Check out what is already available, particularly the solids, run an HPLC sugar analysis to determine type of sweetener used (sucrose or type of glucose syrup), also if a preservative is used.

The sweeteners used in these types of syrup are usually a blend of sucrose, invert, and glucose syrup – 35, 42 or 63 DE. With the availability of HFGS, this can be used to replace the invert and sucrose. Sucrose refiners tend to use a blend of sucrose and invert. Other manufacturers might use blends of different sugars. The aim is to produce a syrup which is sweet, does not crystallise and can be squeezed out of a plastic container.

Additionally, the syrup must be microbiologically stable, which means that the solids are frequently about 70–80%. If the solids are less than 70%, consider including a preservative. Sorbates or benzoates are frequently used, but always check the legislation to see which preservatives are permitted.

Where fruit juices are used for either flavour or sweetness, it is important that the final solids of the topping are in the range of 70–80% range, ideally nearer 80%. The addition of other sweeteners can be used to increase the solids. Because fruit juices contain dextrose, this should be taken into account so as to avoid the possibility of dextrose crystallisation.

As most topping syrups have a fruit flavour, the pH is typically in the range 2.5–3.0, which will improve the microbiological stability, but at this low pH, there is the risk of sucrose being inverted, which will add to the risk of dextrose crystallisation. Providing that there is no risk of dextrose crystallising, the presence of low molecular sugars such as dextrose and fructose will be beneficial in helping to improve the microbiological stability, because low molecular sugars have a higher osmotic pressure, which will act as a way of reducing microbiological spoilage.

The viscosity of a topping syrup can always be increased by the addition of a thickener such as pectin or xanthan gum, usually about 1.0%. Modified waxy starches have also proved to be successful thickeners.

Possible starting recipes could be as follows.

13.13.1 A simple economy topping syrup

Use a 50–55 DE glucose syrup, at 82% solids, and add flavour, colour and possibly citric acid. If a 42 DE syrup is used, the viscosity might make it difficult to squeeze from a plastic container. If a 63 DE glucose syrup is used, whilst it will be sweeter than a 50–55 DE syrup, there is the risk that the syrup could crystallise in the container; therefore, a 50–55 DE glucose syrup is the compromise sweetener.

Adjust solids to 80%.

13.13.2 Fruit-flavoured topping syrup

HFGS (42% fructose, at 70% solids)	470 grams
Sucrose	165 grams
42 DE Glucose Syrup (82% solids)	200 grams
Fruit juice concentrate (60% solids)	165 grams
Citric acid	As required
Flavour	As required
Colour	As required

Adjust the solids to 80%.

13.13.3 All syrup fruit-flavoured topping syrup

HFGS (42% fructose, at 70% solids)	700 grams
42 DE Glucose Syrup (82% solids)	200 grams
Fruit juice concentrate (60% solids)	165 grams
Citric acid	As required
Flavour	As required
Colour	As required

Adjust the solids to 80%.

13.13.4 Chocolate topping

HFGS (42% fructose, 70% solids)	470 grams
Sucrose	165 grams
42 DE Glucose Syrup (82% solids)	200 grams
Fat-reduced cocoa powder	165 grams
Flavour	As required

Adjust solids to 82%.

13.13.5 All syrup chocolate topping

HFGS (42% fructose, 70% solids)	650 grams
42 DE Glucose Syrup (82% solids)	190 grams
Fat-reduced chocolate powder	160 grams
Flavour	As required

13.13.6 Caramel topping

42 DE Glucose Syrup (82% solids)	337 grams
Sweetened condensed milk	277 grams
Sugar	228 grams
Water	100 grams
Flavour	As required
Colour	As required

In order to get the typical caramel colour and flavour, it is necessary to heat the mix to about 100°C (212°F), and then adjust the solids to about 80%. If heating is not possible, then the addition of colour and flavour will give the desired results. Both colour and flavour can be improved if brown sugar is used.

13.13.7 All syrup caramel topping syrup

42 DE Glucose Syrup (82%)	337 grams
HFGS (42% fructose, 70% solids)	326 grams
Sweetened condensed milk	277 grams
Water	100 grams
Flavour	As required
Colour	As required

In order to get the typical caramel flavour and colour, it is necessary to heat the mix to about 100°C (212°F), and then adjust the solids to about 80%. If heating is not possible, then the addition of colour and flavour will give the desired results. By using HFGS, because it contains the reducing sugars dextrose and fructose, these will readily react with the proteins in the milk (Maillard reaction) to form both colour and flavour, so minimal heating might be required.

13.14 Reduced calorie products

For a product to be called an ice cream, its composition must meet the legal requirements. An ice cream with reduced calories would not meet these requirements, and therefore could not legally be called an ice cream. It could however be called a 'frozen dessert'.

The calories of an ice cream come from the sugar, fat and milk. By replacing some or part of these ingredients, a calorie reduction can be achieved, but care must be taken when reformulating the product, otherwise both the texture and sweetness will be affected. Consider the following options.

First, the sugar. Apart from contributing sweetness to an ice cream, the molecular weight of the sugar will influence the freezing point, and hence the texture. Whilst an artificial sweetener can replace any lost sweetness, to obtain the same texture, it will be necessary to use a product which has the same molecular weight as the original sugar. By using the sugar alcohol maltitol, most of these requirements can be met.

- Maltitol has a sweetness value of 90 compared to 100 for sucrose.
- The molecular weight of maltitol is 344, compared to 342 for sucrose.
- The calorific value of maltitol is only 2.4 per gram, compared to 4.0 for sucrose.

Fats contribute mouthfeel to an ice cream, act as a flavour carrier, and are also very calorific. A possible fat replacement could be a low DE maltodextrin, which with its high viscosity has a texture similar to a fat, but is bland and lacks the characteristic flavour of a fat. Fat flavours are oil soluble, but maltodextrins are only water soluble.

These drawbacks mean that in practice a maltodextrin is only suitable as a partial fat replacement. However, a 50% fat replacement with a low DE maltodextrin can produce a thick stable emulsion, which has both an acceptable mouthfeel and balance of both water and oil flavours comparable to that of the original fat. With maltodextrins being carbohydrates, they have a calorific value of 4.0 per gram compared to 9.0 for a fat; therefore, any fat replacement with a maltodextrin will reduce the amount of calories. By using a combination of a maltodextrin and a bulking agent such as polydextrose, products containing zero fat can be produced.

The following examples (Table 13.4) illustrates the possible calorie reduction which could be achieved using a combination of maltitol and a 5 DE maltodextrin in an 'ice cream'. All replacements are on a dry weight for dry weight basis. When major ingredients are changed, expect to make minor changes to other ingredients.

Table 13.4 Possible calorie reduction per gram of ice cream solids.

	Control	Sugar-free	Fifty per cent fat reduced	Sugar-free and 50% fat reduced
Skimmed milk powder	10.0 g	10 g	10 g	10 g
Vegetable fat	8.0	8.0	4.0	4.0
Sucrose	13.4	–	13.4	–
Maltitol powder	–	13.4	–	13.4
5 DE Maltodextrin	–	–	1.0	1.0
Emulsifier / Stabiliser	0.6	0.6	0.6	0.6
Solids	32.0	32.0	28.4	28.4
FPD	13.3	13.3	13.4	13.5
Relative sweetness	13.3	12.0	13.5	12.1
Total kilocalories per 100 g	161.6	140.0	129.6	108.0
Kilocalories per gram of solids	5.05	4.38	4.56	3.80
Percentage of calorie reduction per gram of solids	–	13.2	9.7	24.8

In theory, since there are no hard and fast laws as to what a reduced calorie dessert may or may not contain, the possibilities of which ingredients to use are limitless. It is down to the imagination and skill of the technologist.

Chapter 14
Glucose syrups in jams

14.1 Introduction

Jam and similar products (e.g. marmalade and jelly jams) are made by boiling fruit with a sugar solution, and was originally used as a way of preserving fruit before the days of canning and refrigeration. Today, some or all of the sugar can be replaced with a glucose syrup. The use of glucose syrups can frequently offer the manufacturer additional advantages, not obtainable when using only sugar (sucrose). Typical advantages to a product could be as follows:

- Glucose syrups are not inverted – the sugar spectrum is stable under normal jam processing conditions.
- Glucose syrups, because of their high osmotic pressure, have excellent preserving properties.
- Glucose syrups can help to control crystallisation.
- Glucose syrups can be used to tailor the sweetness profile of a product.
- Glucose syrups can enhance the flavour.
- Glucose syrups can be used to change textural properties enabling higher final solids to be obtained without crystallisation problems, for example biscuit jams.
- Glucose syrups are a source of nutritional soluble solids. (A minimum soluble solids of about 55% is required to form a gel depending upon the amount and type of pectin, typically a low methoxy pectin).

14.2 Effects of boiling

The process of making jam consists of boiling fruit with a sugar solution. During this process, a very important chemical reaction takes place. The acid of the fruit converts some, or possibly all, of the sucrose into dextrose and fructose. This chemical reaction is known as 'process inversion' and it occurs frequently in the food processing industry where sucrose is heated in the presence of an acid. This mixture of dextrose and fructose is known as 'invert sugar', but with the use of high-pressure liquid chromatography (HPLC) for analysing sugars, it is now more convenient to look at the individual sugars present, rather than considering the proportion of invert sugar produced to understand the importance of process inversion to the jam manufacturer. Sometimes, the jam manufacturer will refer to invert as 'the reducing sugars', because both dextrose and fructose are reducing sugars, whilst sucrose is a non-reducing sugar (Fig. 14.1).

$$\text{Sucrose + Water} \xrightarrow{\text{Heat + Acid}} \text{Dextrose + Fructose}$$

Figure 14.1 Process inversion.

To the jam manufacturer, the amount of process inversion which has taken place, and hence the amount of dextrose and fructose produced, is very important for two reasons.

The first reason is that the solubility of sucrose, dextrose and fructose are all different and can result in the jam having a gritty texture due to either the dextrose or the sucrose crystallising in the jam. In a jam, sucrose crystals are usually larger, harder and more glassy in appearance than dextrose crystals, and usually form a crystalline crust on the surface. Dextrose crystals, on the other hand, tend to be smaller, and are usually found in the body of the jam, giving the jam a honey like texture. Fructose being a very soluble sugar does not easily crystallise, and so crystal formation due to fructose rarely occurs. By inverting some of the sucrose, the concentration of sucrose is effectively reduced, thereby reducing the problem of sucrose crystallisation, but if too much inversion takes place, dextrose crystallisation becomes the problem. In jams, which have a low solids, crystallisation is less of a problem, but there could be a problem with microbiological stability.

Table 14.1 Solubility of different sugars.

Sugar	20°C	55°C
Sucrose	67%	73%
Dextrose	48%	73%
Fructose	79%	88%

The second reason is that the presence of dextrose and fructose, because they have a lower molecular weight than sucrose, will increase the osmotic pressure of the jam. This increase in osmotic pressure improves the microbiological stability of the jam. The reason for this is that the jam having a higher osmotic pressure than the moulds, yeasts and bacteria tries to dilute itself by removing water from the moulds, yeast and bacteria, thereby dehydrating the cells, which results in their death. One of the reasons why increasing the osmotic pressure of a jam is becoming more important is because of the trend to reduce jam solids. In the 1950s, the final soluble solids of a jam would have been about 70%, but the average soluble solids today are in the range of 60–62%. Jams with a low solids, that is less than 50% soluble solids, will have to be stored in a refrigerator and will require a preservative, such as benzoate or sorbate, but check local legislation.

Table 14.2 shows the difference in osmotic pressures for a 10% solution of sucrose, 63 DE syrup, dextrose and fructose.

The amount and speed of sucrose process inversion will depend upon the amount of acid and the type of cooker used.

Table 14.2 Osmotic pressures for different sugars.

Sugar	Molecular weight	Osmotic pressure in atmospheres
10% Sucrose solution	342	7.19
10% 63 DE syrup solution	285	8.63
10% Dextrose solution	180	13.67
10% Fructose solution	180	13.67

With open pan cookers which operate at higher temperatures, process inversion is more difficult to control, with too much dextrose and fructose being produced. This results in grainy jam due to the dextrose crystallising, because dextrose has a lower solubility than either sugar or fructose.

The industry then moved away from open pan cookers to vacuum pans, with larger jam producers moving to continuous cookers for products such as industrial bakery jams where fruit texture is less important. Whilst both vacuum pans and continuous cookers are more gentle, this resulted in the process inversion being reduced, and so sucrose crystallisation could become the problem. To overcome this problem, invert syrup, that is a mixture of dextrose and fructose, would be added, but this meant another process, namely making invert. A solution to this problem came in 1958, with the introduction of the 63 DE syrup, which could also replace some of the sucrose as well as all the invert.

14.3 Use of glucose syrups

Glucose syrup had been used in jams as early as 1886, for economic reasons, but possibly due to a combination of the poor quality of the glucose syrup – particularly the colour and the lack of sweetness – its use was discontinued. However, during the 1939–1945 war, because of the acute sugar shortage, 42 DE was used in jam at about a 10% inclusion rate, but with its lack of sweetness and gummy texture, its use was subsequently discontinued with the ending of hostilities.

The sugar spectrum of a 63 DE syrup is as follows:

$$\begin{aligned} \text{Dextrose} &= 34\% \\ \text{Maltose} &= 33\% \\ \text{Maltotriose} &= 10\% \\ \text{Higher sugars} &= 23\% \end{aligned}$$

A 63 DE syrup was used because it was sweeter and being less viscous was more easily handled than a 42 DE syrup. The sweetness of a 42 DE syrup is about 50, whilst a 63 DE syrup has a sweetness of about 70, compared to sucrose which has a sweetness of 100. Additionally, a 63 DE syrup does not readily crystallise. It must also be appreciated that glucose syrups do not invert, and their sugar spectrum does not change when heated, so the dextrose content of the 63 DE syrup remains constant at about 34%. A typical 63 DE inclusion rate in 1967 was about 20–30%, to give a final reducing sugars content of about 20–30% in the jam. However, in the 1970s, there was a world sugar shortage, and the industry had no choice but to either increase the 63 DE syrup inclusion or stop making jam. Therefore, the inclusion rate increased from 40% to 60%, and in some instances up to 80%. At the same time, the solids of the jam was reduced from 67% to 65%. The current solids of UK jam is about 63%, but the solids of some European jams is as low as 60%. The limiting factors for higher syrup inclusion rates at that time were as follows:

- Because of the higher sugars in the syrup, too large an inclusion rate resulted in a gummy mouthfeel to the jam. Additionally, these higher sugars also masked any synergistic effect between the sucrose and the fruit flavours.
- Because a 63 DE syrup is less sweet than sucrose, there was a slight reduction in sweetness.
- Since a 63 DE syrup contained dextrose, there was the concern that this extra dextrose could cause dextrose crystallisation.

The total dextrose content of a jam should not exceed 40%. By aiming for 35%, there is a safety margin but remember that the higher the final solids of the jam, the dextrose content should be reduced, so as to avoid dextrose crystallisation. The reason for this is that at higher solids, there is less free water in the jam to dissolve the dextrose, and so the concentration of the dextrose in solution within the jam will become supersaturated, with the possibility of the dextrose coming out of solution, to give a gritty jam. It is for this reason that some producers of bakery jams which have a solids of 70–80% depending upon the application prefer to use a 42 DE syrup. With the dextrose content of a 42 DE syrup being typically only 19%, the possibility of problems due to dextrose crystallisation is reduced at these higher solids. Additionally, in the case of a bakery jam for use in baked tarts, a 42 DE syrup is more viscous than a 63 DE syrup, and this increase in viscosity, together with a small addition of a food grade anti-foam, can help to reduce the possibility of the jam boiling out of the pastry casing.

As a general rule, where a large proportion of the sucrose has been replaced by a glucose syrup, there will be a higher acid to sucrose ratio, which will result in greater process inversion and at a quicker rate resulting in more of the sucrose being inverted in a shorter time. Where a de-mineralised syrup has been used, there is virtually no buffering action from the syrup, so process inversion will be even quicker and greater. However, process inversion can be reduced by reducing the amount of acid and buffering salts (acidity regulator) in the recipe.

With the availability of high fructose glucose syrups (HFGS), it is now possible to blend different types of glucose syrups so as to totally replace the sucrose. This not only overcomes the problems associated with process inversion and sugars crystallising, leading to grainy jams, but they also offer the possibility of cost savings. Some authorities will also maintain that whilst the jams might be less sweet, this reduction in sweetness allows the flavour of the fruit to come through. Also, the jam spreads better, and there is an improved gloss and sheen, making the end product more appealing to the consumer.

In blending glucose syrups for use in jams, the aim must be for sweetness, with minimal 'gummy mouthfeel', and at a commercially acceptable price.

Table 14.3 Approximate sweetness values of different sweeteners.

Sugar	Sweetness
Sucrose	100
42 DE Glucose Syrup	50
63 DE Glucose Syrup	70
Maltose Syrup – 50% maltose	50
Maltose Syrup – 70% maltose	0.5
Dextrose	80
HFGS – 42% fructose	95 (90–100)
HFGS – 55% fructose	105 (100–110)
Fructose	150 (120–170)

The sweetness will come from the fructose, but that is expensive, so to reduce the cost some dextrose can be used. However, too much dextrose cannot be used, because of the risk of dextrose crystallisation, therefore the dextrose would have to be limited to about 30%, with the rest of the dextrose coming from other syrups in the blend. The balance of the blend would then have to come from the next sweetest syrup, but one which contains minimal higher sugars, so as to avoid the 'gummy mouthfeel' which higher sugars would impart to the blend, therefore a high maltose (70%) containing syrup would be acceptable. The next consideration would be the availability of different syrups.

According to British Patent No. 2273642 B, the following sugar spectrum is suitable:

Sugar	% Range
Fructose	18–25
Dextrose	29–33
Maltose	24–37
Maltotriose	9–10
Higher sugars	6–11

The above sugar spectrum could be made using blends of HFGS 42, and maltose syrups containing 70% maltose with or without a 63 DE syrup. The HFGS would supply sweetness, and some dextrose, whilst the 63 DE could be used, depending on whether a 50% or 70% maltose syrup is available.

Where HFGS containing 55% fructose is available (or syrup blends, containing 55% fructose), then the blending of syrups for total sucrose replacement becomes a lot more simple.

When replacing sucrose with a glucose syrup, always make the replacement on a dry weight for dry weight basis, and where possible, match the sweetness of the syrup blend to that of sucrose.

When using an all syrup sweetener system, slight adjustments might have to be made to the pectin and calcium additions. The temperature at which the pectin sets might increase, but this can be corrected by using a slower set pectin, or by slightly increasing the pH.

14.4 Domestic jam

Whilst in the following examples, a 63 DE syrup has been used to replace sucrose, if a blend of a 63 DE syrup with a fructose syrup had been used, then more sucrose could be replaced, depending upon the amount of fructose in the blend.

(a) Sucrose replaced with a 63 DE syrup

Ingredients	100% Sucrose	75% Sucrose 25% 63 DE	50% Sucrose 50% 63 DE	25% Sucrose 75% 63 DE
Fruit (1)	350 grams	350 grams	350 grams	350 grams
Sucrose	632 grams	474 grams	316 grams	161 grams
Water	489 grams	455 grams	420 grams	382 grams
63 DE Glucose Syrup (2)	–	193 grams	381 grams	574 grams
Raid set pectin. 150 SAG (3)	85 grams	85 grams	85 grams	85 grams
Citric acid solution (50% w/w)	3.8 ml	3.8 ml	3.8 ml	3.8 ml

For all the mixes, the fruit solids are 35%, and the final solids are 67%.

Notes

1. Approximate solids of the fruit = 10%.
2. Solids of 63 DE syrup   = 82.0%.
3. The pectin is sometimes blended with sucrose for ease of dispersion.

N.B. The amount of pectin required will depend on the pectin content of the fruit.

Method

Fruit, sucrose, water and syrup are mixed and gently boiled to the approximate end solids. The pectin solution and acid are added and the solids of the mix are adjusted to a refractometric solids of 67% solids. The mix is cooled to about 85°C, before filling into clean sterile jars, and then capped.

(b) 100% Syrup

Maltose Syrup (70% maltose and 80% RI solids)	392 grams
HFGS (42% fructose and 70% RI solids)	364 grams
Dextrose monohydrate	70 grams
Water	275 grams
Fruit (approximately 10% solids)	350 grams
4% W/W Rapid Set Pectin 150 SAG solution (containing 20% sucrose)	85 grams
50% w/w Citric acid solution	3.8 ml

Method

As per the previous example.

(c) Marmalades

Marmalades are essentially jams made with citrus fruits, typically oranges or lemons. The peel of citrus fruits contains some very bitter tasting chemicals, and therefore when the peel is incorporated into the marmalade, it might be necessary to add extra sweetness to compensate for this bitterness. Additionally, marmalades contain less fruit than jams, which is another reason why extra sweetness is required. How much 'extra sweetness' to add will depend upon the bitterness required in the end product.

14.5 Jelly jams

This type of jam can be regarded as an ordinary jam, with their characteristic clarity being achieved by the removal of any insoluble fruit pieces. Typical jelly jams would be blackberry, red currant, mint and cranberry. The same sweetener system can be used in jelly jams as used for conventional domestic jams, with a rapid set pectin, and a pH of 3.1–3.2. The solids are about 63%.

For mint jelly, pieces of mint are added, together with flavouring and colouring (e.g. copper chlorophyll complex, riboflavin), and the solids are typically about 67–71%.

14.6 Honey type spread

Mention has already been made of the grainy texture which can arise due to dextrose crystallisation. This effect can be used to advantage to produce a honey type spread. Basically, this type of product is like a grainy fondant, except it does not contain any sucrose. A dextrose rich syrup will over time crystallise to form a solid lump, but by incorporating a quantity of non-crystallising 42 DE syrup, the increase in viscosity due to the 42 DE syrup will prevent the dextrose crystals from joining together to form a solid lump. Additional dextrose is added to act as a seed to start the process of dextrose crystallisation. The finer the dextrose, the smoother will be the final texture. Since the end product does not contain honey, it cannot be labelled honey. A possible starting recipe could be as follows:

95 DE Glucose Syrup (at 75% solids)	700 grams
42 DE Glucose Syrup (at 82% solids)	300 grams
Dextrose monohydrate	40 grams
Colour	As required
Flavour	As required

Method

1. Warm the 95 DE syrup to melt any crystals.
2. Add the 42 DE syrup, with constant stirring, to produce a homogeneous mix.
3. Allow the mix to cool to about 60°C.
4. Slowly add the dextrose, with constant stirring.
5. Add colour and flavour, with constant stirring.
6. Pour into containers, and allow to set.

The final solids should be about 75–77%. It will take about 48 hours for the mix to set. A softer texture can be obtained by using a lower DE syrup, for example 85–90 DE instead of a 95 DE syrup or by having lower final solids. Conversely, a harder texture can be achieved by increasing the dextrose content, or by increasing the solids.

14.7 Chocolate spread

The following is a starting recipe for a chocolate spread using syrups as the carrier for the cocoa powder. When formulating this type of spread, there are two factors to bear in mind. The solids must be sufficiently high so as to prevent microbial spoilage and the viscosity must allow the end product to be easily spread and also to keep the chocolate powder in suspension. For microbiological stability, the solids have to be 75% or higher. If 100% sucrose is used, there is the risk of sucrose crystallisation, and if a high DE syrup is used, then dextrose crystallisation becomes the problem. If a 100% 42 HFGS is used, this would provide sweetness, but the viscosity would be too low to keep the cocoa powder in suspension as well as the possibility of dextrose crystallisation. The answer therefore is a syrup/sucrose blend, with sucrose and HFGS supplying sweetness and a 42 DE glucose syrup supplying the viscosity. If the solids are increased by increasing the amount of cocoa powder, this would increase the unit cost and the increase in viscosity would reduce the spreadability of the spread. Where a 55 HFGS is available, then all the sweetness could come from the HFGS allowing the sucrose to be omitted. Additionally, by using HFGS, the microbiological stability would be improved due to the increase in the osmotic pressure which it would impart to the spread.

Ingredients	42% HFGS	55% HFGS
HFGS (42% fructose at 70% solids)	210 grams	–
HFGS (55% fructose at 77% solids)	–	318 grams
Sucrose	100 grams	–
42 DE Glucose Syrup (82% solids)	100 grams	100 grams
Fat-reduced cocoa powder	80 grams	82 grams
Water	10 grams	–
Vanilla flavour	As required	As required

Method

1. The syrups, sucrose and water are heated to about 50°C until the sucrose is dissolved. Where a 55 HFGS has been used, still heat the syrup to 50°C. By heating the syrup, the viscosity is reduced, making it easier to disperse the cocoa powder.
2. The cocoa is slowly added to the warm syrup, with constant stirring, until a smooth homogeneous mix is obtained.
3. Add vanilla flavour, if required.
4. Check solids and adjust to 77–78%.
5. Fill into sterile containers and seal.

A variation of the above is to replace about half of the cocoa powder with a hazelnut paste or a similar nut paste. About 0.5% of lecithin can be added to improve the stability of the final spread.

14.8 Peanut spread

Peanut spreads do not contain any water, therefore syrups cannot be used to replace the sucrose, but syrup solids could be used. A possible starting recipe could be as follows:

Roasted peanuts	770 grams
Crystalline fructose	27 grams
35 DE Glucose Syrup solids	23 grams
Palm oil	10 grams
Salt	5 grams
Lecithin	5 grams

The sweetness can be slightly reduced by replacing some of the fructose with a spray-dried 95 DE Syrup.

14.9 Industrial jams

Bakery products are probably the biggest users of industrial jams covering a large range of different applications, and are technically very challenging, because each application has its own special requirement. When formulating these types of jams, there are several points to bear in mind.

- The jams should be pumpable because they are usually deposited by machinery.
- The fruit should be finely sieved, so as to prevent blocking of the depositing heads. In some applications, the fruit is replaced by a fruit puree or concentrate.
- If the jam is to be deposited, there should be no tailing.
- If the jam is to be spread, it should spread easily, and evenly without damaging the surface on which it is to be spread.

- It must not soak into the surface. Ideally, the water activity (Aw) of the jam should be the same as that of the surface so as to reduce moisture migration.
- The jam must be microbiological stable.

14.9.1 Bake-stable jams

These are jams which are used in tarts, and therefore must withstand the baking conditions, and will not boil out of the pastry casing during baking. For this type of jam, a 42 DE glucose syrup is ideal, because the syrup will increase the viscosity of the jam, thereby reducing the possibility of the jam from boiling out of the casing. A typical starting recipe could be a domestic jam, but using a 42 DE syrup, and a medium set pectin, with the final jam solids being 70%.

14.9.2 Biscuit jams

The same recipe as previously, except a slow set pectin, should be used with the final solids increased to 75–80%.

If the jam is to be used in a Jaffa Cake, then the setting time can be rapidly increased, by the addition of citric acid just prior to depositing. The initial pH of the jam should be in the range 4.1–4.2. With the addition of the citric acid, the pH drops to 3.2–3.3, and a gel will form within 5 seconds. The solids should be about 78%. If a softer gel is required, then a 63 DE syrup should be used to replace the 42 DE syrup.

14.9.3 Spreadable jams

This type of jam is used in covering sponges and Swiss Rolls. A recipe similar to a domestic jam can be used, with a medium set pectin, and pH 3.2. The final solids should be 65%. Since these jams have to be spread, the fruit must be sieved, or a fruit puree used.

14.9.4 Jam fillings

These types of jam are typically used in the filling of doughnuts, and can use a 50:50 blend of sucrose and a 63 DE syrup, or a total syrup blend, for example HFGS and HM50, with a slow set pectin, based on the domestic jam recipe. Since this type of jam is injected into the doughnut, it is important that sieved fruit is used. The final solids should be 76–78%. A cheaper fruit filling can be made using equal parts by weight of sucrose, 63 DE syrup (82% solids), and apple pulp with colour and flavour, cooked to 78% solids.

14.9.5 Flan jellies

These types of products are usually poured over fruit which is in a pastry case, therefore it is important that the gel is slow to set. As with the Jaffa Cake type filling, citric acid is used to control the rate of setting, and depending upon the type of gel required, either a 42 or 63 DE syrup can be used with sucrose. Alternatively, syrups can be used to totally replace the sucrose.

Ingredients	42 DE syrup	63 DE syrup	All syrup
Buffered Slow Set Pectin (1)	0.35 Kg	0.35 Kg	0.35 Kg
Sucrose (2)	25 Kg	25 Kg	0.50 Kg
42 DE Glucose Syrup (82% solids)	10.5 Kg	–	10.5 Kg
63 DE Glucose Syrup (82% solids)	–	10.5 Kg	
HFGS (42% fructose and 70% solids)	–	–	35.7 Kg
Water	20 Kg	20 Kg	–
Citric acid (3)	See note (3)	See note (3)	See note (3)
Colour	As required	As required	As required
Flavour	As required	As required	As required

Final solids = 68%.

Notes

1. Buffered pectin is pectin to which buffered salts have been added, thereby saving the user the trouble of making their own addition. The reason why buffered salts are added is to prevent premature gelling of the pectin. The salts most frequently used are sodium citrate and tetrasodium pyrophosphate, at an addition rate of 0.2–0.5%.
2. The addition of sucrose in an 'all-syrup' recipe is purely to aid the dispersion of the pectin, and reduce the possibility of the pectin forming lumps when being made into a solution.
3. The pH of the flan jelly should be about 4.0 before adding any citric acid. The addition of the citric acid will reduce the pH, thereby starting the gelling process. Depending upon how much citric acid is used will depend upon how quick a gel is formed. In the above recipe, at 68% solids, the jelly is pourable. The addition of 110 grams of citric acid monohydrate will cause the liquid jelly to form a gel within three minutes. If only 55 grams is used, then the gelling time is extended to about ten minutes. The normal addition procedure is to dissolve the citric acid in an equal amount of water, and then rapidly stir in the citric acid solution into the jelly, prior to it being quickly poured into the flan where it quickly sets.

Method

1. Dry blend the pectin with a small amount of sucrose.
2. Heat water to about 70°C, and with constant stirring, slowly add the pectin to the water, making sure that all the pectin is completely dispersed.

3. Bring to the boil and add the remainder of the sucrose and syrup.
4. Boil to a solids of 68%, as measured on a refractometer.
5. Citric acid is added as discussed in footnote (c) above.

14.9.6 Fruit and pie fillings

These types of products can use either pectin or a modified acid stable waxy starch. These types of modified starch are freeze thaw stable, that is they will remain unaffected when frozen then thawed, and being acid resistant, the starch will not be broken down by any acid present in the recipe, for example acid from the fruit or additional acid, for example citric acid. Either whole fruit or a fruit puree can be used. A possible all-syrup recipe using a modified starch could be as follows:

Fruit	340 grams
Maltose Syrup (70% maltose at 80% solids)	50 grams
HFGS (42% fructose, at 70% solids)	46 grams
Dextrose monohydrate	9 grams
Water	55 grams
Modified acid stable waxy starch	15 grams
Citric acid	2.5 grams
Potassium sorbate	0.3 grams
Flavour	As required

Method

1. The starch is slurried with the water.
2. The syrups, fruit and dextrose are blended together and cooked until boiling.
3. The starch slurry is added to the boiling syrup, with constant stirring, and brought back to the boil. The solids are adjusted to 35%.
4. The citric acid and potassium sorbate are dissolved in an equivalent weight of water, then added to the mix with constant stirring, when the temperature is about 88°C.
5. The temperature of the mix should be 88°C, when it is put into suitable containers and immediately sealed.

14.9.7 Tablet jellies

These products are a very English type of dessert, and consists of gelatin, 'sugars', acid, flavouring and colour. Originally, the 'sugar' portion was composed of sucrose and invert syrup. When 63 DE glucose syrup became available, it was used to replace the invert. With the availability of HFGS, it is now possible to produce an 'all syrup' tablet jelly. One advantage of using an all syrup sweetener system is that there is no process inversion – the sugar profile will be unaffected by the citric acid. A possible starting recipe could be as follows:

Part 1
High Maltose Syrup (70% maltose and 80% solids)   216 grams
HFGS (42% fructose and 70% solids)   200 grams
Dextrose monohydrate   38 grams

Part 2
Gelatine (240 Bloom) (See note 1)   35 grams
Water   90 grams

Part 3
Citric acid   5 grams
Colour   As required
Flavour   As required

Note

1. The exact amount of gelatin used will depend upon the Bloom strength of the gelatine.

Method

1. The gelatine in Part 2 is hydrated in the water at 60°C for 2 hours.
2. The syrups and dextrose are heated to 122°C, to give a solids of 90%, and allowed to cool to 90°C.
3. The hydrated gelatine is added with stirring to Part 1, and the mix is held at 80°C for about 30 minutes. During this time, any skin which forms is removed.
4. The ingredients in Part 3 are now added. The citric acid is dissolved in an equal amount of water before being added.
5. The mass is now deposited into a suitable mould, and allowed to set, then wrapped. When set, the jelly has a tough rubbery texture.

To use a tablet jelly for a dessert, 143 grams of jelly is added to hot water, which is made up to 568 ml. The jelly will dissolve in about two minutes. This diluted jelly is then poured into a dish to cool and set before serving.

14.9.8 Mincemeat

Eighty per cent of this product is composed of sieved apples, sugar and vine fruits. The glucose content is only about 10%, and is the only liquid addition. Generally, a 42 DE syrup is used, although at this low inclusion rate, a 63 DE syrup could be used. There are three major reasons why only 10% glucose syrup is used.

1. Dextrose from the vine fruits can crystallise, therefore any extra dextrose will only make the problem worse.
2. Mincemeat is made by a cold process, that is there is no heating involved and it has to have a minimal final solids of 65–72%, so as to make it self-preserving. By increasing the syrup content, extra water would be added, which could not be removed. This extra water would therefore reduce the final solids and allow microbiological growth.

3. Because there is a legal requirement that mincemeat should contain no less than 30% sucrose, there is little opportunity to replace any of the dry sugar with crystalline fructose due to costs, and the addition of extra dextrose would only exacerbate the dextrose crystallisation problem.

Where modern bakeries use mincemeat, it is pumped automatically for depositing into the pastry cases. Because the mincemeat has to be pumped to the depositor, the size of the fruit pieces has to be reduced. This in turn can result in the solids separating from the liquid, resulting in blockages. However, if the viscosity of the mincemeat is increased, this problem can be avoided. Typical thickeners could be cellulose gums, locust bean gum, guar gum, xanthan gum or a low DE maltodextrin.

An advantage of using a maltodextrin is that it would bind the water, increase the viscosity, but not affect the dextrose crystallisation problem, and would contribute nutritive solids to the product.

14.9.9 Fruit curds

These are spreadable gels, and are made in a way similar to a jam. The most common flavour is lemon. The main ingredients are sucrose and glucose syrup, but it is possible to make a fruit curd using only glucose syrup. The advantage of using only glucose syrup is that glucose syrups are unaffected by the acid from the fruit, and will not be inverted like sucrose. The sugar spectrum of a glucose syrup is totally stable. A possible starting recipe for an all syrup lemon curd could be as follows:

Part 1

High Maltose Syrup (80% RI solids)	188 grams
HFGS (42% fructose and 70% solids)	174 grams
Dextrose monohydrate	33 grams
Unsalted margarine	30 grams

Part 2

Pectin solution, slow set (12% w/w)	35 grams
Pre-gelled acid stable, modified waxy starch	25 grams
Liquid egg (or equivalent in other form)	30 grams

Part 3

Lemon juice (or sieved lemon dummy pulp)	5 grams
Sodium citrate	1.25 grams
Citric acid	1.0 grams
Lemon oil	As required
Colour	As required

Method

1. Heat the syrups and dextrose to 60°C.
2. Add the margarine and mix.
3. Add ingredients of Part 2 with constant stirring, and heat to 88°C.
4. Adjust solids to 66% refractometric solids.
5. Add ingredients of Part 3 with constant stirring, and fill into pre-sterilised containers.

14.10 Diabetic and reduced calorie products

Jams with reduced calories can be made by replacing both the sucrose and glucose syrup with sorbitol. Sorbitol is used because like other polyols it is only partially metabolised by the body and therefore it is less calorific than sucrose or glucose syrups. Additionally, it has minimal effect on sugar levels in the blood stream. As the molecular weight of sorbitol is less than that of sucrose, the osmotic pressure is greater than when sucrose is used, which together with a solids of 65% ensures a product with both good microbiological and textural stability. Since the sweetness of sorbitol is half that of sucrose, there would be a loss of sweetness, which could be compensated for by the addition of a heat stable high intensity sweetener.

An alternative to sorbitol could be maltitol. This has an advantage over sorbitol because it has 90% of the sweetness of sucrose, but because the molecular weight of maltitol is nearly the same as sucrose, any jams with a solids of less than 65% should contain a preservative.

Any product made with a sugar alcohol such as sorbitol or maltitol will be laxative. An alternative sweetener system for a diabetic jam instead of using a polyol is to use a combination of a glucose syrup and fructose. Because fructose is 50% more sweet than sucrose, less is required to obtain a comparable sweetness to that of a 100% sucrose jam. The advantage of this approach is that the jam is not laxative. The disadvantage, however, is that whilst the calories are reduced, both glucose syrups and fructose will ultimately affect the sugar level in the blood stream. The calories 'per serving' could be further reduced by reducing the solids. This reduction in solids would however mean that the jam would have to be stored in a refrigerator so as to avoid microbiological spoilage.

Another approach could be a blend of a polyol with fructose and a glucose syrup. All sucrose replacements must be on a dry weight for dry weight basis.

14.11 How to calculate a recipe?

There are several different methods for calculating a jam recipe and the following example is just one of the methods, but before working out a jam formulation, always check the local legislation.

It is proposed to make a jam with the following finished properties:

Soluble solids	= 63%
Fruit content	= 35%
pH on a 50% w/w solution	= 3.0–3.3
Batch size	= 50 Kg
Texture	Medium to firm set

Proposed ingredients

- Fruit at 12% moisture
- Sweetener initially sucrose
- Rapid set pectin – 150 SAG
- Citric acid

(1) Calculate the total weight of soluble solids from all ingredients.

$$\text{Soluble solids} = \frac{1}{100} \times \text{Soluble solids of finished jam} \times \text{Batch size}$$

$$= \frac{1}{100} \times 63 \times 50$$

$$= 31.5 \text{ Kg}$$

(2) Calculate the soluble solids from the fruit.

$$\text{Soluble solids} = \frac{1}{100} \times \% \text{ Soluble solids from fruit} \times \text{Weight of fruit}$$

$$= \frac{1}{100} \times 12 \times \frac{1}{100} \times 35 \times 50$$

$$= 2.1 \text{ Kg}$$

(3) Calculate the amount of sweetener solids required in the finished product.

N.B.   (i)   In this example, it has been assumed that there is no moisture in the pectin. Commercially, the pectin would normally be added as a solution, or blended with sucrose. Any extra water addition would have to be taken into account.
       (ii)   In this example, citric acid is being added as a 50% w/w solution.
       (iii)   The amount of pectin to add will depend upon the type of fruit, its pectin content and the type of set required. In this example, 90 grams is being added.
       (iv)   Commercially, a small scale trial should be used to check the amount of pectin and citric acid to add.

Sweetener required = Total weight of soluble solids − (Soluble solids in the fruit
+ Soluble solids in pectin
+ Soluble solids the acid)

$$= 31.5 - (2.1 + 0.0 + .15) \text{ Kg}$$

$$= 29.25 \text{ Kg}$$

(4) Therefore, the starting recipe would be as follows:

Fruit	= 17.50 Kg
Sweetener (sucrose)	= 29.25 Kg
Pectin (see note 1)	= 0.090 Kg
	(or 3.0 Kg as a 3% solution)
Citric acid	= 0.300 Kg
(as a 50% w/w solution)	

Note

1. If the pectin is added as a 3% solution, the pectin would be blended with sucrose as follows. The water used to make the solution would have to be taken into account in calculation (3).

Pectin 150 SAG	= 90 grams
Sugar	= 360 grams
Water	= 2550 grams
Total	**= 3000 grams**

So as to ensure that all the sucrose dissolves and to compensate for loss of water during boiling, additional water will have to be added, typically 5–10 Kg.

Where a glucose syrup is used to replace the sucrose, the replacement must be on a dry weight for dry weight basis. Because glucose syrups contain water, the addition of any extra water could be reduced.

Chapter 15

Glucose syrups in tomato products and other types of dressings and sauces

15.1 Introduction

Glucose syrups can be used in both tomato based products and salad type dressings. Typical tomato based products would be ketchup, sauce and puree. Other types of dressings and sauces would be salad cream, mayonnaise, salad dressings and vegetable mixes such as sandwich spread. All these products contain sucrose which can be replaced wholly or partly with a glucose syrup frequently offering both a cost saving and a product enhancement by improving both the texture and viscosity of these products. Before undertaking any product reformulation, always check that any changes comply with the current legislation.

15.2 Which glucose syrup to use?

Technically glucose syrups, particularly high fructose glucose syrups (HFGS), can be used to replace the sucrose in all of these types of products. By using a glucose syrup, the flavour balance between sweetness and acidity can be subtly changed, allowing background flavours to be enhanced. Additionally, glucose syrups can improve both the texture and gloss of the product making it more visually appealing.

Generally, there will be no noticeable change when 25% of the sucrose is replaced by a glucose syrup. When 50% of the sucrose is replaced with either a 42 or 63 DE syrup, there will be a slight reduction in sweetness, especially with the 42 DE syrup. This reduction in sweetness allows both the acidic and spice flavours to become more pronounced. When 100% of the sucrose is replaced with HFGS, there is minimal change in either sweetness or flavour. The reason why HFGS does not affect either the sweetness or flavour is because in these types of products the acetic acid in the vinegar inverts the sucrose to give equal amounts of dextrose and fructose, which is the same sugar composition as present in HFGS. Depending upon how the product has been made, this sucrose inversion will be either immediate if heat is involved, or it will occur over a period of time. Therefore, tasting the product immediately after the product has been made might not give an accurate flavour profile.

When dextrose is used as a 100% sucrose replacement, there will be a slight reduction in sweetness but this reduction in sweetness allows the subtle flavours of the other ingredients to come through which would otherwise be smothered by the sweetness of the sucrose. The reasons why both the sweetness and flavour are not so pronounced when a 42 or 63 DE syrup are used are twofold. Firstly, neither of these two syrups are as sweet as sucrose or HFGS. Secondly, both 42 and 63 DE syrups contain higher sugars which alter the sweetness and flavour response by increasing the viscosity. This increase in viscosity slows down the time for both the sweetness and flavour to reach the taste receptors compared to the less viscous sucrose or HFGS.

The following glucose syrups are used in both tomato products and dressings for salads.

Table 15.1 Sugar composition of glucose syrups suitable for use in tomato and other types of dressings and sauces.

	42 DE	63 DE	Dextrose	42 HFGS	55 HFGS
Dextrose	19%	34%	99.9%	52%	41%
Fructose	–		–	42%	55%
Maltose	14%	33%	–	4%	2%
Maltotriose	11%	10%	–	1%	1%
Higher sugars	56%	23%	–	1%	1%
Sweetness (1)	50	70	80	95	105

(1) These sweetness values are relative to sucrose, which has been taken as 100.

15.3 Tomato products

There are standards for tomato products, which will vary from country to country, therefore check the local legislation before making any changes to a recipe.

Typical sucrose contents of UK tomato products would be as follows:

Tomato ketchup	23.5%
Tomato sauce	5.0% Minimum
Tomato puree	13.4%

The sugar content of tomato ketchup can, however, vary from 10% to 25%.

The term 'tomato sauce' refers here to the sauce used in canned products and prepared foods. Typical solids of this type of sauce are usually about 20%.

The maximum sugar content of a tomato puree will be in the range of 45–50%, but it will be dependent upon the solids of the puree. The solids of a puree can vary from 12% to 55%. For a puree with a solids of 28–30%, the sugar content will be about 13.4%.

Commercially, there are three grades of tomato puree based on the solids. The solids for these three grades are usually in the range of 18–20%, 28–30% and 38–40%. Some producers will describe the solids of a tomato puree as '4 fold', '6 fold' or '8 fold', whilst other producers might describe it as being '4X', '6X', or '8X'. Both systems mean that the puree can be diluted 4, 6 or 8 times with water to give the same solids as the original tomato juice.

To obtain a stable tomato product, it is important that the original tomato pulp has been heat treated, technically referred to as 'hot break'. This is important because the viscosity of the tomato pulp is due to the pectin content. By heating the pulp, the enzyme pectinase is destroyed, which would otherwise break down the pectin causing a loss in viscosity.

Commercially, because of the heat sensitivity of tomato products, they are usually prepared in stainless steel vacuum cookers.

The following recipes are examples of a partial and total sucrose replacement with glucose syrups in a tomato ketchup. A similar approach can be used for tomato sauces or purees (Table 15.2).

Table 15.2 Tomato ketchup recipes using sucrose and glucose syrups.

Ingredient	100% Sucrose	50% Sucrose, 50% 63 DE Syrup	100% HFGS 42
Red wine vinegar (1)	77.5 grams	77.5 grams	77.5 grams
Modified starch (2)	13.7 grams	13.7 grams	13.7 grams
Tomato puree (3)	216.0 grams	216.0 grams	216.0 grams
Sucrose (4)	125.0 grams	63.0 grams	–
63 DE Glucose Syrup (82% solids)	–	77.0 grams	–
HFGS 42% fructose (71% solids)	–	–	176.0 grams
Salt	7.0 grams	7.0 grams	7.0 grams
Onions (5)	7.5 grams	7.5 grams	7.5 grams
Water (6)	116.0 grams	89.0 grams	52.0 grams
Flavours and spices (7)	As required	As required	As required

Notes

1. An alternative to red wine vinegar could be spirit vinegar or malt vinegar, but adjustments to the amount added might have to be made depending upon the amount of acetic acid in the alternative vinegar.
2. The modified starch acts as a thickener and because tomato ketchup is acidic, the modified starch should be a pre-gelled acid stable waxy maize starch. An alternative to a modified starch could be gum tragacanth. If this is used, it should be dispersed in the wine using a suitable mixer, before being added to the rest of the ingredients. Some manufacturers do not use a thickener, relying on the viscosity of the tomato paste to act as the thickener.
3. In these recipes, the tomato puree is a double concentrate, that is 28–30% solids. If a tomato puree of a different concentration is used, then adjustments will have to be made to the addition rate.
4. The sucrose in these recipes is granulated. Where liquid sugar is used, then adjustments will have to be made to the water addition.
5. The onions should be minced, or alternatively a good quality onion powder can be used.
6. The amount of water added can vary. The mix should be cooked to 80°C and have a final refractive solids of 38%. Further boiling might be required or water might have to be added. Cooking should be carried out in stainless steel equipment.
7. The flavours and spices should be in the form of a concentrated extract on a salt or dextrose carrier for easy mixing and should be added at the end of the cooking. If they

are added during the cooking, there is the possibility that some of the flavours might be lost to the atmosphere.

15.4 Other dressings

Products such as salad cream, mayonnaise and salad dressings are essentially oil and water emulsions. In the United Kingdom, the difference between salad cream and mayonnaise is not always fully understood. The main difference is in the oil content. Salad cream will contain about 25–40% oil, whilst mayonnaise will contain about 70–85%. By increasing the oil content of the emulsion, the viscosity is increased. The preferred vegetable oil to use is cottonseed oil.

The sugar content of a salad cream can be as high as 25%, but the sugar content of a mayonnaise can be about 5%. Because of the high oil content in mayonnaise, it is difficult to increase the sweetener content without producing either a gummy or gritty product (Table 15.3).

Table 15.3 Typical sugar contents of salad dressings.

Salad cream	8–16%
Mayonnaise	2.0%
Flavoured dressing	4.6%
Thousand island	5.6%
French dressing	12.0%

The glucose syrups which are used in tomato ketchup can be used equally well in these types of products to replace sucrose, although HFGS is possibly the most suitable.

As with tomato products, there are legal compositional requirements to be met with these products, therefore before commencing any recipe changes, but always check that any proposed changes will comply with the legislation.

When making either salad cream or mayonnaise, there are some general points to bear in mind.

1. The eggs act as the emulsifying agent, but some recipes, particularly salad cream, incorporate an acid-resistant pre-gelled waxy maize starch as an additional stabiliser. Other stabilisers could be gums such as gum tragacanth or xanthan.
2. The eggs also impart colour to the product. If a darker colour is required, then a suitable edible colour can be added.
3. Because of the light colour of these types of products, it is important to use either distilled vinegar or spirit vinegar.
4. It is the acidity of the vinegar which acts as a preservative. Depending upon the product, the acidity in the end product should be about 1.5–3.5% measured as acetic acid, therefore the amount of vinegar to add will depend upon its acidity.
5. When adding the oil, always add it in a continuous steady stream.
6. For maximum stability, it is important to produce a stable emulsion, which can usually be achieved by passing the mix through an in-line homogeniser or colloid mill. The correct

operating pressure is important to produce the correct particle size oil globules, typically about 0.5–1.0 microns. The viscosity stability of both salad cream and mayonnaise should be checked 24 hours after manufacture.

A possible starting recipe for a salad cream could be as follows:

Vegetable oil	25.0%
HFGS 42% fructose (71% solids)	22.5%
Mustard	2.0%
Salt	2.0%
Egg yolks	3.0%
Modified starch	2.0%
Spirit vinegar	3.0%
Water	As required

Typical water content of a salad cream is 45–50%.

A possible starting recipe for a mayonnaise using HFGS to replace sucrose could be as follows:

Vegetable oil	75.0%
Egg yoke	8.0%
Spirit vinegar	3.0%
Water	8.5%
HFGS 42% fructose (71% solids)	5.0%
Mustard	0.5%
Salt	As required

Typical water content of a mayonnaise is about 18–20%.

15.5 Other sauces, marinades and pickles

Glucose syrups can also be used to replace the sucrose in products such as tartare sauce, which contain 15–20% sucrose, in horseradish sauce which contains 8–10% sucrose, and in sandwich spread and similar sandwich fillings which contain 20–25% sucrose. Other products could be English or French mustard.

A typical brown sauce might contain about 23% 'sugars', being a mixture of sucrose and molasses. Whilst the sucrose can be replaced by HFGS, or a glucose fructose syrup blend, the molasses contribute both colour and flavour to the product and therefore could be more difficult to be totally replaced. A 25–30% replacement of the molasses with HFGS would be possible without compromising the flavour. Any loss in flavour can usually be overcome by a slight increase in the spices and other flavouring ingredients and caramel can be added to increase the colour. The solids of a brown sauce are about 35–40%.

The following brown sauce recipe based on British Patent 2273642 B, the sucrose has been totally replaced by a blend of HFGS 42, high maltose syrup (70% maltose) and dextrose.

Malt vinegar	213 grams
Water	71 grams
Tomato paste	30 grams
High Maltose Syrup (70% solids)	12 grams
HFGS 42% fructose (71% solids)	10 grams
Dextrose	2 grams
Molasses	19 grams
Dates and raisins (finely chopped)	13 grams
Onions (finely chopped)	5 grams
Modified pre-gelled starch	5 grams
Soy sauce	8 grams
Salt	2 grams
Native maize starch	2 grams
Xanthan gum	0.65 grams
Ground nutmeg	0.065 grams
Ground coriander	0.025 grams
Caramel colour	As required

For a budget price brown sauce, some of the tomato puree, dates and raisins could be replaced by apples.

Depending upon the required sweetness, glucose syrups can also be used in marinades and similar products. A marinade for a sweet and sour dish can be made using a blend of a 63 DE syrup and HFGS. In this type of product, the sweetness must not over power the sour and spicy flavour notes from the other ingredients. Blending a 63 DE syrup with HFGS moderates the overall sweetness level, whilst the higher sugars from the 63 DE syrup provide viscosity, body and mouthfeel to the marinade, thereby making the product more visually appealing, as well as improving the organoleptic qualities. The viscosity can be further increased by the addition of a suitably modified starch.

Similarly in dark pickles, all the sucrose and part of the molasses could be replaced with the same provisos as for a brown sauce.

Light coloured pickles such as piccalilli contain about 4–7% sucrose, which could be totally replaced with HFGS 42.

For the replacement of sucrose in mint jelly, red currant jelly or cranberry jelly, see Chapter 14, 'Glucose syrups in jams'.

15.6 Reduced calorie products

A possible way of reducing the calorie content of the products considered in this chapter would be to replace either the sucrose or glucose syrup, with the sugar alcohol maltitol, which has approximately half the calories of sucrose, with about 90% of the sweetness of sucrose.

Chapter 16
Glucose syrups in soft drinks

16.1 Introduction

Several years ago, there was in the United Kingdom a legal requirement that all soft drinks must contain a minimum of 10% carbohydrate sweetener, when consumed. This requirement covered both ready to consume drinks such as carbonated drinks and dilutable drinks such as squashes and cordials. The carbohydrate sweetener was invariably sucrose, because that was the only sweetener suitable – conventional glucose syrups such as 42 DE were not sweet enough and 63 DE syrup was not produced in the United Kingdom until 1958, whilst fructose syrups were still waiting to be discovered. When 63 and 95 DE syrups became available, some soft drink producers started to use a blend of sucrose and a glucose syrup to meet the required 10% minimum carbohydrate requirement. Additionally, glucose syrups were cheaper than sucrose, so there was a cost incentive for using glucose syrups, but as these blends were not as sweet as 100% sucrose, the sweetness was boosted with saccharine. There was, however, a legal maximum on the amount of saccharin which could be added.

Since those early days, there have been many changes within the soft drink industry, both legal and technical mainly due to the availability of high-fructose glucose syrups (HFGS), and the arrival of new artificial sweeteners which are considered to be superior to saccharin. This has resulted in many more different types of soft drinks being produced, and the legislation has been updated to reflect this new diversity of drinks.

The major change has been to reduce the minimum carbohydrate requirement from 10% to 4.5%, which has opened the way for cheaper drinks. In the case of a squash (a very British drink), which has to be diluted, the carbohydrate content of the concentrate should be 22.5%. This assumes that one part of concentrate will be diluted with four parts of water to give a solids of 4.5% when consumed.

On the positive side, the new legislation now ties in the composition of a soft drink with the product description and claims. For example, if a soft drink is sold as 'low calorie', then it must not contain more than 5% carbohydrate, and if the drink claims to be a 'diet' drink, then it will not contain any carbohydrate sweetener. In a diet drink, the sweetness comes only from the artificial sweetener. Interestingly, if a drink is sweetened only with artificial sweeteners, the producers do not have to make any claim that it is a 'diet drink'.

Low-energy drinks contain less than 20 kcal/100 ml.
Energy reduced contain 30% less energy than that in the original drink.
Energy free contain less than 4 kcal/100 ml.
'Lite' drinks are the same as reduced drinks.

16.2 Ingredients

Retail soft drinks usually contain some or all of the following, depending upon the type of drink:

Water	up to 98% weight volume
Bulk sweetener	5–12% weight volume
Fruit juice	1–10%
Carbon dioxide	0.4–1.0% weight volume
Acidulant	0.05–0.6% weight volume
Preservative	300 ppm
Flavours	0.1% weight volume
Natural colours	0.0–0.2%
Synthetic colours	0.0–0.1%
Acidity regulator	0.0–100 ppm
Artificial sweetener	

Additionally, other ingredients could be emulsions, anti-oxidants, stabilisers, clouding agents, vitamins and minerals.

The bulk sweeteners are usually sucrose or HFGS.

The significance of the fruit juice is that most fruit juices contain fructose, dextrose with lesser amounts of sucrose and therefore have the potential to be a sweetener within their own right. Typical juices which are used for their sweetening properties are apple, grape and pear. Additionally, by using a fruit juice as the sole sweetener, it allows the manufacturer to make a 'No added sugar' claim.

The reason why an artificial sweetener is used in conjunction with a conventional bulk sweetener is frequently for commercial / price reasons. In non calorie drinks, they are present because they contribute sweetness but zero calories.

16.3 Effect of process inversion

Whilst the acid is present to enhance the flavour of a drink, it will also react with the sucrose, and over a period of time, the sucrose will be inverted. That is, it will be changed into equal amounts of dextrose and fructose. This is known as 'process inversion' (see figure 16.1), and is a common reaction in food manufacturing, where both acid and sucrose occur in a product.

$$C_{12}H_{22}O_{11} + H_2O \xrightarrow{Acid} C_6H_{12}O_6 + C_6H_{12}O_6$$

$$\text{Sucrose} + \text{Water} \longrightarrow \text{Dextrose} + \text{Fructose}$$

$$\text{Molecular weight} \quad 342 + 18 \longrightarrow 180 + 180$$

Figure 16.1 Process inversion of sucrose.

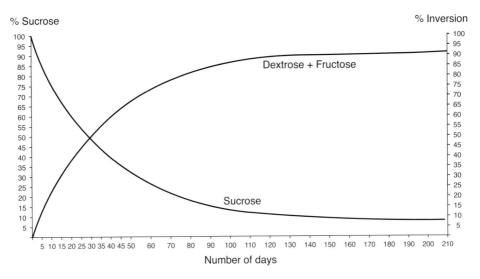

Figure 16.2 Inversion of a cola drink – pH 3.0, held at room temperature for 210 days.

The above graph in figure 16.2 illustrates the inversion of sucrose in a typical cola type drink (pH 3.0) over a storage period of 210 days. In summary, after 24 hours, 5% of the sucrose has been inverted. After 10 days, over 20% of the sucrose has been inverted, and after 30 days, 50% has been inverted. However, after 150 days, the inversion rate has levelled out at about 90% inversion, and after a further 60 days, the sucrose inversion has only increased by a further 0.6%. This slowdown in the rate of inversion is in line with most chemical reactions as they approach completion. When process inversion occurs, there is an increase in the solids or Brix reading, due to the inclusion of water. This increase in solids is known as the hydrolysis gain.

Days	% Sucrose	% Dextrose	% Fructose	% Sucrose inversion
0	100	0	0	0
1	95	2.5	2.5	5.0
2	92	4.0	4.0	8.0
3	88	6.0	6.0	12.0
4	87	6.5	6.5	13.0
5	85	7.5	7.5	15.0
6	84	8.0	8.0	16.0
7	82	9.0	9.0	18.0
8	81	9.5	9.5	19.0
9	80	10.0	10.0	20.0
10	76	12.0	12.0	24.0
15	70	15.0	15.0	30.0
20	64	18.0	18.0	36.0
25	56	22.0	22.0	44.0
30	50	25.0	25.0	50.0

35	45	27.5	27.5	55.0
40	40	30.0	30.0	60.0
45	36	32.0	32.0	64.0
50	32	34.0	34.0	68.0
60	28	36.0	36.0	72.0
70	22	39.0	39.0	78.0
80	18	41.0	41.0	82.0
90	16	42.0	42.0	84.0
100	14	43.0	43.0	86.0
110	12	44.0	44.0	88.0
120	11	44.5	44.5	89.0
130	10.2	44.9	44.9	89.8
140	10.0	45.0	45.0	90.0
150	10.0	45.0	45.0	90.0
160	9.8	45.1	45.1	90.2
170	9.8	45.1	45.1	90.2
180	9.6	45.2	45.2	90.4
190	9.6	45.2	45.2	90.4
200	9.4	45.3	45.3	90.6
210	9.4	45.3	45.3	90.6

The rate at which process inversion occurs will depend upon many factors, but particularly

- the concentration of acid used,
- the hydrolytic or inverting power of the acid,
- the temperature at which the drink is stored,
- the length of time the drink is stored at that temperature,
- the presences of any potential buffering salts.

All of the above factors will contribute to the profile of the inversion curve. High concentrations of acids, which have strong hydrolytic (or inverting) power together with high temperatures, will all contribute to a rapid rate of inversion to give a steep inversion curve. The process of inversion occurs when the hydrogen ion of the acid is released when the acid dissociates.

With the sucrose in the drink being inverted, both the sweetness and flavour profiles of the drink will change because dextrose, fructose and sucrose have different individual sweetness responses. Fructose has a very high intense sweetness, which only lasts for a short length of time. Sucrose on the other hand has a lower sweetness intensity, but lasts for a longer time. The sweetness intensity of dextrose is less than that of either fructose or dextrose, and lasts for a shorter time.

Comparing the sweetness response profiles, as shown in figure 16.3, for dextrose, fructose and sucrose with the inversion graph of the cola drink, one can see how the changes

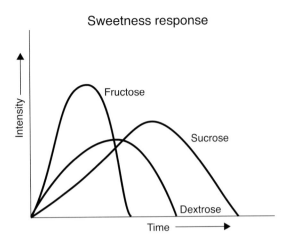

Figure 16.3 Sweetness response profiles for fructose, sucrose and dextrose.

in the relative proportions of dextrose, fructose and sucrose due to inversion of the sucrose will affect the perceived sweetness response of the drink. Some of the perceived changes in the sweetness might also be due to possible synergy between the three different sugars.

Similarly, as the sugar profile changes, so the perceived flavour profile will change. This is because different flavours react differently with different sugars, for example fructose enhances citrus flavours.

When the high-intensity sweetener aspartame is used in a drink, there is a loss of sweetness over a period of time because at low pH, as in a soft drink, aspartame is broken down in to its component amino acids – aspartic acid and phenylalanine. This loss of sweetness is accelerated with increase in temperature.

16.4 Use of glucose syrups

It is because of these different changes which occur in a soft drink that the use of glucose syrups, particularly HFGS are ideally suited for soft drinks – the sugar profile of glucose syrups is totally stable being unaffected by pH or temperature, and therefore will not change during storage resulting in both the sweetness and flavour profiles remaining unchanged.

The following glucose syrups are typically used in soft drinks.

Sugar.	63 DE Syrup.	42% HFGS.	55% HFGS.
Fructose	–	42%	55%
Dextrose	34%	52%	41%
Maltose	33%	4%	2%
Maltotriose	10%	1%	1%
Higher sugars	23%	1%	1%

To match the sweetness of a sucrose based drink, use an HFGS containing 55% fructose. Where only a HFGS containing 42% fructose is available, then the syrup addition rate might have to be increased by about 5% on a solid for solid basis. HFGS due to their method of manufacture, and particularly the many refining stages which they undergo, have a very clean, even a 'clinical' like taste profile. This taste / sweetness profile can be modified by blending a HFGS with a 63 DE glucose syrup.

Because a 63 DE syrup contains a proportion of higher sugars, these higher sugars effectively increase the viscosity of the drink, thereby slowing down the time for the fructose to reach the taste receptors, and is responsible for giving the blend a sweetness intensity / time profile similar to that of sucrose. This slowing down of the sweetness release is sometimes referred to as 'sweetness drag'. Additionally, glucose syrup will also increase the overall molecular weight of the 'sweetener system'.

Due to the restrictions brought about by the Isoglucose Production Quota which operates within the EU, so as to increase the availability of fructose containing syrups to as many users as possible, blends of a 63 DE syrup with HFGS 42 are frequently used in soft drinks. As the sweetness of a 63 DE syrup is less than that of a HFGS, the downside of blending a 63 DE syrup with a high fructose syrup is that there will be a reduction in sweetness, but this reduction in sweetness allows the flavour to come through. If the reduction in sweetness is too critical, then an artificial sweetener can be added. The relative sweetness of 63 DE syrup, 42% and 55% HFGS, compared to sucrose is as follows:

Sugar	Approximate sweetness
Sucrose	100
Fructose	150 (120–170)
Dextrose	80
HFGS 42% FX	100 (95–105)
HFGS 55% FX	105 (100–110)
63 DE Syrup	70

A possible alternative to using a 63 DE syrup to modify the perceived sweetness profile could be to use a maltodextrin. Maltodextrins contain mainly higher sugars, therefore a lesser amount of maltodextrin would be required to provide a comparable amount of higher sugars from a 63 DE syrup. Because the addition of maltodextrin would be small, there would be a correspondingly less reduction in the overall sweetness. Some solutions of maltodextrin do not have the same clarity as a 63 DE syrup, therefore the use of a maltodextrin might affect the clarity of the end product. However, where a clouding agent is required, then a maltodextrin could be a useful addition to the formulation.

When a high intensity sweetener is used in a soft drink, the drink will usually lack mouthfeel and body, resulting in the flavour quickly disappearing. These problems can be overcome by adding a small amount of glucose syrup. Once again, the higher sugars of the glucose syrup will improve both the mouthfeel and body, with the increase in viscosity helping to retain the flavour in the mouth, and increasing the sweetness drag, thereby making the drink more acceptable. Obviously, this approach cannot be used when formulating a drink which is claimed to be a diet or low calorie drink.

Frequently artificial sweeteners, particularly saccharin, can give a drink an unpleasant after taste. This unpleasant after taste can often be masked by the addition of a 63 DE syrup to the formulation. The higher sugars of the 63 DE syrup increase the viscosity of the drink. This increase in viscosity allows the after taste to be slowly released and in so doing effectively dilutes the after taste making it less obvious.

Much is said about soft drinks being 'empty calories'. The term 'empty calories' refers to the fact that these drinks only contain 'sugars' without any other nutrients, such as minerals and vitamins. It must be understood that the 'sugars' in a drink will ultimately be broken down by the body to provide dextrose, which is the fuel for the body – it is dextrose which circulates in the blood stream. Therefore, to think of soft drinks as the popular press would have us to believe, as having no nutritional value, is totally misleading. However, when a high intensity sweetener has been used as the sole sweetener in a drink, then the description 'empty calories' for that type of drink, meaning that it does not contain any calories, would be a correct description.

16.5 Quality considerations

One of many important considerations when using any bulk sweetener in a soft drink is its microbiological quality. Most soft drink producers will insist that the syrup meets The American Bottlers Standard – an American industry standard. A summary of this standard for liquid sugar, but equally applicable to glucose syrups, is as follows:

Mesophilic bacteria	100 organisms or less per 10 grams on a dry solids
Yeast	10 organisms or less per 10 grams on a dry solids
Moulds	10 organisms or less per 10 grams on a dry solids

Another quality consideration for sweeteners which are to be used in clear drinks such as a lemonade is the clarity of the sweetener in use and if it will form a floc when in solution. The sucrose industry produces a special grade for use in soft drinks – Mineral Water grade which when dissolved produces a solution of exceptional clarity. Glucose syrups which are used in soft drinks, and particularly syrups such as HFGS, because of their method of production, have equal clarity to that of Mineral Water grade sucrose and will easily pass the Spreckles test for clarity. (This test was developed by The Spreckles Sugar Company of San Francisco, and is now an industry standard.)

16.6 Laboratory evaluation of glucose syrups in soft drinks

When carrying out a laboratory evaluation of any glucose sweetener in a soft drink, there are several basic rules to follow. It must be stressed, however, that laboratory results are only indicative as to what can be expected from a production run, and remember that production runs are frequently able to improve upon laboratory results.

16.6.1 Water

Water is the major ingredient in any soft drink. If the quality of the water is unacceptable, then the final drink will be unacceptable. It is also important to always use water from the same source for both the control drink and all trial drinks. In a laboratory situation where acceptable water is not readily available, then the best way to ensure that the quality of the water is consistently acceptable is to buy a reputable brand of bottled water from a supermarket, and always use the same brand of water for any future work. By using this approach, it is possible to make either still or carbonated drinks by buying either still or carbonated water. As the amount of carbonation in a drink can affect the taste profile of the drink, in buying carbonated water from a supermarket, the level of carbonation will be consistent, from one bottle to another – something which cannot be guaranteed by hand carbonation, unless an expensive carbonator is available. For concentrated drinks, these can be made using still water. The finished drinks can then be made by diluting the concentrate with either carbonated water for a carbonated drink, or still water for a still drink.

16.6.2 Sweeteners

1. When replacing one sweetener with another, so as to find an approximate acceptable replacement level, make the initial replacements in large stages, typically at 25%, 50%, 75% and 100% replacements. This will give a quick indication of the likely achievable replacement levels. Having established the approximate replacement level, the next stage should be to make further replacements but at 5% replacement levels within this approximate achievable range.

   All replacements must be on a dry weight for dry weight basis.

2. When replacing crystalline sucrose with a syrup, always convert everything to a dry basis, and replace dry for dry. The same approach must also be used when using syrups which have a different solids content.

3. Some syrups are deionised, some are not. This difference could influence the taste of the end product in two ways. First, some ions such as chlorides and sulphates can affect the taste. Secondly, because deionised syrups do not contain any minerals, they have no buffering capacity, which means that the amount of acidulant to be added could be different. Therefore, always check whether or not the syrup has been deionised.

4. Some drinks might use a combination of sweeteners usually to obtain the required sweetness for the best price (Table 16.1).

Table 16.1 Comparative sweetness values[a]

Sweetener	% Inclusion	Relative sweetness	% Sucrose equivalent
Sucrose	10%	1.0	10
Fructose	6.7%	1.5	10
Saccharin	0.02%	450	10

[a] Further sweetness values can be found in Appendix C, Sugar Data, page 321.

For example, if the sweetness of sucrose is 100, and it is necessary to replace some of the sucrose with fructose, without changing the overall sweetness, this is the calculation:

The sweetness of fructose is 150, compared to sucrose at 100.

$$\text{Equivalent sweetness} = \frac{\text{Sucrose sweetness}}{\text{Fructose sweetness}} = \frac{100}{150} = 0.67$$

Therefore, 67 parts of fructose will have the equivalent sweetness of 100 parts of sucrose.

Whilst the overall sweetness of the drink will be the same, it is likely that both the perceived sweetness and flavour profiles will be different. The addition of a small amount of a 63 DE syrup could bring the profiles more in line with that of sucrose by changing the sweetness drag.

16.6.3 Acidulants

1. Be prepared to experiment with both the amount and type of acidulant, as both can have an affect on the flavour and perceived 'sharpness' of the drink as well as the sweetness profile.

   From Table 16.2, it will be seen that, for an equal pH drop, malic acid will give the most acidic taste, followed by citric, lactic, tartaric and with phosphoric giving the least acidic taste. In practice, when formulating a drink, the amount of acid added depends upon the required 'sharpness' in the end product.

Table 16.2 Approximate relationship between acidic taste and pH of different acids.

	Parts by weight to give	
	Equal pH drop	Equal acidic taste
Citric acid	1.0	1.0
Malic acid	1.0	0.80
Lactic acid	1.0	1.25
Tartaric acid	0.56	1.00
Phosphoric acid	0.23	0.90

2. A mention has already been made of the hydrolytic or inverting power of different acids. Table 16.3 shows the hydrolytic or inverting power of some common acids. The reason why some acids have greater hydrolytic or inverting power than others is because of their molecular structure and the ease in which they dissociate.

A recipe contains 10 grams of phosphoric acid. To replace the phosphoric acid with citric acid, and maintain the same inverting power, then the calculation is as follows:

Table 16.3 The hydrolytic or inverting power of different acids.

Acid	Hydrolytic or inverting power
Hydrochloric	100.0
Sulphuric	53.6
Phosphoric	6.21
Citric	1.72
Malic	1.55
Tartaric	3.0
Lactic	1.07

The hydrolytic or inverting power in the recipe is as follows:

$$\text{Weight of acidulant} \times \text{hydrolytic or inverting power}$$
$$= 10 \times 6.21$$
$$= 62.1$$

The hydrolytic or inverting power of citric acid is 1.72. Therefore, obtaining the equivalent hydrolytic or inverting power in the recipe will require

$$\frac{\text{Hydrolytic power of present acid} \times \text{amount}}{\text{Hydrolytic power of alternative acidulant}} = \frac{6.21 \times 10}{1.72}$$
$$= 36.1 \text{ grams}$$

Therefore, 36.1 grams of citric acid will have the equivalent hydrolytic power as 10.0 grams of phosphoric acid.

16.7 Soft drink recipes

The following recipes should be taken only as a guide for further experimentation. In these recipes, all the sucrose have been replaced and the amount of HFGS 42 syrup solids has been increased by about 5%, compared to HFGS 55 so as to compensate for the reduced sweetness of HFGS 42 compared to sucrose. In some drinks, this might not be necessary, and whilst these examples are for a 100% sucrose replacement, a partial replacement might be just as acceptable.

Using the previously mentioned calculations for determining sweetness, further changes to the sweetener system can be achieved by replacing any of the sweeteners with any other syrup or high intensity sweetener or a combination of both syrup and high intensity sweetener(s) and still end up with a product which has the same perceived sweetness as the original drink. This approach can be useful when costs are a consideration. However, it is important that the sweetness of the final drink matches the sweetness and flavour expectations of the consumer, and each country has its own ideas as to what is sweet and what is not, as well as their ideas about the flavour.

Similarly, the perceived sweetness and flavour can be changed by using different acids.

16.7.1 Carbonated drinks, for example lemonade

Ingredients	Sucrose control	HFGS 42% fructose	HFGS 55% fructose
Sucrose	540.00 grams	–	–
HFGS 42% fructose (71% solids)	–	800.00 grams	–
HFGS 55% fructose (71% solids)	–	–	760.00 grams
Potassium sorbate	1.25 grams	1.25 grams	1.25 grams
Citric acid anhydrous	13.50 grams	13.50 grams	13.50 grams
Lemon/lime flavour	6.00 ml	6.00 ml	6.00 ml
Water to make	1000 ml	1000 ml	1000 ml

One part of this concentrate is diluted with five parts of carbonated water and then bottled. (This procedure of diluting the concentrate with water, prior to bottling, is known technically as the 'Throw').

16.7.2 Dilutable drinks, for example orange squash

Ingredients	Sucrose control	HFGS 42% fructose	HFGS 55% fructose
Sucrose	100 Kg	–	–
HFGS 42% fructose (71% solids)	–	148 Kg	–
HFGS 55% fructose (71% solids)	–	–	141 Kg
Sodium saccharin	0.5 Kg	0.5 Kg	0.5 Kg
Aspartame	0.5 Kg	0.5 Kg	0.5 Kg
Orange comminute (40 Brix)	30 litres	30 litres	30 litres
Citric acid	10 Kg	10 Kg	10 Kg
Potassium sorbate	1.2 Kg	1.2 Kg	1.2 Kg
Flavouring	1.0 Kg	1.0 Kg	1.0 Kg
β-Carotene	0.5 Kg	0.5 Kg	0.5 Kg
Ascorbic acid	0.25 Kg	0.25 Kg	0.25 Kg
SCC[1]	0.5 Kg	0.5 Kg	0.5 Kg
Water to make	1000 litres	1000 litres	1000 litres

Note

1. Sodium carboxymethyl cellulose.

The above concentrate is pasteurised before bottling. To drink, one part of the concentrate is diluted with four parts of water, to give 2% whole orange comminute when consumed.

Where a compound flavour is used, it will usually contain a preservative and citric acid. The solids of a typical compound flavour is usually about 60–65%.

If concentrated fruit juices are used, they are described as '4, 5 or 6 fold'. This means that one volume of the concentrate can be diluted with 4, 5 or 6 volumes of water to give a fruit juice with the same solids as was in the original juice. The amount a concentrate can be diluted will vary from one juice to another.

16.8   Powdered drinks

A typical recipe could be as follows:

Spray dried fruit juice	2.00 grams
Dextrose monohydrate	20.00 grams
Citric acid anhydrous	0.60 grams
Sodium saccharin	0.035 grams
Free-flow agent, for example silicon dioxide	0.2 grams
Spray dried flavour	0.02 grams
Colour (where applicable)	

The above mix is packed into a moisture-proof sachet, and to make a drink, the whole sachet is added to 200 ml of water.

Whilst a blend of dextrose and saccharin has been used in the above recipe, any powdered sweetener could be used – sucrose, fructose, maltodextrin, powdered polyol, a high intensity sweetener or a combination. Which ingredients are used will depend upon the type of drink to be produced and the cost involved.

When producing a powdered drink, consideration should be given to the particle size of the ingredients. Ideally, all the ingredients should have approximately the same particle size. If there is too big a difference in the particle size distribution of the ingredients, there is the possibility of the ingredients separating during handling, with the smaller particles gravitating to the bottom of the container leaving the larger particles at the top. This separation would result in the drink having a variable flavour, and is especially important if the powdered drink is to be used in a vending machine. To reduce the problem of separation and to ensure that a consistent flavour is produced, the ingredients can be either co-spray dried or dry blended. If dry blending is used, then it is useful to agglomerate the ingredients and then reduce the agglomerated particles to a consistent particle size.

16.9   Reduced calorie drinks

One of the easiest ways to reduce the calories in a drink and maintain the same sweetness level, without using an artificial sweetener, for example in a drink which is sweetened with HFGS, is to use fructose. If the sweetness value of HFGS is 100 and the sweetness value of fructose is 150, then to match the same sweetness, 100 parts of HFGS solids could be replaced with only 67 parts of fructose. Since the caloric value is the same for both HFGS and fructose, by using fructose, a 33% calorie reduction could be achieved and therefore legally it could be claimed that there has been a calorie reduction.

Chapter 17
Glucose syrups in health and sports drinks

Nomenclature

In this chapter, dextrose will be used to refer to the medical ingredient powdered glucose.
Glucose syrup will refer to the syrup containing a mixture of dextrose, maltose, maltotriose, etc.

17.1 Introduction

The main object of a health or sports drink is to supply the body with a ready source of energy, and glucose syrups are the ideal ingredients since they are easily broken down by the body. There are also two other reasons for drinking these types of drink, namely to replace electrolytes, for example sodium chloride and water, lost through perspiration and to maintain or enhance an athlete's performance.

Carbohydrates, such as starch and sugars, are the major energy sources for the human body, but it is the simple sugar dextrose (known medically as glucose) which circulates in the bloodstream, and it is dextrose which is the 'fuel' powering the human body. Therefore, one of the functions of the digestive system is to convert complex carbohydrates into dextrose. Whilst dextrose will be absorbed immediately by the body, it takes longer for the body to convert the more complex carbohydrates into dextrose, and for their subsequent utilisation by the body.

17.2 The energy source

Starch based glucose syrups such as a 55 or 63 DE syrup, maltodextrins and fructose are the ideal ingredients for inclusion in a sports or health drink, and because of their broad sugar spectrum, they can over a period of time be easily broken down to dextrose, thereby ensuring a steady release of energy which athletes require.

A typical 63 DE glucose syrup will contain 36% dextrose, 33% maltose, 31% maltotriose and higher sugars, so when this syrup is used in a sports drink, the dextrose will be immediately available, and whilst it is being used by the body, the maltose is being converted into dextrose to replace the original 36% of dextrose. Similarly, whilst the dextrose from the maltose is being consumed, the remaining higher sugars are being broken down first to maltose, and then to dextrose, thereby ensuring a continuous supply of energy over a period of time.

Maltodextrins, because they are made up of higher sugars, are more slowly broken down releasing energy over a longer period, making them ideal for an endurance athlete.

Since maltodextrins typically have a molecular weight of over 1000, they have a greater energy potential on a weight for weight basis than other glucose polymers, without affecting the osmotic pressure. As low DE maltodextrins form a cloudy solution in water, this can be a limiting factor for their inclusion in sports and health drinks which are normally crystal clear. Higher DE maltodextrins which produce near-clear solutions are preferred.

Fructose, because it has to pass first to the liver to be converted into dextrose, there is a delay from the time that the fructose is digested to the time when its energy becomes available. This time lag in the availability of energy can make fructose a useful sugar where a delayed energy release is acceptable.

Therefore, by using a blend of different starch based sweeteners, it is possible to formulate a sports or health drink to match the energy requirements of the user (Table 17.1).

Table 17.1 Typical sugar spectrum of different starch-based sweeteners frequently used in sports and health drinks.

	15 DE Maltodextrin	55 DE Glucose Syrup	63 DE Glucose Syrup	HFGS 42% fructose
Fructose	–	–	–	42.0%
Dextrose	1.3%	30.0%	34.0%	52.0%
Maltose	4.1%	24.0%	33.0%	4.0%
Maltotriose	6.0%	10.0%	10.0%	1.0%
Higher sugars	88.6%	36.0%	23.0%	1.0%

Whilst in theory, the process of breaking down the large molecules into smaller molecules allows dextrose to be slowly released into the bloodstream to give a continuous supply of energy, it is the hormones insulin and glucagons produced by the pancreas, which actually regulate the amount of dextrose in the bloodstream, with any excess dextrose being stored as glycogen.

Sucrose, on the other hand, for reasons which will be explained later, is not an ideal sugar for use in sports and health drinks.

During strenuous exercise, the body loses water through perspiration, and in this perspiration are salts (electrolytes). Both water and these salts are essential to the body, and must be replaced, particularly the sodium ions which maintain the osmotic drive to drink and to facilitate fluid absorption. Sodium ions are also involved in the transport of dextrose across the intestinal wall. Therefore, another function of a sports drink is not only to supply energy, but also to replace both the lost water and the electrolytes. The inclusion of fructose in a sports or health drink is not only useful as a potential source of energy, but being a very sweet sugar – approximately 50% more sweet than sucrose, this sweetness also masks the salt flavour of the electrolytes, making the drink more palatable.

17.3 Classification of health drinks

When formulating sports or health drinks, care must be taken to ensure that the osmotic pressure of the drink does not upset the osmotic pressure of the body or interferes with any bodily functions. The osmolality of body plasma is 290 mOsm/Kg. It is for this reason that sports drinks are conventionally divided into three main types depending upon their osmotic pressure or osmolality.

1. ISOTONIC DRINKS, which have the same concentration as body fluids and are usually based on glucose syrups. They are designed to give a supply of energy before, during and after exercise. Isotonic drinks will typically have an osmolality of 270–330 mOsm/Kg. It should be realised that the theoretical osmolality of body plasma, based on known constituents, does not always agree with actual measurements due to some electrolytes not being completely dissociated whilst others are bound to proteins, hence the range of 270–330 mOsm/Kg.
2. HYPERTONIC DRINKS, are more concentrated than body fluids. They also contain glucose and are taken before and after exercise. Being more concentrated than the body fluids, hypertonic drinks are gradually absorbed into the bloodstream over a longer period of time, thereby allowing a prolonged period of activity. The osmolality of a hypertonic drink is greater than 330 mOsm/Kg.
3. HYPOTONIC DRINKS, are less concentrated than the body fluids, and because they are less concentrated, they are quickly absorbed and rapidly replace water and salts, lost through excessive perspiration while exercising. If, however, the object of the exercise is to lose weight, there is some debate if hypotonic drinks should contain any calories, but if it contains a few calories in the form of a glucose syrup to make the drink more palatable, then calories are acceptable. The osmolality of a hypotonic drink will be less than 270 mOsm/Kg.

17.4   Osmotic pressure of health drinks

Osmosis is the passage of a liquid, for example water, through a semipermeable membrane from a weak to a more concentrated solution and the osmotic pressure is the pressure required to prevent this movement of liquid and in medical applications is usually expressed as either osmolality, which is the number of moles per kilogram of water, or as osmolarity, which is the number of moles per litre of solution. As the volume of a liquid changes with temperature, osmolarity is more difficult to measure, therefore the preferred osmotic pressure measurement for sports and health drinks is osmolality, with the units being milliosmols (mOsm/Kg).

$$\text{Osmolality} = KnM \text{ (Osmol/Kg)}$$

where
K = osmotic coefficient (see note 1)
n = number of particles into which the molecule dissociates (see note 2)
M = molality (see note 3)

Notes

1. The osmotic coefficient is taken as being 1 when there is total dissociation with no water binding.

2. Dissociation is the process whereby a molecule is split into simpler fragments which may be smaller molecules, atoms, free molecules or ions. The dissociation number, therefore, is the number of smaller molecules into which the molecule can split. Because sugars do not dissociate, the number of particles for dextrose, sucrose, fructose, etc, would be 1.

   For substances which dissociate such as sodium chloride (NaCl), it would be 2, namely, sodium and chloride. For potassium, sulphate (K_2SO_4), it would be 3, namely, two molecules of potassium and one molecule of sulphate, and so on.

3. A molal solution contains the gram molecular weight of a substance in 1000 grams of water to give a weight/weight solution. Therefore, a molal solution of dextrose would contain 180 grams of dextrose dissolved in 1000 grams of water.

The above osmolality formula can be simplified as follows:

$$\text{Osmolality (mOsm)} = \frac{\text{Weight of solids in 1000 grams} \times 1000}{\text{Molecular weight} \times \text{dissociation number}}$$

Example 1

Osmolality of a non-ionic solution, for example 8% dextrose solution.

Molecular formula of dextrose = $C_6H_{12}O_6$
Molecular weight of dextrose = 180
The dissociation number = 1
Osmotic coefficient assumed = 1

Since 180 grams of dextrose + 1000 grams of water = 1000 mOsm,
An 8% dextrose solution = 80 grams + 1000 grams water.

$$\text{Therefore, osmolality} = \frac{80 \times 1000}{180 \times 1}$$

$$= 444 \text{ mOsm}$$

N.B. In the above calculation, anhydrous dextrose has been used. Commercially, dextrose monohydrate would be used, which typically contains between 7.0% and 9.5% water. Therefore, when using dextrose monohydrate, this water would have to be taken into account.

Example 2

Osmolality of an ionic solution, for example 5% sodium chloride solution.

Molecular formula of sodium chloride = NaCl
Molecular weight of sodium chloride = 58
N, the dissociation number = 2
Osmotic coefficient assumed = 1

Since 58 grams of sodium chloride + 1000 grams of water = 1000 mOsm,
A 5% sodium chloride solution = 50 grams + 1000 grams water.

$$\text{Therefore, osmolality} = \frac{50 \times 1000}{58 \times 2}$$

$$= 430 \text{ mOsm}$$

17.5 Sucrose in sports or health drinks

It has already been mentioned that sucrose is not an ideal sugar for inclusion in a sports or health drink. One reason is that sports and health drinks are formulated to a specific osmolality, and because these types of drinks frequently contain an acid, over a period of time, the sucrose will be inverted to give dextrose and fructose which have a different molecular weight to that of sucrose, with the result that the osmolality of the drink will change. The following example demonstrates this change in osmolality. Initially, a drink containing 8% sucrose would have the following osmolality.

Molecular formula of sucrose = $C_{12}H_{22}O_{11}$
Molecular weight of sucrose = 342
The dissociation number = 1
Osmotic coefficient = 1

Since 342 grams of sucrose + 1000 grams of water = 1000 mOsm,
An 8% solution of sucrose = 80 grams sucrose + 1000 grams of water.

$$\text{Osmolality} = \frac{80 \times 1000}{342 \times 1}$$
$$= 234 \text{ mOsm}$$

If the 8% sucrose is totally inverted (which is unlikely), there would now be 4% dextrose and 4% fructose, and as the molecular weight of both dextrose and fructose is 180, the new osmolality of the drink would now be as follows:

Molecular formula of dextrose and fructose = $C_6H_{11}O_6$
Molecular weight of dextrose and fructose = 180
The dissociation number for both dextrose and fructose = 1
Osmotic coefficient for both dextrose and fructose = 1

$$\text{Osmolality} = \frac{8 \times 1000}{180 \times 1}$$
$$= 444 \text{ mOsm}$$

Therefore, by using sucrose, the osmolality of the drink will have increased during storage from 234 to 444 mOsm. In practice, the amount of sucrose which is inverted will depend upon the type and amount of acid used in the drink, together with the storage conditions of the drink with high temperatures speeding up the rate of inversion. For a weak acid such as citric acid, the sucrose inversion would be considerably less. This change in the osmolality of drinks sweetened with sucrose means that a drink which was originally an isotonic drink could become a hypertonic. Because these types of drinks contain sugars, minerals and other ingredients, the osmolality of the drink cannot always be accurately calculated, therefore commercially, the osmolality will usually be measured using an instrument called an osmometer. The anion and cation content of the drinks can be determined using selective ion electrodes.

Both of these types of instrument are ideal for QC applications, being both quick and simple to use.

Table 17.2 Approximate osmolality in mOsm/Kg, of different starch-based sweeteners with different DEs and at different concentrations.

	5 DE	10 DE	15 DE	20 DE	28 DE	35 DE	42 DE	63 DE	72 DE	100 DE (1)
	Molecular weight (2)									
	3600	1800	1200	900	643	514	429	286	250	180
1%	2.8	5.6	8.3	11.1	15.6	19.5	23.3	35.0	40.0	55.6
2%	5.6	11.1	16.7	22.2	31.1	38.9	46.6	69.9	80.0	111
3%	8.3	16.7	25.0	33.3	46.7	58.4	69.9	105	120	167
4%	11.1	22.2	33.3	44.4	62.2	77.8	93.2	140	160	222
5%	13.9	27.8	41.7	55.6	77.8	97.3	117	175	200	277
10%	27.8	55.6	83.3	111	153	195	233	350	400	556
15%	41.7	83.3	125	167	233	292	350	525	600	833
20%	55.6	111	167	222	311	389	466	699	800	1110
25%	69.4	139	208	278	389	486	583	874	1000	1390
30%	83.3	167	250	333	467	584	699	1050	1200	1670
35%	97.2	194	292	389	544	681	816	1220	1400	1940
40%	111	222	333	444	622	778	932	1400	1600	2220

Notes

1. This is anhydrous dextrose. In a commercial situation, dextrose monohydrate would be used, which typically has a moisture content of 7.0–9.5%, and this moisture should be taken into account.
2. Molecular weight $= \frac{18000}{DE}$.

17.6 Formulating a sports drink

Physiological and sensory research suggests the following guidelines when formulating a sports drink:

- They should not be carbonate, because carbonation can cause gastric discomfort.
- For optimum fluid absorption and energy delivery, the carbohydrate concentration should be less than 8%.

- The preferred carbohydrate should be glucose or glucose polymers (maltodextrin), to encourage fluid absorption and energy uptake by muscles.
- Sodium salts should be included so as to maintain the osmotic drive to drink and to facilitate fluid absorption.
- The drink should have an appealing taste, to encourage a greater liquid intake, and hence a more complete rehydration.

Warning! Drinking excessive amounts of water can cause hyponatraemic encephalopathy in athletes. This is a potentially fatal condition caused by a sever lack of salt in the blood due to the excessive intake of water diluting the sodium chloride level in the blood. It is for this reason that any water drunk after exercising should contain sodium chloride.

A basic recipe for a sports drink, could be as follows:

55 DE Glucose Syrup (81% solids)	= 55.0 Kg
Citric acid anhydrous	= 900.0 grams
Sodium citrate	= 96.0 grams
Potassium chloride	= 72.0 grams
Sodium chloride	= 144.0 grams
Ascorbic acid	= 24.0 grams
Flavour	= As required
Water to make	= 100.0 litres

As the drink does not contain a preservative, it should be either fully pasteurised or aseptic packaged. If either of these options is not possible, then potassium sorbate would be the preferred preservative at about 300 mg/L.

This concentrate is then diluted to give a carbohydrate content in the final drink of about 6–8%.

1. **55 DE Glucose Syrup**: This supplies the 'energy' part of the drink. The typical sugar spectrum of a 55 DE syrup is:-

Dextrose	= 30%
Maltose	= 24%
Maltosetriose	= 10%
Higher sugars	= 36%

    Therefore, the original dextrose in the syrup supplies the initial energy, followed by dextrose produced by the breakdown of the maltose, followed by the breakdown of the maltotriose and eventually the higher sugars. This breaking down of the sugars results in a continuous supply of dextrose to give a continuous supply of energy – ideal for an athlete.

    By changing the balance of soluble carbohydrates in a drink, that is by using a glucose syrup which has a different sugar spectrum, it is possible to change the rate at which energy becomes available to the athlete. For example, by replacing the 55 DE syrup with dextrose, the amount of 'instantly available energy' could be increased, which could be of benefit to a short distance runner. Similarly, by replacing the 55 DE syrup with say a maltodextrin, then the 'long term available energy' could be increased, at the expense of the 'instant energy', and would benefit the endurance athlete who requires a release of energy over a longer time period.

The higher sugars will also give mouthfeel and body to the drink helping to make the drink more palatable.

2. The citric acid gives acidity or sharpness to the drink, with the sodium citrate acting as a pH buffer. The sodium citrate also acts as a supply of sodium ions. If the drink contains a citrus flavour to make the drink more acceptable, then the citric acid will enhance this fruit flavour.
3. The potassium chloride and sodium chloride supply the electrolytes so essential for maintaining normal heart rhythm and regulating the water balance in the body amongst many activities. The absorption of dextrose is linked with the absorption of sodium ions. *N.B.* The absorption of fructose, if present, is independent of sodium ions.
4. Ascorbic acid has two functions – a source of vitamin C and as an antioxidant.
5. The flavour makes the drink more palatable by masking the taste of the salts.
6. Additionally, the drink might also contain a stabiliser and vitamins.

The above drink could be produced in a powdered form, replacing the 55 DE glucose syrup with a dry blend of dextrose, and spray-dried 35 DE glucose syrup, and a spray-dried flavour.

A similar approach of dry blending dextrose, fructose, spray dried glucose syrup and maltodextrins can be used to produce drinks of any omolality, and all that has to be done is to mix the powder with the correct amount of water.

17.7 Energy values

The energy value of drinks is expressed in either kilojoules or kilocalories, and can be calculated by multiplying the carbohydrate content by 17.0 to give a value in kilojoules per gram, or multiplying by 4.0 to give a value in kilocalories per gram.

A typical isotonic sports drink will contain about 6.0–6.5% carbohydrate, and could be a 50/50 blend of a 55 DE glucose syrup and a 15 DE maltodextrin. Such a blend provides 'instant energy' from the dextrose in the glucose syrup, whilst the maltodextrin provides the 'long-term energy' so needed by an athlete. The energy value of the drink would be about 110 kJ (25 kcal) per 100 ml of drink.

An energy drink as expected would contain a higher amount of carbohydrate, typically between 15% and 20%. This carbohydrate would be in a form which would make it easily assimilated and typically would be based on a high fructose glucose syrup. High fructose glucose syrups are composed mainly of dextrose and fructose. The dextrose will be quickly assimilated, whilst there would be a delay before the potential energy from the fructose is released. The fructose will also add sweetness and enhance the palatability to the drink. The energy value of these types of drink will be about 290 kJ (68 kcal) per 100 ml of drink.

Typically, a low-energy drink will contain less than 20 kcal/100ml. Energy-reduced drinks will contain 70% less calories than the original drink, and energy-free drinks will contain only 4 kcal/100ml or less.

17.8 Oral rehydration

The need for oral rehydration is often due to excessive loss of fluids, typically caused by diarrhoea. With this loss of fluids, there will also be a loss of electrolytes, both of which must be replaced if life is to continue.

In formulating a suitable rehydration solution, the following guidelines should be considered:

- The solution must increase the rate at which water and electrolytes are absorbed.
- The correct amount of electrolytes are replaced in a safe way. The solution should be slightly hypo-osmolar (about 250 mmol/l to prevent the possible induction of osmotic diarrhoea).
- The solution could contain an alkalising agent (sodium bicarbonate) to prevent acidosis.
- The solution must be simple to use.
- It must be palatable, especially to children. Some formulations might contain flavouring, such as blackcurrant or lemon.
- It must be readily available.

As this is a worldwide problem, the World Health Organisation (WHO) has recommended the following recipe:

Sodium chloride	= 3.5 grams
Potassium chloride	= 1.5 grams
Sodium citrate	= 2.9 grams
Anhydrous dextrose	= 20.0 grams
Water – sufficient to make 1 litre	

The water should be freshly boiled, and then cooled. It should be discarded 1 hour after being made up, unless it is stored in a refrigerator, in which case it can be stored for up to 24 hours.

This solution would provide the following:

Sodium	= 90 mmol/l
Potassium	= 20 mmol/l
Chloride	= 80 mmol/l
Citrate	= 10 mmol/l
Dextrose	= 110 mmol/l

The dose rate will depend upon the fluid loss, but as a guide 200–400 ml of solution for an adult, after every loose motion; 200 ml for a child, and for infants 1–1.5 times the usual feed volume.

UNICEF suggests the following dose rate:

> Infants   = 1 litres/24 hours
> Children = 1 litres/8–24 hours
> Adult   = Drink freely

As with athletes, so it is for people suffering from diarrhoea, if excessive amounts of water are drunk it will dilute the electrolytes in the bloodstream which could then cause hyponatraemic encephalopathy – a potentially fatal brain condition. Both sodium and potassium chloride are essential for the human body's well being. They are involved in food absorption, regulating the body's water balance and the rhythm of the heart.

In preparing this section on oral rehydration, acknowledgement is made to The British National Formulary for permission to reproduce the above information.

17.9 Geriatric drinks and liquid foods

Drinks for geriatrics and invalids are basically the same as for other sports and health drinks, but have a different appearance. Whereas sports and health drinks look like any other soft drink, drinks designed for geriatrics and invalids are designed to supply energy, but at a slower rate. This slower rate is achieved by using maltodextrin, which noticeably increase the viscosity, making the drinks look more like a soup!

Maltodextrins, like a conventional glucose syrup, is a carbohydrate, and therefore is a source of energy, however maltodextrins are composed mainly of higher sugars, which are slowly broken down by the body, as opposed to the glucose syrup, which is absorbed relatively quickly.

An advantage of using maltodextrins as an energy source is that with a molecular weight, frequently over 1000, they offer a greater energy loading, with very little change in the overall osmotic pressure. This is something which is important, because any increase in the osmotic pressure is likely to cause diarrhoea.

Some geriatric drinks might contain a protein in the form of a protein hydrolysate which can also be readily digested. The drink will also contain minerals, vitamins and supplements, as well as other carbohydrates. Depending upon the nutritional requirements, the types of carbohydrates used will vary, as will the other ingredients. Lactose is frequently used in geriatric foods, but for people who are lactose intolerant, this could result in acute diarrhoea. By including a maltodextrin in the formulation, the viscosity of the drink is increased, and this increase in viscosity not only increases the body and mouthfeel of the drink, but also helps to keep in suspension any particulate material as well as acting as an emulsion stabiliser for any oils in the formulation.

Where a geriatric drink is in the form of a dry powder, this can be made by either dry blending, or co-spray drying the ingredients. Where dry blending is used then maltodextrins, and spray dried low DE glucose syrups (35 DE) can be used. If dextrose or fructose is used, then care must be taken to make sure that the osmotic pressure of the drink is acceptable.

It is important that all powdered drinks are easily dispersed, and wetted out when added to water, without the formation of any lumps.

17.10   Slimming foods

The popular image of these types of foods is that they satisfy a person's hunger, without increasing their weight. Since carbohydrates can increase a sedentary person's weight, the inclusion of glucose syrups in these types of food is therefore limited. However, maltodextrins, fructose and dextrose are often included at relatively low levels, for example 2%, so as to increase the body and mouthfeel as well as increasing the sweetness, thereby making the drink more palatable and satisfying.

Chapter 18
Carbohydrate metabolism and caloric values

Nomenclature

In this chapter, the word 'dextrose' will be used to refer to the medically used word 'glucose' as in blood glucose levels.

18.1   Introduction

Whilst all carbohydrates have a caloric value of 17 kJ or 4 cal/g regardless of the type of carbohydrate, not all can be utilised by humans. Cellulose, hemi-cellulose, pectin and gums such as alginates are typical carbohydrates which the human body cannot break down. However, whilst they cannot be a source of energy, they nevertheless form a very valuable function in the digestive process by providing fibrous bulk or roughage. Of the carbohydrates which the human body can utilise, not all are absorbed by the same metabolic route due to their different chemical structures.

It is the sugar dextrose which is the normal sugar circulating in the bloodstream, supplying energy to all parts of the body – irrespective of the type of carbohydrate which has been digested. Typically, there will be 3.3–7.0 grams of dextrose circulating in the bloodstream of an average adult, assuming that the body has a blood volume of five litres.

Since the body can only metabolise simple sugars, that is monosaccharides, one of the functions of the digestive tract is to convert the complex starch carbohydrates present in our foods into simple sugars. These simple sugars are then absorbed and converted into dextrose.

Generally, sugars which are easily absorbed by the body all have several chemical features in common (see figure 18.1). They all have a six-membered ring; they all have one or more carbon atoms attached to carbon 5, and they all have an hydroxyl group

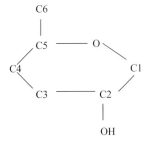

Figure 18.1   Molecular structure of an easily absorbed sugar, for example dextrose.

at carbon 2, with the same stereo configuration which occurs in dextrose (d-glucose) (Textbook of Physiology, by Bell, Emslie-Smith and Patterson).

18.2 Human digestive system

In discussing the human digestive system, only the fate of digestible carbohydrates will be considered.

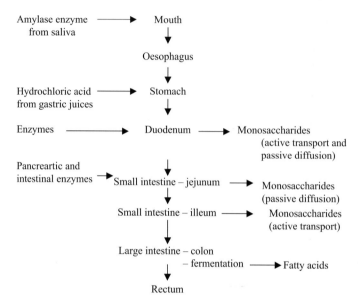

Figure 18.2   Simplified human digestive system.

A carbohydrate will enter the human digestion system (see figure 18.2) through the mouth, where it is mechanically broken down into smaller pieces by the teeth and mixed with saliva. During this process, the surface area of the carbohydrate is increased, which allows the amylase enzyme present in the saliva to start the breaking down or hydrolysis of the carbohydrate into the sugar maltose.

The carbohydrate now passes via the oesophagus to the stomach, where it will remain for about two to four hours. The churning action of the stomach further mechanically breaks down the carbohydrate, allowing the amylase to continue hydrolysing the carbohydrate until the hydrochloric acid in the gastric juices stops the reaction. Whilst in the stomach, some of the disaccharides are broken down into monosaccharides by acid hydrolysis.

The partially digested carbohydrates which now have a semi-liquid consistency pass into the duodenum by a process of peristalis – that is a wave like motion caused by muscles of the digestive system contracting and relaxing. In the duodenum, pancreatic amylase is added, together with other enzymes and the hormones insulin and glucagon. It is here where the monosaccharides which have already been formed are absorbed. The remainder passes into the small intestine, where it remains for between one and six hours. It is the small intestine where the main conversion of the carbohydrates into sugars takes place. Enzymes are secreted from the brush border lining of the small intestine – invertase to hydrolyse sucrose into dextrose and fructose, maltase to hydrolyse maltose into dextrose and lactase to hydrolyse lactose from milk into dextrose and galactose. The sugars are now absorbed mainly in the jejunum, with minor amounts being absorbed by the ileum and pass into the bloodstream where they are transported via the hepatic portal vein to the liver.

18.3 Carbohydrate absorption

It is important to understand that carbohydrates are absorbed by two main processes. The first is by 'passive diffusion'. That is, the concentration in the intestine is greater than that in the blood, and this mainly occurs in the duodenum. The second process is by 'active transport'. Here, the sugar molecules attach themselves to a protein carrier, which in turn connects to a sodium ion. The sugar is now transported through the intestinal wall as a sugar–protein–sodium complex into the bloodstream, which transports it to the liver.

Fructose is only absorbed by passive diffusion, but dextrose and galactose are absorbed by both passive and active transport. Interestingly, dextrose is more quickly absorbed whilst galactose, which has the same molecular configuration as dextrose is absorbed at a slightly slower rate. The reason for this difference is possibly because dextrose has a greater affinity for the carrier than galactose. Fructose, on the other hand, takes twice as long to be absorbed.

Whilst all parts of the digestive system play an important part in the breakdown of carbohydrates, there are, however, two particular organs which are exceptionally important, namely the pancreas and the liver.

The pancreas produces the digestive enzymes which break down the carbohydrates and produces the hormones insulin and glucagon. Insulin controls the amount of dextrose in the blood and its production is controlled by the brain. Glucagon is involved in (1) the conversion of excess dextrose into glycogen, (2) the storage of glycogen in the liver and (3) the subsequent conversion of the stored glycogen back into dextrose for maintaining the normal blood dextrose levels.

The liver is the largest internal organ of the body, and is important for the body's utilisation of carbohydrates, amongst its many other functions. Think of the liver as being a sugar refinery – it receives a feedstock of several different sugars, which comes from the stomach, small intestine and part of the large intestine via the hepatic portal vein. The liver then converts this feedstock into dextrose, which it distributes to other parts of the body for use as energy. The liver maintains the dextrose level in the bloodstream at about 100 mg per 100 ml of blood. Excess dextrose is converted into glycogen (sometimes called animal starch) and stored either in the liver or the muscles. Some of the excess dextrose is also converted into triglyceride fats. When the digestive system starts to absorb sugars, the liver stops circulating dextrose, and starts to convert it into glycogen. (This conversion of dextrose to glycogen is controlled by insulin.) When the supply of sugars to the liver stops, the glycogen is then converted back into dextrose, which is then circulated in the bloodstream.

The liver also converts fructose into dextrose. This conversion of fructose into dextrose results in a time delay from when fructose is digested to when it becomes available as dextrose, and means that there is not an immediate increase in the blood sugar levels as is the case with dextrose. This 'time delay' on becoming available, together with fructose being approximately 50% more sweet than sucrose, makes fructose a useful ingredient in diabetic foods.

As the body cannot directly utilise galactose, another function of the liver is to convert it into dextrose, with unconverted galactose being excreted (see notes on galactosaemia).

18.4 Summary of carbohydrate metabolism

18.4.1 Sugars

Dextrose (glucose)

This is the most readily absorbed and utilised sugar. It is absorbed by the small intestine, and passes directly into the bloodstream, but it is the hormone insulin which controls how much dextrose is circulated for use by the body's metabolism. Excess dextrose is stored either as glycogen or is converted and stored as fat.

Fructose

This is absorbed by the small intestine and passes into the bloodstream. The bloodstream carries the fructose to the liver, where it is converted into dextrose which is then either circulated around the body in the bloodstream, or is stored either as glycogen or as fat. Because fructose has to be converted into dextrose before it can be utilised, there is a time delay from when it is digested to when it becomes available as dextrose. This delay means that there is not an immediate increase in blood sugar levels as is the case with dextrose (see notes on fructose and diabetes below).

Sucrose

This is broken down in the intestine by the enzyme invertase into dextrose and fructose, which are then absorbed with the dextrose as described previously for dextrose and fructose.

Maltose

This is broken down during the digestive process by the amylase enzyme into two dextrose molecules, which are then absorbed as described previously for dextrose.

Maltodextrin and higher sugars

These are broken down by the amylase enzymes in the small intestine, in the same way as starch is converted into dextrose, which then follows the normal dextrose pathway. Being larger molecules than maltose, they take longer to be broken down, which means that the dextrose is not immediately available, but becomes available over a period of time.

Trealose

> Being a disaccharide consisting of two dextrose molecules, similar to maltose, it is broken down in the small intestine by the enzyme trehalase into dextrose, which is then absorbed by the normal pathway.

Galactose

> Before galactose can be utilised by the body, it has to be converted by the liver into dextrose. Unconverted galactose is excreted via the kidneys (see notes on galactosaemia).

Lactose

> This is hydrolysed into dextrose and galactose. The dextrose is absorbed by the normal metabolic pathway, whilst the galactose will either be excreted or possibly converted into dextrose.

Other sugars

> Only simple sugars can be absorbed from the small intestine. Sugars which are not absorbed pass into the colon, where they might be converted into volatile fatty acids, by the colon's bacteria. Some of these fatty acids might be used by the body, whilst the remainder are lost in the faeces together with other undigested material.

18.4.2 Starch

> Starch is a polymer of dextrose and is therefore hydrolysed by the enzyme amylase first into higher sugars, then into maltose and finally into dextrose, which is then absorbed by the normal pathway.

Resistant starch

> This is not broken down in the small intestine, but might be available for fermentation in the colon by bacteria, to produce short chain fatty acids such as butyrates, acetates and propionates. Some of these fatty acids might be used by the body whilst the bulk will be lost in the faeces.

18.4.3 Fibre (that is roughage) and non-starch polysaccharides (that is cellulose)

> Unlike herbivores, humans are unable to utilise these types of products and therefore they act as bulking agents. Whilst their nutritional contribution is minimal, they are a useful bulking agent, which encourages water retention and increased microbial action in the faecal biomass.

Inulin (fructo-oligosaccharide)

As there are no enzymes in the human digestive system able to hydrolyse inulin, about 88% is not absorbed by the body and passes unchanged through the digestive tract like any other non-starch polysaccharide. However, approximately 12% is fermented by bacteria in the colon, which suggests an energy value of about 1–1.5 kcal/g.

Polydextrose

For the most part, polydextrose is not absorbed by the body. Because it is not fully absorbed by the body, it tends to act like roughage, with the resultant increase in faecal matter. However, bacteria in the lower bowel can under certain circumstances use it to produce volatile fatty acids, which can be utilised by the body. Because of the production of these volatile fatty acids, its caloric value is considered to be 1 kcal/g.

18.4.4 Polyols

As a general rule, all polyols are only partially metabolised by the body and are frequently used in diabetic and low calorie products, sometimes as bulking agents, but because they are only partially metabolised by the body, they are a potential laxative. Due to this laxative effect, there are recommended daily consumption levels of foods and sweets containing polyols. The laxative effect of polyols is an osmotic phenomenon.

Erythritol

Erythritol is unique among polyols. Due to its low molecular weight, about 90% is rapidly absorbed from the small intestine, but it is not metabolised, and is subsequently excreted unchanged in the urine. It is considered to be non-caloric.

Maltitol

Some small amounts of maltitol can enter the bloodstream by two possible routes – diffusion and enzyme hydrolysis. Diffusion is very slow due to the large size of the maltitol molecule. Enzymes can hydrolyse maltitol into sorbitol and then dextrose. However, the bulk of the maltitol and sorbitol will pass undigested to the colon where it is fermented into volatile fatty acids.

Mannitol

A very small amount of mannitol is absorbed in the small intestine by passive absorption, which is then converted into fructose. All the fructose passes to the liver where it is converted into dextrose. The bulk of the mannitol passes to the colon where some is fermented into volatile fatty acids. As mannitol is possibly one of the least tolerated polyols, most will pass through the body unchanged.

Sorbitol

A significant amount of sorbitol is not absorbed by the body, and passes to the colon unchanged where it is fermented to produce volatile fatty acids. Some sorbitol is very slowly absorbed by the small intestine into the bloodstream and passes to the liver, where it is converted initially into fructose and then into dextrose. The acceptable daily intake (ADI) is 30 g/day.

18.5 Carbohydrate metabolic problems

Unfortunately, carbohydrate metabolism, like most processes, sometimes can go wrong. The most common problem is diabetes. Other problems include lactose intolerance, galactosaemia, and coeliac disease. Since these conditions will impinge upon glucose applications, personnel working within the industry should be aware of these potential problems.

18.5.1 Diabetes mellitus

Diabetes occurs when too little insulin is produced by the pancreas and results in too much dextrose in the bloodstream. In a healthy person, the dextrose level is controlled by the hormone insulin. Too high a level of dextrose in the bloodstream will result in long-term health problems, such as poor circulation and blindness, whilst too low a level will result in fainting.

Diabetes occurs in two forms:

- *Insulin dependent diabetes mellitus (Type 1)*, occurs when the pancreas fails to produce sufficient insulin, but this condition can be treated by regular insulin injections.
- *Non-insulin-dependent diabetes mellitus (Type 2)* occurs when the pancreas produces insulin, but it is not recognised by the body's cells. This results in the dextrose (glucose) not being absorbed by the body and so the dextrose (glucose) level in the bloodstream continues to increase.

Non-insulin-dependent diabetes mellitus can be controlled by dietary means, typically by eating complex carbohydrates such as starchy foods. These foods are slowly broken down into dextrose, producing a slow but steady release of dextrose (glucose) into the bloodstream over a long period of time, thereby keeping the blood dextrose level low by not overloading the bloodstream.

18.5.2 Fructose and diabetes

The main virtue of using fructose in diabetic foods is that fructose is approximately 50% more sweet than sucrose; therefore, less fructose is required for an equivalent sweetness.

Additionally, because fructose has to be converted into dextrose (glucose) by the liver before it can be utilised by the body, the rate and peak of sugar concentration in the blood is much lower with fructose, than with an equivalent weight of dextrose. It must also be appreciated that since fructose will ultimately be broken down into dextrose, fructose should only be used in moderation in foods for people with mild or well controlled diabetes.

18.5.3 Lactose intolerance

Whilst lactose is not a product of the glucose industry, it does impinge upon glucose applications, particularly confectionery and dairy applications. Lactose intolerance is the inability to digest lactose, which is the main sugar present in milk. This condition is most common in older children and adults who originate from Africa, Asia and South America, and in malnourished children. Caucasian people seem to be less susceptible to this problem.

With this condition, the body is unable to absorb lactose. Lactose has to be broken down into dextrose and galactose by the enzyme lactase before it can be absorbed. It is the absence of the enzyme lactase, which is responsible for the medical complaint 'lactose intolerance'.

When there is insufficient lactase, the lactose is not broken down, and passes into the large intestine, where it is fermented by bacteria into lactic acid, which results in diarrhoea.

Whilst galactose is absorbed by the body, it cannot be utilised by the body and is either converted into dextrose or is excreted.

In some cases, lactose intolerance can be due to problems associated with the lining of the intestine.

Milk and milk products which contain lactose are frequently used in foodstuffs, because they are a potential source of calcium, which is essential for good bone structure.

18.5.4 Fructose intolerance

This is a rare congenital condition. The body is unable to metabolise sucrose or fructose, due to the lack of the B isoenzyme of fructose-1-phosphate aldolase, which is required to break down fructose, and the subsequent conversion of glycogen into dextrose.

18.5.5 Galactosaemia

This is a rare genetic disorder in which the body is unable to convert galactose to dextrose. Whilst galactose is again not a product of the glucose industry, it is an end product of the hydrolysis of milk, and therefore could be a problem with confectionery and dairy applications, similar to lactose intolerance.

Galactosaemia can cause diarrhoea, vomiting and jaundice, and if not treated, will result in liver disease (enlargement), cataract, blindness, poor growth and mental retardation.

18.5.6 Gaucher's disease

This is an inherited genetic disorder, caused by the lack of the enzyme glucocerebrosidase, which affects the body's ability to digest sugars, and results in an accumulation of fatty lipids in the liver, and other parts of the body.

18.5.7 Coeliac disease (also known as gluten intolerance)

This is a disorder of the small intestine, caused by sensitivity of the lining to the protein in wheat and rye gluten, and in some cases even barley. This protein damages the intestinal lining, which causes poor absorption of foods, resulting in weight loss, anaemia and skin problems. Additionally, because the food is not totally digested, the stool are bulky and foul smelling. Interestingly, in Africa and Asia, coeliacs disease is not as common as in Western populations.

The significance of this condition to the glucose industry is that some glucose syrups are made from wheat, and despite the efficient refining of the syrup, sufferers are concerned that some of the protein could still be present. However, starch and syrup made from maize are generally considered safe, because the protein composition of maize gluten is different from that of wheat and rye. If wheat starch is used to make a syrup, the total protein content of the starch would be less than 0.25%, and the total protein content in the final syrup would be less than 0.025%, and with modern refining technology, considerably less.

The ill effects of gluten intolerance can be greatly reduced by keeping to a wheat free diet, and avoiding bread, pasta, and any manufactured foods which contain wheat based products.

18.5.8 Phenylketonuria and aspartame

This is an inherited metabolic defect, in which the body is unable to fully metabolise the amino acid phenylalanine. The by-product of this incomplete metabolism – phenylketones – can in turn affect the brain, causing mental deficiency.

Whilst this condition is not directly related to glucose syrups, it is included for two reasons. The first is that phenylalanine is an essential amino acid, required for body proteins, and is present in both maize and wheat, but fortunately, with modern methods of glucose syrup manufacture, total protein levels in syrup are less than 0.025% d.b.

The second reason for its inclusion is that phenylalanine is present in the artificial sweetener aspartame, which is sometimes used in conjunction with glucose syrups in different foods and drinks – the syrup supplies mouthfeel, whilst the aspartame provides the sweetness. This combination of a glucose syrup with aspartame is frequently used when formulating a product where minimal cost and maximum sweetness are required. By using the artificial sweetener aspartame to supply the sweetness, the amount of carbohydrate sweetener, that is sucrose and / or glucose syrup, can be reduced, thereby reducing the amount of calories in the product.

18.6 Caloric values

When calculating the caloric value of foods, the following values should be used:

Energy per gram d.b.	Kilojoules	Calories
Carbohydrates	17	4
Protein	17	4
Fat	37	9
Polydextrose	4.7	1.1
Polyols – other than erythritol	10	2.4
Erythritol*	0.85*	0.2*
Ethanol	29	7
Organic Acid	13	3

*Because the energy available to the body from erythritol is less than 5%, some countries feel that erythritol should be regarded as having a zero energy value.

N.B. Because of the way in which different polyols are metabolised, different countries have given different calorific values for different polyols. Within the EU, under the EC Directive 90/496/EEC, all polyols are given a value of 10 kJ/g (2.4 cal/g). Whilst this makes for easy calculations, especially where blends of polyols have been used, the results might not be totally accurate especially where blends of polyols have been used.

To convert joules into calories, multiply by 0.2389.
To convert calories into joules, multiply by 4.1868.

Chapter 19

Caramel – the colouring

19.1 Introduction

Caramel colouring – E150 – is made by the controlled heating of sugars, particularly high DE (dextrose equivalent) glucose syrups. Straightforward heating of a sugar will produce a plain caramel which can be used more for its flavouring properties rather than for its dark colouring powers, but by heating different sugars in the presence of a catalyst, darker coloured products are obtained, and by using different sugars with different catalysts, caramels can be produced which have different properties.

There are four main types of caramel approved for food use in the EU and covered by Directive 94/36/EC, Part 1, Annex V, and the four caramels are:-

Plain caramel	E150 (a)
Caustic sulphite caramel	E150 (b)
Ammonia caramel	E150 (c)
Sulphite ammonia caramel	E150 (d)

The method of manufacture for each of the above four caramels is slightly different, which gives each caramel slightly different properties, thereby making each one suitable for a specific application.

Whilst caramels are generally recognised as being safe, there has been some concern raised in the past over the safety of ammonia caramel (E150 c) due to the possible production of carcinogens such as nitrosamines during its manufacture. The amount of nitrosamine formed during cooking can be reduced if the ammonia is drip feed into the reaction vessel during cooking.

19.2 Process

The basic process for making caramel involves the heating of a dilute glucose syrup in a stainless steel reaction vessel with or without an inorganic catalyst to 120–160°C for 2 to 7 hours depending upon the type of caramel being produced. When the required depth of colour (tinctorial power) has been reached, the batch is filtered then cooled. The colour is standardised and the viscosity is adjusted by the addition of water or a 42 DE glucose syrup. It is then packed or spray dried. Because metals can adversely affect the colour, it is important that the caramel is stored in stainless steel or plastic containers.

The glucose syrups used to make caramel should contain a minimum of 75% reducing sugars, but preferably higher.

Typical inorganic catalysts are ammonium hydroxide, which is used to produce a caramel with a positive charge or ammonium metabisulphite to produce a negatively charged caramel. The chemistry of the reaction is not fully understood but involves the reaction of the reducing sugars of the glucose syrup to produce a melanoidin-type pigment.

When sucrose is used to make a caramel, the chemical reaction inverts the sucrose to give an equal mix of dextrose and fructose. As fructose is chemically a more reactive sugar than dextrose, the colour of the resulting caramel has a slight red colour compared to a caramel made only from dextrose. Therefore, by using blends of different sugars, in different proportions, it is possible to produce a range of caramels with each having a different colour, tinctorial power and functional property.

19.3 Properties

When caramel is dissolved in water, the colloidal particles will carry an electrical charge – either a positive or negative charge – depending upon which catalyst has been used in its production. These differences in the electrical charge will have a considerable effect on the properties of the caramel, and hence on their application. When positive and negatively charged particles are brought together, the charges are neutralised and the particles are precipitated out of solution. This is known as the isoelectric point of the caramel and it is pH dependent – when the pH is above the isoelectric point, the particles become negatively charged, and when the pH is below the isoelectric point, the particles become positively charged.

Therefore, so as to avoid the caramel being precipitated out of solution, it is important that the correct caramel is used for the correct pH of the application.

The isoelectric point of a positively charged caramel is pH 4–7 with a colour of 60 000 European Brewing Convention (EBC) units. The isoelectric point of a negatively charged caramel is pH 3.0 or lower and would have a colour of 25 000 EBC units.

A spirit caramel is made from sucrose using sodium hydroxide as the catalyst. During the process, the sucrose is inverted to give dextrose and fructose. Spirit caramel will typically have a very weak or no charge at all, and have a colour of 20 000 EBC units. Spirit caramel, as its name implies, is stable and soluble in ethyl alcohol, propanol and propylene glycol.

19.4 Applications

Whilst the primary application for caramels is in colouring foods and drinks, the skill is to match the isoelectric point of the caramel for the correct application.

Typical applications would be in gravy browning and soups, bakery and biscuits, vinegar, malt and non-brewed vinegars, cider vinegar, confectionery, canned goods, soft drinks, beer and cider, whilst spray-dried caramel is used in dry mixes.

As a general rule, positively charged caramels, that is with an isoelectric point in the pH range of 6.0–7.0, can be used for general purpose colouring. Whilst dark-coloured barleys are frequently used to give a dark colour to beer, positively charged grades of caramel can also be used for additional colouring. Negatively charged caramels, that is with an isoelectric point in the pH range of 2.5–3.0, can be used in vinegar and pickled vegetables. Caramels with an isoelectric point of less than 1.0 can be used in soft drinks, particularly colas, cider and cider vinegar. Spirit caramel with its weak charge and good stability in alcohols is used in products containing proof spirit.

Glossary

The primary object of this glossary is to explain the terminology as used within the starch and glucose industry, including its specifications and the process – it is not, however, intended to be a definitive dictionary of scientific terms.

Because the products of the glucose industry are very versatile, and with the syrups being used in many industries, the secondary purpose of this glossary is to explain some of the terminology which is used in those different applications.

Absolute temperature	It is the temperature at which a perfect gas would occupy zero volume, if it could be cooled without turning into a liquid or solid. In calculations, absolute zero is taken as $-273.16°C$, or $0°K$. Therefore, $27°C$ is equivalent to $300°K$. A typical calculation in which absolute temperature is used would be the determination of osmotic pressure.
Achrodextrin	A name sometimes used to describe maltodextrin.
ADI	Acceptable daily intake.
Aflatoxins	See **Mycotoxins**.
Akrah	See **Palm syrup**.
Aldose sugars	These are sugars containing an aldehyde group. They reduce Fehling's solution (alkaline copper sulphate) and Tollens' reagent (ammonical solution of silver oxide). They also reduce and decolourise bromine water. Since fructose will not decolourise bromine water, this can be used as a confirmatory test that the sugar is aldose sugar.
American Bottlers Standard	This is an American microbiological standard for sugars which are used in soft drinks.
Amioca	An alternative name for waxy maize. *See also* **Waxy maize**.
Amyloglucosidase	An enzyme which will detach glucose units from a chain of amylopectin units. Also known as 'glucoamylase' or 'AMG'. Normally, the amylopectin has to be gelled for this reaction to occur. This enzyme is used in the making of high DE syrups.
Amylopectin	One of the fractions which make up starch – the other being amylose. Amylopectin is a highly branched structure of dextrose units. Typically, maize starch contains approximately 74% amylopectin. *See also* **Starch**.
Amylose	One of the fractions which make up starch. Amylose is made up of dextrose units arranged in long straight chains. Maize starch will contain about 26% amylose. *See also* **Starch**.
Animal starch	See **Glycogen**.

Anion	A negatively charged atom, for example the chloride ion in a solution of sodium chloride.
Anti-caking agents	These are substances which are added to powders such as icing sugar to prevent the particles from sticking together and to improve the flow of the powder. Typical anti-caking agents are tricalcium phosphate, silicon dioxide, sodium alumina silicate and maize starch.
Artificial sweetener	These are substances, other than sugars, capable of producing a sweet taste. Some are man made, such as saccharin (discovered in 1879), aspartame, acesulfame K and sucralose; others such as thaumatin and dihydrochalcones are derived from plants. Artificial sweeteners are characterised by being intensely sweet substances, with a sweetness value several hundreds or even thousands times more sweet than sucrose. Because of their intense sweetness, they are used in very small quantities, and as they are non-calorific, they are frequently used in low-calorie foods, drinks and diabetic foods.
Aw	*See* **Water activity**.
Azide dextrose	Azide dextrose broth is a liquid medium used for the selective detection of streptococci. The medium is composed of beef extract, tryptone, dextrose, sodium chloride and sodium azide.
Bacteriophage	It is any one of a large number of viruses which can infect bacteria. Some types can destroy starter organisms.
Balling	A hydrometer scale originally constructed to read direct percentages of sucrose in a solution. This scale contained a slight inaccuracy, which was subsequently corrected by Dr Brix to give the Brix scale. Basically, one degree is equivalent to 1% of solids. The measurements are temperature dependent. *See* **Brix**.
Bastard sugar	An American term for a grade of soft brown sugar.
Baumé	A hydrometer scale used to measure the specific gravity of a solution, and particularly of glucose syrups. It is frequently abbreviated to 'Bé'. Baumé readings are temperature dependent.
Biotin	Also known as vitamin H, it is essential for the growth of yeast and several species of bacteria. In humans, it is normally being constantly produced by intestinal bacteria.
Birch sugar	An old name for xylitol. So called because xylitol was originally made from the xylose obtained from Finnish birch trees.
Black strap	The American name for molasses. *See* **Molasses**.
Black syrup	A by-product from the refining of sugar.
Bland apple syrup	A sugar syrup originally produced by treating apple juice with lime to precipitate the acids as their calcium salt. The protein comes out of solution by denaturation. The product is then filtered and concentrated.
Bloom	A greyish colouration on the surface of chocolate confectionery. If the bloom has a white tint, and is not removed when scratched, it is probably a sugar bloom, and if it has a bluish tint, and

	feels greasy and is removed when scratched, it is a fat bloom. *See* **Fat Bloom** and **Sugar Bloom**.
Bloom strength	The measurement of the firmness of a gel, for example of gelatine, using a Bloom gelometer. In this instrument, a plunger (diameter 127 mm) is depressed into the surface of a gel by the addition of lead shot to a depth of 4 mm. The weight of shot used in grams to achieve this depression is referred to as the bloom strength.
Bob syrup	This term has several different meanings. (1) A dilute syrup into which a product is dipped, with the object of either sealing the surface of the product or to build up the existing surface layer. Where the bob syrup is being used in wet crystallisation, it is usually a 74% sucrose solution containing 0.2% sodium citrate buffer, and the reducing sugars should not exceed 4%. (2) A syrup which is used in confectionery production as a processing aid, for example, the thinning down of fondant or soft centres.
Bolting cloth	A cloth frequently made of nylon and used for the sieving and grading of flour. Because of the very small mesh size of the weave, it is sometimes used for sieving pulverised sugar. Where a powder has been sieved using a bolting cloth, the resulting powder is generally identified by a number followed by the letter 'X'. The larger the mesh size number, the finer the powder, for example a flour sieved using a 10X bolting cloth is finer than that sieved using 4X.
Brabender	An instrument to measure the changes in viscosity of a starch slurry when it is being heated, and stirred under controlled conditions. A typical test cycle would be to heat the slurry to 95°C, at a rate of 1.5°C/minutes, with constant stirring. Holding at 95°C for 15 minutes and then cooling to 20°C, at the rate of 1.5°C/minutes, where it is held for a further 20 minutes. As the starch is heated, it gels and the viscosity increases. During the cooling from 95°C to 20°C, the viscosity continues to increase. These changes in viscosity are continually measured, and recorded by the Brabender on either a chart or a computer.
Brewer's Extract	The soluble solids obtained from malt, sugars and glucose syrups.
Brix	A hydrometer scale which measures the percentage by weight of sucrose in an aqueous solution. Brix readings are temperature dependent and only accurate for sucrose solutions. Degrees Brix = degrees Plato = % sucrose solids W / W. (rounded up)
Browning	*See* **Maillard reaction**.
Brown sugar	White sucrose crystals coated with clarified molasses.
Buffered acid	An acid to which a buffer salt has been added. Typical example would be buffered lactic, which is lactic acid with added sodium lactate. *See also* **Buffers**.

Buffers	These are chemicals which can resist considerable changes in the acid/alkali balance in a formulation. They are a convenient way of maintaining a fixed pH. They are usually salts of strong alkalis and weak acids, for example sodium citrate or potassium citrate. Typical applications are in confectionery and jam making where they can either prevent pregelation or assist in gel formation.
Buffer salts	See **Buffers**.
Caloric value	Starches, glucose syrups, maltodextrins, dextrose, fructose and sucrose are all carbohydrates, and therefore **all have the same caloric value**, on a dry basis, of 17 kJ or 4 kcal/grams.
Calvin Cycle	This is the photosynthetic route used by root crops, such as chicory, beet and potatoes to fix atmospheric carbon dioxide. It is sometimes referred to as the 'C3' route. This is also the route used by fruits and trees.
Capillaire	The name given to a sucrose syrup used to sweeten wines and liqueurs.
Caramel	(1) A dark brown substance resulting from the heating of sugars. It is used as a colouring agent – E150. Because caramel can be made by different processes, this has resulted in several different types of caramel (150a–150d inclusive), with each type having different properties. (2) A piece of confectionery similar to toffee. The main difference is that caramel contains more fat, milk products and moisture than toffee. The bite is usually softer than toffee.
Carbohydrates	Carbohydrates are compounds of carbon, hydrogen and oxygen, in which the ratio of hydrogen to oxygen is 2:1. The term 'carbohydrate' was first used in 1862 by the German chemist Ernst Wagner.
Cassava	The Southeast Asia and African name for the manioc plant (botanical name *Manihot esculenta* or *M. utilissima*), which grows in tropical countries, and from which tapioca starch is produced. It should be noted that the manioc root contains goitrogen and linamarin, which produce cyanide due to enzymes in the root.
Cassonade	A French Canadian name for brown or soft sugar (derived from the French word which means moist sugar).
Caster sugar	A grade of sucrose with an approximate particle size of 150–650 microns.
Catabolite regression	See **Crabtree effect**.
Cation	Positively charged atoms, for example the sodium ion in a solution of sodium chloride. Metals are cations.
Chip Sugar	See **Corn Chip Sugar**.
Chromatography	An analytical method for separating and identifying different substances in a mixture. Chromatography can take many forms – paper, thin layer (TLC), gas–liquid (GLC), or

	high pressure liquid chromatography (HPLC). The word is derived from the Greek, meaning 'colour writing'. A German chemist Runge was the first person to use paper chromatography to separate dye colours in 1850.
CIP	The abbreviation for 'clean in place'. This procedure enables the cleaning of entire machines, storage tanks, production lines or other components without the need for the equipment to be dismantled.
Colligative properties	Properties of solutions such as vapour pressure, osmotic pressure, boiling point and freezing point, which depend only upon the number of particles (molecules, atoms or ions) present in a solution, and are independent of the solute.
Confectioners Glucose	The name used by confectioners to describe the grade of glucose which they use, and is usually taken as meaning a 42 DE regular glucose syrup.
Copper Sugar	A sugar which is added at the brewing copper (kettle) as a source of soluble carbohydrate, and did not originate from the mash tun. Originally in a solid form and derived from sucrose, such as solid invert, or it can be based on solidified glucose syrup, such as chip sugar. The term is now used to refer to any sugar or syrup which is added at the copper. *See* **Corn Chip Sugar**.
Co-products	An alternative name for by-products.
Corn	The name used in the United States for maize (botanical name *zea mays*). There are several different types of corn. *See* **Dent Corn, Flint Corn, Flour Corn, Popcorn** and **Sweet Corn**.
Corn Chip Sugar	A solid form of glucose syrup. Sometimes referred to as 'crude sugar', it is usually made from a high DE syrup, and allowed to crystallise to form a hard solid product. Being in a solid form, it was frequently used by brewers, and other industrial users who did not have a liquid storage facility.
Cornflour	This is the name most commonly used in the UK domestic circles for maize starch.
Corn flour	(1) This was the name used by John Polson of Brown and Polson, of Paisley, Scotland, in 1850's for starch which they had extracted from maize, using alkaline steeping, as opposed to acid steeping, which uses sulphur dioxide. Steeping with sulphur dioxide is now the normal steeping process for maize. (2) In the United States, corn flour refers to the flour produced by the dry milling of maize, after the removal of grits, germ, bran and meal.
Corn meal	It is the end product of grinding whole maize, including the germ. As the corn meal contains the germ, the shelf life of the corn meal is limited, as over a period of time, the oil in the germ is broken down by the enzymes in the original maize, resulting in the development of a bitter taste.

Corn oil	Also known as maize oil. *See* **Maize oil**.
Corn steep liquor	Evaporated steep water. It is usually added to the maize fibres to make maize gluten feed. Corn steep liquor is also referred to as 'CSL'.
Corn syrup	The name used in the United States for glucose syrups. Most glucose syrups in the United States are made from corn starch, hence the name 'corn syrup'.
Corn syrup solids	The US term for dried glucose syrup.
Coulis	A thin fruit or vegetable puree used as a sauce. Typical solids; 25–30% solids. Because of the low solids, the product has to be kept in a refrigerator after opening, despite containing a preservative, typically potassium sorbate.
Coupling sugar	A range of sugars produced by the action of the enzyme cyclomaltodextrin glucanotransferase on a mixture of starch and sucrose.
Crabtree Effect	A phenomenon usually associated with brewing, when a high concentration of dextrose affects yeast metabolism, and usually results in a 'hanging fermentation', that is the fermentation stops before all the maltose and trisaccharides have been utilised. This phenomenon can also occur in other fermentations.
Cryoscope	This is an instrument which measures the freezing point of a solution. Because glucose syrups will alter the freezing point of a solution, a cryoscope can be used to measure these changes, and by reference to tables, the DE of the syrup can be determined. A more accurate description for a cryoscope would be a 'freezing point osmometer'.
Cryoscopic constant	It is the molar depression of the freezing point, also known as the freezing point depression constant. In the case of water, the constant is 1.86 K kg/mol.
Crystal starch	It was originally made using wet starch from a filter press or centrifuge cake, which was pressed into small cubes and allowed to dry. When dried, the lumps are broken into smaller pieces. The modern method of production is to pass dry starch through compacting rolls, with the resulting solid starch board being broken up into dust free lumps.
Crystalline glucose	Usually taken to mean dextrose monohydrate.
CSU	Corn syrup unmixed or corn syrup unrefined. Typical DE range would be 40–45 DE.
Cyclodextrins	These are enzymatically produced oligosaccharides, which at the molecular level have a closed circular structure, which has been likened to a doughnut. This cyclic structure consists of six to eight glucose units linked together by α-1,4 bonds. These are sometimes referred to as Schardinger dextrins or cycloamylases. Cyclodextrins are totally different from the dextrin which is used in adhesives.

D-sugar	The letter 'D' does not refer to the right-handed polarisation of light through a solution of the sugar, but to the right-handed arrangement of the hydroxyl group on the penultimate carbon atom.

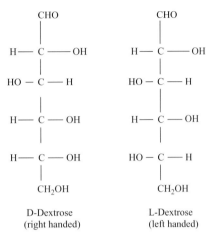

The body can only absorb right-handed sugars. Virtually all naturally occurring monosaccharides are D-sugars. As the body cannot absorb left-handed sugars, they are treated in the same way as an indigestible bulking agent, such as fibre.

Date sugar	See **Palm sugar**.
d.b.	Dry basis.
DE	Dextrose equivalent. It is the measurement of the total reducing sugars, calculated as dextrose, and expressed as a percentage of the dry solids. Starch has a DE of zero. Dextrose, the ultimate product of starch hydrolysis (breakdown), has a DE of 100. Therefore, DE can be regarded as an indication of the degree of hydrolysis, which the product has undergone. Whilst DE is frequently used in specifications to characterise a particular starch hydrolysate, a more precise method is a sugar spectrum, determined by HPLC.
De-ionised	A syrup is said to be 'de-ionised' when it has had the minerals removed by de-ionising resins. These resins are similar to those used for removing minerals from water. How well a syrup has been de-ionised can be determined by measuring its conductivity. De-ionisation is beneficial in improving the storage life of the syrups, particularly colour development and microbiological stability. The process of de-ionisation is also referred to de-mineralisation or ion exchange.
Demerara	A type of brown sugar which was named after the Demerara region of Guyana where it is grown.
De-min	It is the process of removing minerals from a liquid in this case glucose syrup. Such syrups are referred to as 'de-min syrups'. See also '**De-ionised**'.

Dent corn	A type of maize which, when dried, can be characterised by a dent across the crown of the kernel. During the drying of the kernel, there is a loss of moisture from the floury endosperm, which then shrinks and forms the dent. Dent corn grows well in the hot sunny conditions of the American Midwest.
Dew point	It is the temperature below which the moisture content of saturated air will be deposited as water. This is a very important factor in the correct operation of chocolate coolers so as to prevent 'sugar bloom', as well as in the correct handling and storage of confectionery, sugar and dextrose.
Dextran	A complex polysaccharide made by certain bacteria. It is manufactured by the fermentation of sucrose, and is used as a blood plasma extender, also as an emulsifying and thickening agent.
Dextrin	A polysaccharide produced by the heating or roasting of starch, sometimes in the presence of an acid, for example hydrochloric acid. Sometimes referred to as 'pyrodextrin' because of the heating process involved in their manufacture. By altering the processing conditions, several different types of dextrins can be produced – white dextrin, yellow or canary dextrin and British gums. Dextrins are used in remoistenable adhesives, and must not be confused with the dextrins associated with glucose syrups, which are now more accurately referred to as 'higher sugars'.
Dextrorotatory	A substance is said to be dextrorotatory when in solution it rotates the plane of polarised light to the right. Dextrorotatory substances are given the prefix of 'd' (or +) to distinguish them from laevorotatory 'l' (or −) isomers. Solutions of dextrose and sucrose will rotate a beam of polarised light to the right. Fructose on the other hand rotates light to the left. (The term 'dextrorotatory' is derived from a Latin word 'dexter' which means 'to the right'.)
Dextrose monohydrate	The purified and crystallised d-glucose containing one molecule of water of crystallisation, in each molecule of d-glucose. Traditionally, within the industry, 'dextrose' is taken to mean the white crystalline powder. See also **Glucose**.
Diatomaceous earth	A filter aid consisting of the skeletal remains of unicellular marine creatures (diatoms), and is frequently used in the filtration of beer. Also known as Kieselguhr.
Dibs	Date syrup, produced by pressing dates, and collecting the resultant syrup.
Disaccharide	A carbohydrate containing two sugar molecules chemically bound together. Examples of disaccharide sugars would be sucrose (dextrose + fructose), maltose (dextrose + dextrose), lactose (dextrose + galactose).
Doctor	A substance which will prevent crystallisation. 42 DE glucose syrup is used as a 'doctor' in boiled sweets to prevent sucrose

	crystallising. Invert is another 'doctor'. Acids and cream of tartar are also considered as being 'doctors', because they will convert sucrose into invert.
Dollo	See **Palm syrup**.
DP	The degree of polymerisation. Often used to refer to the degree of breakdown of starch after hydrolysis, and hence it is used to describe the composition of a glucose syrup – DP1 refers to dextrose, DP2 refers to maltose, DP3 refers to maltotriose, etc.
Dragee pan	Rotating drum used to coat sweets and tablets.
Dry solids (d.s.)	The percentage of dry substance in a syrup (see also moisture content).
Dry solids basis (d.b.)	The reporting of calculations, based on dry substance.
Dulloah	See **Palm syrup**.
EBC colour	European Brewing Convention Colour. This is a method for measuring the colour of beers or liquid adjuncts, such as glucose syrups, and involves the comparison of a sample against a set of standard coloured glass discs. These coloured discs are made by the Lovibond Tintometer Co.
Ebullioscopic constant	The elevation produced by 1 mol of solute in 1 Kg of water. For water, the ebullioscopic constant is 0.52 K·°C/(mol/Kg).
Empty calories	This term is used by the popular press to refer to a food or drink which only supplies calories, and no other nutrients such as mineral and vitamins.
Engrossing	The process of building up layers on a piece of confectionery in a dragee pan.
Enzyme	Proteins which catalyse reactions with a high degree of specificity and efficiency. They are obtained mainly from micro-organisms, but also from plants and animals. They usually react within a very narrow pH and temperature range. Some enzymes but not all will require a non-protein co-factor for stability which will vary from enzyme to enzyme. Typical enzymes which the glucose industry uses are α- and β-amylase, amyloglucosadase (AMG), malt extract, pullulanase and isomerase.
Equilibrium relative humidity	It is a point at which a product neither gains nor loses moisture to the atmosphere. It is very dependent upon the product, its moisture content and the relative humidity of the surrounding atmosphere. See also **Water activity**.
Ergot sugar	See **Trehalose**.
ERH	Abbreviation for equilibrium relative humidity. See **Water activity**.
Erythritol	It is a sugar alcohol, containing four carbon atoms and four hydroxyl groups, and is produced by the fermentation of dextrose or sucrose. It is a tetrahydric polyol.
Excipient	An inactive substance which acts as a carrier for other ingredients.
Extract	The soluble constituents from malt, including sugars and protein. See **Brewer's extract**.

Farina	An alternative name for potato starch, derived from the latin 'farina', meaning meal or flour.
Fat bloom	A grey colouration on the surface of chocolates, which when touched is removed from the surface and is greasy to the touch. It must not be confused with sugar bloom. See **Sugar bloom**.
Feed cane molasses	Molasses derived from sugar cane. Typical solids 74%; total sucrose content 33% d.b., reducing sugars 15% d.b.
Fehling's solution	An alkaline copper sulphate solution developed by von Fehling in 1848 for measuring reducing sugars by titration. It was subsequently modified by Lane and Eynon, who introduced the addition of methylene blue as an indicator for the end point of the titration. Von Fehling's observations were also the basis for the methods developed by Soxhlet (1878), Pavy (1894), and Main (1932).
Fermentable sugars	Sugars which can be fermented by micro-organisms, particularly by yeast. As a general rule, dextrose and maltose are readily fermented by yeast; maltotriose is slowly fermented, whilst maltotetrose and higher sugars are not fermented.
Feuilletine	A praline with a wafer-like texture achieved by blending in crushed baked crepes.
Flaked maize	De-husked and de-germed maize kernels, lightly milled, steamed, rolled and dried. Used in brewing, where it is added to the grist.
Flat Sours	The name given to a group of heat-resistant acid-producing microorganisms, for example *Bacillus macerans*. When the infected ingredient, for example starch, is used in a canned product, the acid produced by the micro-organism will give the product a sour or acidic taste. Generally, flat sours do not produce any gas, however if the inside of the can is not lacquered, then the acid will attack the metal to produce hydrogen gas which can distort the can. See also **Sulphide stinkers** and **Hydrogen swells**.
Flint corn	This is a type of maize, which can be characterised by its smooth round kernel, with no dent across the crown of the kernel. Flint corn is best adapted to the shorter growing seasons, and cooler temperatures found in Europe.
Flour corn	Maize which has a high proportion of floury endosperm.
Fluidity	See **Starch fluidity**.
Fold	This is a term used to describe the concentration of a fruit juice. Typical concentrations are 4, 5 or 6 fold. It means that one volume of juice can be diluted with water to give 4, 5 or 6 volumes of juice, with a concentration equivalent to that of the original juice.
Fondant	A sweet consisting of very fine sucrose crystals held in suspension by a glucose syrup.
Fondant cream	A fondant which has been diluted with a glucose syrup. Fondant creams usually have slightly lower solids than a fondant.

Fondant sugar	This is a dry mix of icing sugar and a maltodextrin. Water is added to make it into a fondant.
Foot sugar	Cane sugar which has been allowed to crystallise. It is claimed to have good flavour and aroma but tends to be dirty.
Ford cup viscometer	A metal cup used to measure the viscosity of liquids.
Fortified sugar	During the Second World War, sucrose was in very short supply. Therefore, to make the available sucrose go further, it was blended 50/50 with saccharine, and the mix was referred to as 'fortified sugar'. After the war, when sucrose became more freely available, the production of fortified sugar stopped.
FOS (fructooligosaccharides)	See **Fructosans**.
Fourths	A grade of brown sugar, used in confectionery to give colour and flavour. So called because it is made from the fourth sucrose crystallisation, hence the brown colour and flavour. It can also be made by adding molasses to white sucrose.
Free-flow agents	See **Anti-caking agents**.
Freeze thaw stable	A product is said to be 'freeze thaw stable' if it remains stable after being frozen and then allowed to reach ambient temperature.
Freezing point depression	The freezing point of a solution is dependent upon the number of molecules in solution – the greater the number of molecules in solution, the greater the freezing point is depressed. This is one of the colligative properties of solutions, the others being boiling point, osmotic pressure and vapour pressure. Sweeteners with a high DE, such as dextrose and fructose, have a greater effect on the freezing point than a low DE sweetener such as maltodextrins. The freezing point depression can be measured using a cryoscope.
Freezing point factor	Molecular weight of sucrose divided by the molecular weight of the alternative sweetener. This factor can be used when formulating an ice cream recipe to ensure the desired texture is obtained. The greater the freezing point factor, the greater the effect on the freezing point depression.
Fructosans	Polysaccharides of fructose, for example inulin. They are not digested, and therefore can be considered as dietary fibre.
Fructose	The ketose sugar D-fructose – sometimes referred to as laevulose or fruit sugar. It is approximately 50% more sweet than sugar (sucrose).
Fruit sugar	The old fashion name for fructose. See **Fructose**.
Fucose	6-Deoxy-*d*-galactose. A sugar found in seaweed, algae, bacteria cell walls and gums such as gum tragacanth.
Galactooligosaccharide (GOS)	A polymer of galactose.
Ganache	A blend of chocolate and cream (or butter), and forms the basis of truffles.
Gaucher's disease	An inherited genetic disorder, caused by the lack of the enzyme glucocerebrosidase, which affects the body's ability to digest

	sugars, and results in an accumulation of fatty lipids in the liver, and other parts of the body.
Gelatinise	This occurs when an aqueous suspension of starch is heated, and the starch granules swell and gelatinise to form a thick viscous gel.
GI	See **Glycaemic index**.
Gianduja	A smooth blend of hazel nuts and chocolate.
GL	See **Glycaemic load**.
Glucagon	A hormone produced by the pancreas and is involved in regulating the level of glucose (dextrose) in the blood. It opposes the action of insulin.
Glucitol (*d*-glucitol)	Another name for sorbitol.
Glucose	The hexose sugar d-glucose (dextrose). Traditionally, within the industry, 'glucose' is taken to mean glucose syrup, whilst 'dextrose' refers to the crystalline material 'dextrose monohydrate'. In medicine, glucose usually refers to dextrose. See **Dextrose**.
Glucose Effect	See **Crabtree effect**.
Glucose syrup solids	Usually, refers to dried glucose syrup which has been spray dried, but it can also refer to the solids contained in a glucose syrup.
Glycaemic index	It is a numerical system of measuring how fast a carbohydrate when ingested triggers a rise in circulating blood sugar – the higher the number, the greater the blood sugar response – 70 or more is high; 56–69 is considered medium, and 55 or less is low. It is a qualitative measurement.
Glycaemic load	The glycaemic load relates the glycaemic index of a food, to the amount of carbohydrate which the food contains. It assesses the impact of carbohydrate consumption, and is a quantitative measurement. It is calculated by multiplying the available carbohydrate content of a food by its glycaemic index.
Glycogen	Also known as 'animal starch'. It is the major carbohydrate reserve of the body and is composed of glucose molecules. Excess glucose in the bloodstream is converted by the liver into glycogen. The glycogen is stored in both the liver and muscles. When glucose is required, the glycogen is converted back into glucose.
Glycosuria	The presence of glucose in the urine. Glucose in the urine need not always be due to diabetes. The condition may be due to failure of the kidneys to reabsorbed glucose from the blood.
GMS	Glyceryl monostearate. An emulsifier, suitable for use in the manufacture of caramels and toffees.
Golden Syrup	A product of the sucrose refiners. It is made from syrup separated from a second or third sucrose crystallisation (the equivalent in the glucose industry would be hydrol from the dextrose crystallisation process), to which invert is then added. It is used mainly for its flavour.

	It was invented in 1884 by Charles Eastick, when he was second chemist at Abram Lyle's Plaistow Sugar Refinery, in East London.
GOS	See **Galacto-oligosaccharide**.
Graining	Crystallisation of sugars, giving the product a gritty texture. This crystallisation can be controlled by using a 'doctor syrup'. By controlling the crystallisation, the texture can range from very smooth, as in fondant, to coarse, as in fudge. See **Doctor**.
Granulated sugar	A grade of sucrose with an approximate particle size of 0.8–2.0 mm.
Granulation	One of two processes used in the production of lozenges. The other process is 'slugging'. See **Slugging**.
Grape sugar	An old name for glucose, that is dextrose; so called because dextrose is present in grape juice.
GRAS	'Generally Recognised As Safe'. This indicates that there are no known health problems involved with the substance.
Greens	See **Hydrol**.
HACCP	The abbreviation for Hazard Analysis Critical Control Point, which is a system for highlighting potential points in a process where hazard, for example contamination, could occur. It is sometimes spelled as HACCAP.
Hatch–Slack Cycle	This is the photosynthesis route, used by grasses (which includes cereals e.g. wheat and maize, also sugar cane), to fix atmospheric carbon dioxide. It is also referred to as the 'C4' route.
HCS	See **Hydrolysed cereal solids**.
Hexose sugars	Sugars containing six carbon atoms, for example dextrose, fructose, galactose and mannose.
HFCS	High-fructose corn syrup. These are glucose syrups made by the isomerisation of a dextrose syrup, and usually contain 42% or 55% fructose. In the United Kingdom, they are usually referred to as HFGS or high-fructose glucose syrups.
HFGS	High fructose glucose syrups. These are glucose syrups which have been made by the isomerisation of a dextrose syrup, and usually contain 42% or 55% fructose. In the United States, they are usually referred to as HFCS, or high fructose corn syrup.
HFS	High fructose syrups.
Higher sugars	Within the glucose industry, higher sugars refer to all sugars in a glucose syrup, other than mono-, di-, and trisaccharides (i.e. dextrose and fructose, maltose and maltotriose). Higher sugars are also referred to as 'oligosaccharides'.
High gravity brewing	Worts of higher than normal extract, or gravity (15–20 Plato), are fermented and then diluted with water prior to packing.
High test molasses	High test molasses are manufactured directly from the sugar cane. Typical solids 75% minimum; sucrose content 20–30% dry basis, and reducing sugars 45–55% d.b. See also **Molasses** and **Feed cane molasses**.

HMF	The abbreviation for 5-hydroxymethylfurfural. This compound is associated with poor colour in glucose syrups. It can also impart a bitter flavour.
Honey	A sweet, sticky yellowish brown fluid, made by bees from flower nectar. Typically, it will contain about 18% water; 41% fructose; 34% dextrose; 2% sucrose; 5% higher sugars, gums, wax, protein. The fructose content is always greater than the dextrose in genuine honey.
HPKO	Hydrogenated palm kernel oil, which is used in confectionery products. The effect of the hydrogenation process is to reduce the unsaturated nature of the fat, and thereby improving its oxidative stability.
HPLC	The abbreviation for high-pressure liquid chromatography. This is an analytical method frequently employed to separate individual sugars from a mixture.
HSH	This can mean either hydrogenated starch hydrolysis, that is the process of hydrogenating a starch-based product, for example a dextrose syrup. Alternatively, it can refer to the end product of hydrogenated starch hydrolysis, namely hydrogenated starch hydrolysis, for example sorbitol.
Humectancy	The property of preventing moisture loss from a food product.
Humectant	A material which, because of its hygroscopic nature, can be used to retain the moisture content of a product. Typical humectants used in the food industry are glycerol, sorbitol and glucose syrups, for example 63 DE syrups.
Hydrogen swells	The name given to clostridium bacteria, which produce hydrogen. The bacteria are usually associated with starch, and when the starch is used in a canned product, the hydrogen gas which it produces distorts the can, hence the term 'hydrogen swells'. Hydrogen gas can also be produced by corrosion, possibly due to the action of acid producing bacteria, with the resulting gas again distorting the can. *See also* **Sulphide Stinkers** and **Flat Sours**.
Hydrogenation	A process in which a substance is reacted with hydrogen under pressure, and in the presence of a catalyst. This process is used in the manufacture of sorbitol from dextrose. It is also used to change the properties of fats and oils, typically palm kernel oil to make the vegetable fat HPKO. This fat is used in confectionery, ice creams and bakery products. Hydrogenation of oils changes their melting point and melting profiles. It also makes the fat more resistant to oxidation.
Hydrol	The mother syrup remaining after the crystallisation of dextrose. Because this syrup will typically have a reducing sugar content of about 75%, the syrup is frequently used as a feedstock for fermentations. This syrup is also referred to as greens.

Hydrolysates	The products resulting from acid or enzymatic hydrolysis of starch.
Hydrolysed cereal solids	A classification sometimes given to maltodextrins with a DE of less than 10.
Hydrolysis	The process of splitting the starch molecule into smaller parts, with the introduction of water into the resultant molecules. This introduction of water is known as 'hydrolysis gain'. Hydrolysis is usually carried out using either acid or enzymes, or a combination of both. This term can also apply to the inversion of sucrose.
Hydrolysis gain	When starch is hydrolysed, there is an increase in molecular weight, due to the introduction of water into the molecule. This increase in molecular weight is known as 'hydrolysis gain', and is similar to the gain in molecular weight during the inversion of sucrose. Examples of hydrolysis gain are as follows:

Trisaccharides

$$C_{18}H_{32}O_{16} + 2H_2O \rightarrow 3C_6H_{12}O_6$$
$$100 \text{ grams} \qquad\qquad 107.14 \text{ grams}$$

Disaccharides

$$C_{12}H_{22}O_{11} + H_2O \rightarrow 2C_6H_{12}O_6$$
$$100 \text{ grams} \qquad\qquad 105.26 \text{ grams}$$

Hydrometer	An instrument for measuring specific gravity. It usually consists of a glass or metal-graduated tube, weighted at one end. The level to which the hydrometer sinks is noted, and from relevant tables, the solids can be determined. The word 'hydrometer' is derived from two Greek words, hudor meaning water, and metron meaning measure.
Hygroscopic	The ability of a product to absorb moisture from its environment.
Hypoglycaemia	This is a medical term, and refers to an abnormally low level of glucose (dextrose) in the blood. The normal level of glucose (dextrose) in the blood is typically 6.0 mmol/L. Hypoglycaemia is defined as a blood glucose (dextrose) concentration of 2.2 mmol/L or less. Causes for hypoglycaemia are many, but typically diabetes, consumption of excessive alcohol and fasting.
Icing sugar	Sucrose which has been milled to a fine powder. Typical particle size 20–30 microns or less. Frequently anti-caking agents, such as tricalcium phosphate, sodium alumina silicate, or maize starch added. Icing sugar is usually made from white sucrose, but brown sugar can be used. There are several different grades, based on particle size, and hence its suitability for different applications. Because of the fine particle size, there is the risk of a dust explosion when icing sugar is being handled.
ICUMSA	These are the initials of the INTERNATIONAL COMMISSION for UNIFORM METHODS of SUGAR ANALYSIS. These

methods of analyses are the 'official referee' methods of the sucrose industry and would be used where there is an analytical dispute. The term is typically used when describing the colour of a sucrose solution, for example 7.5 ICUMSA units, and is normally abbreviated to 'ICU'.

IGIU   The abbreviation for International Glucose Isomerase Units. It is the amount of enzyme which will convert 1 μmol of dextrose into fructose per minute, in a solution containing 2 mol of dextrose per litre, 0.02 mol of magnesium sulphate per litre and 0.01 mol of cobalt chloride per litre, at a pH of 6.84–6.85 (0.2 sodium maleate) and at a temperature of 60°C.

IMO   Abbreviation for isomalto-oligosaccharide.

Indian corn   An old name used to describe corn (maize). It was so called because it was grown by the native North American Indians.

Inositol   A growth factor for animals and micro-organisms. It is obtained commercially from corn steep liquor. It is present in maize as phytic acid (inositol hexaphosphate). As a vitamin B complex, it acts by releasing energy from food. It is sometimes referred to as meat sugar.

Insulin   The hormone in the human body which regulates the sugar level in the bloodstream.

Invert   A mixture of dextrose and fructose, made from sucrose by treating it with either acid and heat, or by the enzyme invertase. Invert is usually a syrup, but under certain conditions, will form a crystalline solid.

Invertase   An enzyme which converts sucrose into dextrose and fructose.

Inverting power   The efficiency of an acid to convert sucrose into dextrose and fructose.

Invert sugar   A mixture of dextrose and fructose obtained by hydrolysing (inverting) sucrose.

Inulin   It is a polymer of fructose and is the chemical form in which members of the Compositae family, such as chicory and Jerusalem artichoke, store fructose in an insoluble form. It is used to make fructose syrups. Inulin is not absorbed by the body.

Iodine number   It is a measure of the unsaturation of fatty acids and fatty esters. Since unsaturated fatty acids are associated with rancidity, it could be said that the iodine number is an indication of the likelihood of the product going rancid.

Ion exchange   See **De-ionisation**.

Isoelectric point   It is the pH at which the positive and negative charges on particles in a solution are zero, and can result in the particles flocculating. This property is used to precipitate proteins and fats out of solution during the manufacture of glucose syrups.

Isoglucose   The European Community name for high-fructose glucose syrup.

Isomalt	A polyol produced from sucrose by enzymatic conversion to isomaltulose, which is then hydrogenated. It is a disaccharide polyol.
Isomaltose	A reversion product of dextrose, produced during processing.
Isomaltulose	A disaccharide isomer of sucrose. It is produced by the enzymatic conversion of sucrose, and is an intermediary product in the manufacture of isomalt. Despite being a sucrose isomer, it is less than half as sweet as sucrose, and the enzymes in the digestive tract take nearly five times longer to breakdown the isomaltulose in to dextrose and fructose.
Isomerisation	The process of converting a dextrose rich syrup into a fructose containing syrup. This process normally uses an immobilised enzyme, which is packed into a vertical column, and as the dextrose-rich syrup passes over the enzyme, it is converted into fructose. In theory, a 50/50 equilibrium mixture of fructose and dextrose can be obtained, but in practice, as the equilibrium is approached, the reaction slows, and so commercially the reaction is stopped earlier and then blended to give a fructose content of 42%. Higher fructose containing syrups are made by chromatographic enrichment.
Isomers	Molecules having the same number and kind of atoms, but arranged differently within the molecules, thereby giving the molecule different properties.
IX	An abbreviation for 'ion exchange'. Sometimes used when referring to 'de-ionised' syrup.
Kafir corn	*See* **Sorghum**.
Kelvin scale	*See* **Absolute temperature**.
Ketose sugar	A sugar containing a ketone group. Fructose is a ketose sugar. Ketose sugars can reduce Fehling's solution (alkaline copper sulphate), and Tollens' reagent (ammoniacal silver oxide), but cannot reduce or decolourise bromine water. The inability to decolourise bromine water is a confirmation test that the sugar is a ketone and not an aldehyde.
Kettle sugar	*See* **Copper sugar**.
Kieselguhr	*See* **Diatomaceous earth**.
Lactitol	Produced by the hydrogenation of lactose.
Lactose	The disaccharide sugar found in milk, and is composed of dextrose and galactose. Sometimes referred to as 'milk sugar'. It is a reducing sugar and the only sugar made by mammals. It does not occur in the vegetable kingdom. It has an approximate sweetness of 40, when compared to sucrose with a sweetness of 100.
Lactose intolerance	The inability of humans to digest lactose. This is usually caused by a deficiency of the enzyme lactase, which is found in the small intestine.

Lactulose	A semi-synthetic disaccharide based on lactose (milk sugar). It is not absorbed in the gastrointestinal tract. It is a laxative medicine used in the treatment of constipation and liver failure.
Laevulose	Another name for fructose, but not frequently used these days. The term is derived from the fact that a solution of fructose will rotate a beam of polarised light to the left – laevorotatory, hence laevulose. Derived from the latin 'lavus' meaning left.
Lane and Eynon's titration	A method for determining reducing sugars using Fehling's solution. See **Fehling's solution**.
Leavening agent	A product which releases carbon dioxide that aerates a product, for example the use of sodium bicarbonate in baked products.
Laevorotatory	A substance which when in solution will rotate the plane of polarised light to the left. Substances which are laevorotatory are given the prefix 'l' (or −) to distinguish them from 'd' (or +) isomers. See also **Laevulose**.
Limit dextrins	Amylopectin contains some 1-6 linkages which are the starting point of the branched chains. These 1-6 linkages points are not attacked by amylase enzymes and therefore they remain intact to become limit dextrins. The 1-6 linkages can however be attacked by de-branching enzymes such as pullulanas.
Lintner starch	A soluble starch used in the determination of the diastatic power of malt. It is sometimes referred to as 'Lintnerised starch'. It is made by treating starch with 7.5% hydrochloric acid, for 7 days at room temperature. This starch is named after its German inventor, C.J. Lintner in 1886.
Liquid sugar	A solution of sucrose in water, typically with a solids content of 67%.
Loaf sugar	See **Sugar loaves**.
Lovibond colour	A method of measuring the colour of a liquid by comparing the colour of a sample with a set of coloured glass standards. These glass standards are made by The Lovibond Tintometer Co. When used for comparing glucose colours, usually only the red and yellow colours are used, and are added together to give the reported colour. This instrument is also used to measure the colour of maize oil.
L-sugars	See **D-sugar**.
Lycasin	The trade name for a range of hydrogenated glucose syrups containing both hydrogenated saccharides and polysaccharides.
Magma	The mixture of dextrose crystals and high DE glucose syrup, produced in the dextrose crystallisers. The syrup phase is frequently referred to as 'the mother liquor', and after the dextrose crystals have been separated out from the syrup, the 'mother liquor' is then called either hydrol or greens.
Maillard reaction	This reaction is sometimes referred to as non-enzymic browning. It was named after a Frenchman, Louis-Camille Maillard, who studied the complex reactions of reducing sugars,

for example fructose and dextrose, with amino acids/proteins, frequently at elevated temperatures, to produce a brown colour and flavour. He published his findings between 1912 and 1916. In the Maillard reaction, the hydroxyl groups of the sugars react with the amide groups of the protein to form melanoidins. The reaction is dependent upon the following:
- Amount and type of protein present.
- The amount and type of reducing sugars present, with fructose being the most reactive.
- Temperature. Generally, the reaction starts at about 115°C (239°F) – the higher the temperature, the greater the colour and flavour development.
- The length of time the mix is held at 115°C (239°F) – the longer the time, the greater the colour and flavour development.
- The reaction is influenced by pH – the more alkaline the mix, the darker the colour and stronger the flavour.

Maize germ It is the embryo of the new maize plant. It contains 40% oil, 40% protein, with the remainder being starch and fibre.

Maize germ meal The residue after the oil has been extracted from maize germ. Typically, it contains 22% protein and 24% starch, and is used in animal feed.

Maize gluten feed This is the fibrous residue obtained after the extraction of maize starch by the wet milling process. Normally, corn steep liquor (i.e. the evaporated steep water) is added to the fibres so as to increase their nutritive value. The fibres are normally dried and sold to animal feed compounders either as dry fibres or as pellets. In some cases, it is sold wet (i.e. without the drying stage), in which case it is put into a silage clamp prior to use. Typically, maize gluten feed contains 20% protein, 4% oil, 20% starch, with the rest being fibres.

Maize gluten meal It is the protein obtained during the extraction of maize starch by the wet milling process. It contains 60% protein. It also contains the yellow pigment xanthophyll, making it useful in poultry rations where the yellow pigment contributes to the yellow colour of the yolk. Unlike wheat gluten, maize gluten is not vital, that is it does not react with water to form a viscous tenacious mass.

Maize oil It is obtained from maize germ by either mechanical or solvent extraction. The oil contains 98% triglycerides of which 61% are polyunsaturated, and 13% saturated. The main uses are in salad and cooking oils. The advantage of using maize oil for frying is its high smoke point of 230°C.

Malbit Trade name for a hydrogenated maltose syrup.

Maltitol A disaccharide polyhydric alcohol, composed of dextrose and sorbitol, made by the hydrogenation of a high maltose syrup.

Maltodextrins	Starch hydrolysates with a DE of less than 20. They are spray dried, non-hygroscopic, bland powders. Frequently used as bulking agents or as carriers for flavours.
Maltose	A disaccharide sugar, composed of two dextrose molecules, chemically joined together.
Malt sugar	Another name for the sugar maltose.
Maltulose	A reversion product, produced during the production of dextrose, and maltose.
Manioc	See **Manioca**.
Manioca	An alternative name for cassava, from which tapioca starch is derived. It should be noted that the manioca root contains goitrogen and linamarin which can produce cyanide due to enzymes in the manioca root. See **Cassava**.
Mannitol	A sugar alcohol made by the hydrogenation of fructose.
Massecuite	The mass of dextrose crystals and syrup, produced in the dextrose crystallisers, which then passes into the dextrose centrifuge, from where the dextrose crystals are recovered. This term is also used in the sucrose industry, during the recovery of crystalline sucrose (French for cooked mass). See also **Magma**.
Meat sugar	An obsolete name for inositol.
Metabolisable carbohydrates	A term used in animal nutrition. Some animals can utilise some carbohydrates, whilst others cannot. It depends upon whether the animals have one stomach, or more than one, like ruminants such as cows.
Microtal	Tate & Lyle's registered trade name for micro-crystalline sucrose made by a process of rapid crystallisation of a concentrated sucrose solution, using a colloid mill.
Millimole	One thousandth of a mole (1 mOsm).
Millivals	A measurement of water hardness per litre. It is calculated by dividing the millimoles per litre by the valency of the element.
Milk sugar	See **Lactose**.
Milo	Popular name for the cereal sorghum. See **Sorghum**.
Modified starch	Starch which has been either physically or chemically changed so as to alter its rheological properties. Some modified starches will be both physically and chemically modified. A physical modification would be to gel the starch, and then dry it so as to make a pregelled starch. This dried starch could then be slurried in cold water to form a gel. Chemical modifications typically change the viscosity, improve the freeze thaw and acid stability. Common chemical modifications are as follows:

- Acid modification
- Oxidised
- Cross-linked
- Hydroxyethyl
- Cationic
- Phosphated

Moisture content	The amount of moisture in a product, and is usually expressed as a percentage on a weight/weight basis or as a weight/volume basis.
Mol	A mol is the SI unit of weight of a compound equal to its molecular weight in grams (see also osmotic pressure).
Molal concentration	Concentration of a solute measured by the number of moles of solute per kilogram of solvent.
Molal solution	A solution containing 1 mol of solute (dissolved substance) per 1000 grams of solvent.
Molar solution	A solution which contains 1 mol of solute (dissolved substance) dissolved in a total volume of 1 litres.
Molasses	A dark-coloured syrup remaining after the crystallisation of sucrose. Usually contains a high proportion of minerals, together with lesser amounts of sucrose, fructose and dextrose. *See also* **High test molasses** and **Feed cane molasses**.
Mole	The mole is the amount of substance of a system which contains as many elementary entities as there are atoms in 0.012 Kg of carbon. Its symbol is 'mol'.
Molecular weight	It is the sum of the atomic weights present in a compound.
Monosaccharide	A substance which contains only-one sugar. Typical monosaccharides are dextrose, fructose and galactose.
Moulding starch	This is the starch used by the confectionery industry for making moulds. An impression is made on the surface of the starch. The impression is then filled with the prepared mix, typically fondants, pectin or gelatin jellies. The starch usually contains about 0.5% mineral oil, which reduces the risk of dust explosions, and helps to retain the mould impression. For successful moulding, the moisture content of the starch should be less than 6%.
MSNF	This term is used in both the dairy and ice cream industries, and stands for Milk Solids Not Fat.
MTHase	Abbreviation for the enzyme malto-oligosyl trehalose trehalohydrolase, which is used in the production of the sugar trehalose, in conjunction with the enzyme MTSae.
MTSase	Abbreviation for the enzyme malto-oligosyl trehalose synthase, which is used in the production of the sugar trehalose, in conjunction with the enzyme MTHase.
Muriatic acid	An old name, often used in United States, for hydrochloric acid.
Muscovado sugar	A soft brown sugar obtained from sugar cane, containing molasses (derived from the Spanish word 'mascababo', meaning 'of lowest quality').
Mushroom sugar	*See* **Trehalose**.
Mycose	*See* **Trehalose**.
Mycotoxins	These are naturally occurring toxic compounds, produced by fungi such as *Fusarium*, *Aspergillus* and *Penicillium* when they infect agricultural crops, particularly cereals. Aflatoxin B1 is the most toxic of the metabolites. Other toxins are Ochratoxin A, and Fusarium toxins such as T-2 Toxin, Deoxynivalenol, Zearalenone

	and Fumonisins. The physiological effects of mycotoxins are as varied as the toxins themselves.
NAS	No added sugar.
Neosugars	A mixture of varying quantities of fructo-furanosyl sucrose oligomers.
NIBS	During the fermentation of raw cocoa beans, the cotyledon takes up moisture and when dried, it breaks into pieces called nibs. *See also* **Sugar nibs**.
Oligofructose	An oligosaccharide made up of fructose molecules. *See also* **Fructosans**.
Oligosaccharide	A carbohydrate containing four to seven sugar molecules chemically joined together. *See also* **Higher sugars**.
Osmolality	The concentration of osmotically active particles in solution, expressed in terms of osmoles of solute per kilogram of solvent. Units are mOsm/Kg.
Osmolarity	The concentration of osmotically active particles in solution, expressed in terms of osmoles of solute per litre of solvent.
Osmole (unit)	It is the amount of osmotically active particles that when dissolved in 22.4 litres of solvent at 0°C exerts an osmotic pressure of 1 atmosphere.
Osmosis	It is the diffusion of a solvent through a semi-permeable membrane.
Osmotic pressure	It is the hydrostatic pressure produced by a difference in concentration between two sides of a semi-permeable membrane.
Ostwald ripening	The name given to the phenomenon where small crystals dissolve at the expense of the large crystals, which grow larger, and is particularly relevant to the change which occurs in the texture of fondant over a period of time.
Overrun	A term used in the ice cream industry, and refers to the amount of air which is whipped into an ice cream so that the volume of solidified ice cream is considerably more than that of the original mix. This increase in volume due to the added air is known as overrun, and is expressed as a percentage of the original volume. Typical overrun would be 120–140%.
Palitinose	A slow release carbohydrate derived from sucrose. It is also called isomaltulose.
Palm sugar	A sugar obtained from date palms, by making incisions in the trunk, and collecting the juice, which is then evaporated. It is composed mainly of sucrose and lesser amounts of invert.
Palm syrup	Syrup made from the juice of date palms. It is composed mainly of sucrose and invert depending upon the type of palm tree and time of consumption.
Partially inverted syrup	When sucrose is inverted and changed into an equal mixture of dextrose and fructose, it is referred to as being fully inverted. With a partially inverted syrup, only part of the sucrose is inverted. Typically, partially inverted syrups will contain either 66%

	sucrose and 33% as equal parts of dextrose and fructose. Alternatively, there will only be 33% sucrose with 66% being an equal mix of dextrose and fructose. *See* **Invert**.
Pearl starch	A term used in America, and refers to native or unmodified maize starch.
Pearson's Square	An easy method for calculating the additions of water or syrup when diluting or concentrating syrups or when making syrup blends.
Pectin sugar	A name sometimes used for the pentose sugar D-arabinose, which is found in hemi-cellulose, and gums such as pectin.
Pentose sugars	Sugar molecules containing five carbon atoms, for example xylose, arabinose.
Perlite	A filter aid produced by the heating of volcanic rock, which is frequently used in the filtration of glucose liquors, prior to colour removal.
Pet bottles	Plastic containers, usually soft drink bottles, which are made of polyethylene terephthalate.
Phage	*See* **Bacteriophage**.
Phenylketonuria	The inability of the body to metabolise the amino acid phenylalanine. *l*-Phenylalanine is a component of the artificial sweetener aspartame. When aspartame is ingested phenylalanine is produced.
Plato	An empirical scale hydrometer used by brewers to measure the gravity of beer. One degree Plato is equivalent to 1 grams of sucrose, dissolved in 1 grams of water. Therefore, 10 degrees Plato has the same specific gravity as a 10% sucrose solution w/w. Degrees Plato = degrees Brix = % solids w/w (rounded up).
Pol	A term used in the sucrose industry, and refers to the reading obtained with a polarimeter, and hence the purity of the sucrose.
Polarimeter	An instrument to measure the rotation of any optically active substance. *See also* **Saccharimeter**.
Polydextrose	A polymeric randomly bonded melt condensation product based on dextrose. It is prepared by melting dextrose, with lesser amounts of a glucose syrup, sorbitol and an acid such as citric or phosphoric. As the molecular structure has been enlarged, it will not be absorbed by the body and is frequently used as a bulking agent.
Polyglycitol	A polyol containing less than 50% maltitol and less than 20% sorbitol, with the balance being hydrogenated gluco-oligosaccharide.
Polyfructose	*See* **Fructosans**.
Polyol	The name given to any of the polyhydric alcohols such as sorbitol.
Polysaccharide	A high molecular weight carbohydrate composed of a large number of simple sugars. Starch is a polysaccharide, being composed of a large number of dextrose molecules.

Popcorn	A type of maize which, when heated, expands due to the entrapped moisture being turned into steam which expands the corn to produce a solidified puffed product when cooled. Frequently, the puffed product will then be coated with a butterscotch or caramel flavoured coating.
Pot ale syrup	The residue after alcohol distillation, particularly from whisky distillation.
Praline	A blend of roasted nuts, usually hazelnuts, but can be walnuts or almonds, with chocolate.
Pregelantinised starch	This is a starch which has been gelled and then dried. When the dried starch is subsequently mixed with cold water, it forms a viscous paste, but both the thickening power and tendency to form a gel are less than that of the original starch. Typical application for these types of starch would be in convenience foods.
Priming sugar	A sugar which is added to the beer after primary fermentation has finished, so as to support secondary fermentation. The secondary fermentation usually occurs in a closed vessel, for example cask or barrel, where the carbon dioxide produced 'carbonates' or conditions the beer. Priming sugars can also be used to give the beer sweetness or flavour.
Process inversion	The sugar sucrose is made up of two sugars, fructose and dextrose, chemically bound together. In a process when sucrose is heated under acidic conditions, this chemical bond is broken to give an equal mixture of fructose and dextrose. The breaking of this chemical bond is known as process inversion.
Ptyalin	An old name for the enzyme amylase, present in human saliva.
Pyrogen free	A substance which does not contain any chemicals released from micro-organisms, such as endotoxins, which will have an adverse effect on the body's heat regulating mechanism, and result in a 'fever'.
Raffinate	Within the glucose industry, the word raffinate refers to the dextrose-rich syrup stream produced when ion exchange chromatography is used to separate fructose and dextrose from a high-fructose glucose syrup.
Raffinose	A trisaccharide composed of galactose, dextrose and fructose, and found in unrefined white beet sucrose.
Rancidity	A characteristic off-flavour which develops in fats, and products which contain fat, for example toffees. Where it is caused by the reaction between unsaturated oils and oxygen, it is referred to as oxidative rancidity. Where it is caused by the hydrolysis of a triglyceride molecule to give glycerol and a fatty acid, it is referred to as hydrolytic rancidity. Oxidative rancidity can be reduced by the addition of an antioxidant to the fat. Typical antioxidants are natural and synthetic tocopherols, and BHA (butylated hydroxyanisole).

Reducing sugar	It is a sugar which is capable of chemically reducing cupric salts in alkaline solution (Fehling's solution) to cuprous oxide. Fructose, dextrose and maltose are all reducing sugars. Sucrose is **not** a reducing sugar.
Refractive index	When a beam of light is passed through a transparent substance, it is defracted. The ratio of the sine of the angle of incidence of the light beam to the sine of the angle of refraction of the light beam is a constant for any particular substance and for any particular colour. This constant is called the refractive index. This physical property is used to measure the solids of a glucose syrup. Solids determined by this method are referred to as 'refractive index solids' or 'RI solids'.
Refractometer	An instrument which can measure soluble solids by refracted light, with the solids being expressed either as a refractive index value or as direct percentage sucrose solids.
Reinheitsgebot	A German purity law, dating from 1487 and originated in Bavaria. This law restricts the raw ingredients for use in beer to malted barley, hops, yeast and water. This law does not apply to German beer which is exported.
Resin refined syrup	This means that the syrup has been treated with ion exchange resin, as opposed to being treated with carbon, to remove colour, etc. *See also* **De-ionised**.
Resistant starch	Typically, a high-amylose starch, which is not broken down in the human small intestine, similar to dietary fibre. On reaching the large intestine, it could be fermented by bacteria to produce short-chain fatty acids, such as butyrate, acetate and propionates.
Retrogradation	The reversion of a cooked starch paste to an insoluble form. The amylose fraction of starch is soluble in hot water, and so when the starch cools, the amylose comes out of solution.
Rework	A term used in the confectionery industry for the re-processing of confectionery which does not conform to the required standard. A typical example would be the reworking of high-boiled sugar confectionery. The normal addition of rework to a recipe would be about 5–10%. Higher additions run the risk of increasing the amount of reducing sugars in the end product. Rework is not frequently used in chocolate confectionery, as it is considered in some countries to be an adulterant, and can only be used in centres.
Ribose	A five carbon monosaccharide, containing an aldehyde group. It is made in the body and found in cells and forms the structural backbone of the co-enzyme ATP (adenosine triphosphate) and DNA (desoxyribose nucleic acid).
Roping	A process for making confectionery products. Instead of depositing the cooked syrup directly into moulds, the mass is manipulated using ribbed conical rollers to form a rope. The diameter of the rope of cooked confectionery is then gradually

	reduced prior to being cut to size by a rotary guillotine. The pieces of confectionery are then wrapped.
Saccharimeter	This is an instrument similar to a polarimeter and measures the optical rotation of a sucrose solution, but unlike a polarimeter, the scale which measures optical rotation is replaced by one which is graduated to measure percentages.
Saccharin	An artificial sweetener, discovered in 1876. It is approximately 300 times sweeter than sucrose. It is made from *o*-sulphamoylbenzoic acid.
Saccharine	In 1876, Alexander Manbre made a glucose syrup for use in brewing, which was called 'saccharine'.
Saccharose	Another name for sucrose, but not frequently used in the United Kingdom.
Saccharum	The name originally given to invert produced by William Garton in 1855. The name was subsequently used by Customs and Excise to describe any type of invert which was to be used in brewing. Abram Lyle in the 1870s produced a fondant-like material for use in brewing, which was also called 'saccharum'.
SAG	This refers to the strength of a pectin gel, and determined by the USA-SAG method, which measures the loss of height or sag of an unsupported jelly. The test method is normally used for unbuffered high methoxyl pectins.
Sanding sugar	A sugar, generally sucrose, which is used to coat confectionery and bakery products to prevent the products from sticking together.
Saturated solution	A solution in which the solute in solution is in equilibrium with the dissolved solute.
Schardinger dextrins	See **Cyclodextrins**.
Scott Viscosity	A method for determining the hot paste viscosity of a gelled starch. The method originated from the oil industry.
Seventies and eighties sugar	An old American term for 70 and 80 DE acid-converted syrups. Typically, syrups made with these high DEs would crystallise when cooled. Nowadays, these high DE syrups are made using either acid enzyme, or enzyme enzyme conversions.
Slugging	One of the two processes used to make lozenges. The other process is 'granulation'. See **Granulation**.
Soluble solids	This is a term frequently used in the jam industry, and refers to the dissolved solids present in a solution. In the case of jam, the 'soluble solids' would consist mainly of dissolved sugars derived from the fruit, sucrose and glucose syrup, if used.
Soluble starch	See **Lintner starch**.
Solute	The substance dissolved in a liquid.
Sorbitol	A sugar alcohol, made by the hydrogenation of dextrose.
Sorghum	A cereal grown in countries which have a limited rainfall such as Africa, India, Pakistan, China and the Southern States of America. The scientific name is *Sorghum vulgare*, and it belongs to the

	economically important family of grasses, the gramineae. It is also known as milo or kafir corn.
SOS	Abbreviation for soya-oligosaccharide.
Spent carbon	Activated carbon which has been used to decolourise syrup.
Spreckels test	This is a qualitative test to determine if an ingredient for a soft drink, such as sucrose or a glucose syrup, will form a floc. The test involves heating a solution of a sugar or glucose syrup under test in a water bath, and allowing it to cool for about 24 hours, and observing if a floc has been formed. The test was developed by The Spreckels Sugar Company of San Francisco, USA.
Starch	It is a polymer of dextrose, and it is the chemical form in which plants such as cassava, maize, potatoes, rice, sago and sorghum store dextrose in an insoluble form. Starch is composed of amylose and amylopectin. The ratio of amylose to amylopectin will vary from plant species to plant species, but typically the amylose content will be 23–27%, and the amylopectin about 73–77%. In amylo maize, the ratio is about 50/50, whilst in waxy starches, the amylopectin will be nearer to 99%.
Starch fluidity	It is a measure of the viscosity of a starch paste, particularly oxidised starches which have a lower viscosity than a conventional unmodified starch. Oxidised starches are used in starch jelly confectionery.
Starch sugar	An alternative name for spray-dried 95 DE syrup. Also known as total sugar.
Starch syrup	An old name for glucose syrup. So named because the syrup was derived from starch.
Steeping	Steeping is the initial stage in the separation of starch from the maize kernel in the wet milling process. It consists of soaking the dry maize kernels in an aqueous solution of sulphur dioxide, until the kernels become soft.
Steep water	Water containing soluble proteins, minerals, vitamins and sugars which have been leached out of the maize during the steeping process. Because of its high nutritional content, steep water is usually added to maize fibres, which are then sold as animal feed. It is also used as a nutritive supplement and yield enhancer in antibiotic fermentations.
Stein Hall Cup	A metal tube used to measure the viscosity of starch based adhesives (glue) used in the manufacture of corrugating boards. Inside the metal tube, there are two pins, projecting from the inside walls of the tube. The time taken for the adhesive (glue) to pass between these two pins is the viscosity.
Stoving	A process used in the confectionery industry to reduce the moisture content of confectionery, typically used for jelly goods. The process involves storing a jelly in a fan oven at 40–50°C for 12–24 hours to reduce the moisture.

Sucralose	A high-intensity sweetener, made by chlorinating sucrose. It is sold under the name of 'Splenda'. The original work was carried out by Dr Hough at Queen Elizabeth College (now part of King's College), University of London, in 1976, and sponsored by Tate & Lyle. Covered by B.P. 1,543,167 (1979).
Sucrase	An old name for the enzyme invertase.
Sucrose equivalence factor	The molecular weight of sugar divided by the molecular weight of the ingredient. This factor represents the degree of freezing point depression relative to an equal weight of sucrose. *See also* **Freezing point factor**.
Sugar alcohols	An alternative name for polyhydric alcohols (also known as polyols). Typical sugar alcohols are sorbitol and maltitol. So named because the starting material is a sugar, and when hydrogen is added, the aldehyde group of the sugar is changed into a hydroxyl or alcohol group.
Sugar bloom	A greyish appearance on chocolate. In severe cases, it will manifest itself as a crystalline deposit. Unlike fat bloom, it is not removed when scratched, and does not feel greasy. Sugar bloom is caused by atmospheric moisture condensing on a chocolate surface, and dissolving the sugar from the chocolate, which subsequently crystallises out on the surface, when the atmosphere becomes drier. *See* **Fat bloom**.
Sugar fungus	A name sometimes used for the yeast genus *Saccharomyces*.
Sugar loaves	Prior to the introduction of the centrifuge, a concentrated sugar solution would have been poured into cone-shaped moulds, where it was allowed to crystallise. The solid sugar produced was referred to as 'sugar loaves', because of the shape of the resulting block of crystallised sugar.
Sugar nibs	Very coarse sugar – typical particle size range is 1.4–3.0 mm.
Sugar of lead	Nothing to do with 'sugars'. It is a name given to lead acetate.
Sulphide stinkers	The name given to hydrogen sulphide producing bacteria such as B*acillus desulphotomaculum nigrifican* and *B. stearothermophilus*. These bacteria also produce acid. These types of organisms are associated with starch, which when used in a canned product, will putrefy the product. Frequently, the container will be distorted due to the pressure of the gas inside the container. *See also* **Flat sours** and **Hydrogen swells**.
Supersaturated solution	It is a solution in which more solute is in solution than is present in a saturated solution of the same substance at the same temperature and pressure. High-boiled sweets are a typical example.
Sweet bread	The 'butchers' or gastronomic name for the pancreas of an animal. The pancreas produces the hormone insulin, which regulates the storage of glucose in the liver, and the amount of glucose in the blood.
Sweet corn	It is distinguished from normal dent corn by the presence of the recessive sugary gene SU 1. This recessive gene changes the

	carbohydrate content of the endosperm, and in the case of sweet corn, results in some of the starch being stored as sucrose. Once the sweet corn has been harvested, the sucrose quickly reverts to starch during storage.
Sweet potato	A tropical root vegetable – *Ipomoea batatas* – which not only contains mainly starch, but it also contains a small amount of sucrose, which gives the vegetable its sweet taste. The sweet potato is not related to the common potato – *Solanum tuberosum*.
Sweetness drag	The slow release of sweetness over a longer period of time than would normally be expected. Typically due to the presence of higher sugars, which reduce an immediate sweetness release to a slow sweetness release over a longer period of time.
Syneresis	This is sometimes referred to as 'weeping', and is the release of water from a starch gel. This release of water is due to the amylose fraction of the starch coming out of solution which then reforms, and in so doing, the amylose linear chain squeezes out the water.
Synergistic effect	The working together of two or more ingredients to produce an effect which is greater than the sum of their individual properties. In the case of sugars, a synergistic effect occurs when two or more 'sugars' are mixed, and the overall sweetness of the mixture appears to be greater than the sum of the individual 'sugars'.
Syrup Fuscus	The name sometimes used for molasses, but not very often in the United Kingdom.
Syrup Simplex BP (Simple Syrup)	A solution of sucrose in water, with a concentration of 66.7% w/w. Specific gravity is 1.315–1.333. It is used by the pharmaceutical industry as a carrier for medicinal ingredients, typically in a cough linctus.
Tagatose	It is the isomeric ketose of galactose, and is obtained from milk by enzymatically splitting lactose into dextrose and galactose. The galactose is then alkaline converted into tagatose, which is a reducing monosaccharide.
Takasaki–Tanabe process	The original Japanese process for making high fructose glucose syrups.
Tapioca	Starch derived from Cassava.
Tempering	A term used in chocolate confectionery to describe the process of heating and cooling chocolate to precise temperatures, and in a particular order. The fat in chocolate can crystallise into six different molecular arrangements, of which only one molecular arrangement – the β-form – has all the desired properties. The object of tempering is to encourage the formation of the β-form, and for it to be evenly dispersed throughout the chocolate mass.
T_g	The glass transition temperature. It is the temperature at which an amorphous glass softens. Typically, high-boiled sweets are non-crystalline glassy products, so the T_g will be the temperature

	at which the sweet exhibits 'cold flow' and starts to deform. Different substances have different T_g values.
Throw	This term is used in the manufacture of soft drinks and refers to the dilution of a concentrate with water, prior to bottling. A throw of one to five would mean the dilution of one part of the concentrate with five parts of water to make the finished drink.
Tinctorial power	Colouring power. Usually used when discussing the depth of colour of caramel or other colouring agents.
Torrified	A substance is said to be torrified when it has been subjected to very hot and dry processing conditions. Typical torrified products are torrified starch – a type of dextrin, used in remoistenable adhesives and torrified barley, which is used in brewing.
Total sugar	Spray-dried 95 DE glucose syrup. Whilst a 95 DE syrup is used to make both dextrose and total sugar, during the manufacture of dextrose, the dextrose is separated out from the maltose, maltotriose and higher sugars by crystallising out the dextrose. In the manufacture of total sugar, all the sugars present in the original 95 DE syrup are all spray dried and are present in the spray-dried powder. No sugars are removed.
Treacle tap	Because glucose syrups are viscous liquids, it is difficult to control their flow from a drum (45 gallon or 200 liters size), therefore a treacle tap can be used. The tap is usually made of cast bronze or gunmetal and consists of a small length of pipe which has a screw thread on one end and a tight fitting flange at the other bend. The bung of the drum is removed and the treacle tap is screwed into its place. Attached to the flange is a leaver. By lifting or lowering the leaver, the pipe is opened or closed. Because the flange is a very tight fit, the flange acts like a guillotine, cutting off the flow of the syrup.
Trehalose	A non-reducing disaccharide sugar, derived from the fungus *amanita muscaria*. It is composed of two dextrose units, but in a 'flipped configuration', with an α-1,1 linkage, which accounts for its non-reducing properties. It is metabolised by the human digestive system. It has half the sweetness of sucrose, and a molecular weight of 342.
Trimoline	A name sometime used for invert.
Trisaccharide	A carbohydrate containing three monosaccharide sugar molecules chemically bound together.
True solids	These are the solids in a glucose syrup, which have been determined by drying in an oven, as opposed to RI solids which have been determined using a refractometer.
Turbinado sugar	Raw sugar which has had some of the surface molasses removed.
Twaddell	A hydrometer scale devised by a Scottish chemist, William Twaddell (1792–1839) to measure the solids of liquids more dense than water, particularly solutions of chemicals, for

	example sodium silicate and sodium hydroxide. Results expressed as 'degrees Twaddell' (abbreviated Tw).
Unavailable carbohydrates	This term refers to pentosans, pectins, hemi-cellulose and cellulose which are not broken down in the human gastrointestinal tract. They are, however, available to ruminants.
Vanilla sugar	Sucrose which has a vanilla flavour. The flavour is added to the sucrose crystals by the addition of either vanilla extract or synthetic vanillin flavour to the sucrose crystals or icing sugar. Typical vanilla sugar would contain 10% vanillin, 90% sucrose.
Vanillin sugar	*See* **Vanilla sugar**.
Virgin carbon	Activated carbon which has not been used to decolourise a glucose syrup.
Vital wheat gluten	The protein derived from wheat containing gliadin and glutenin. It has the ability to form a viscous tenacious mass when mixed with water. If, however, the wheat gluten is dried above 60°C, the vitality is destroyed, and the vital wheat gluten is referred to as being 'non-vital'.
Water activity (Aw)	The susceptibility of a food to microbiological spoilage depends upon the unbound or free water in the product, rather than the moisture content of the product. This unbound or free water is referred to as the water activity (Aw). When the free water and the moisture in the atmosphere in which it is in contact are in equilibrium, the relative humidity of that atmosphere is called the equilibrium relative humidity (ERH). Aw is expressed as a fraction, for example 0.6, whilst ERH is expressed as a percentage and can be obtained by multiplying the Aw by 100. As a general rule, microbes may only grow in substrates having an Aw higher than 0.61. Typically, osmophilic yeasts grow at about 0.60, fungi at 0.72, most yeast at 0.88 and most bacteria at 0.92. By comparison, the Aw of a jam and fondants will be about 0.8, the Aw of gums will be about 0.60, caramels about 0.50 and toffees about 0.30, whilst the Aw of a biscuits will typically be about 0.2. Water activity can be measured by either using concentrated salt solutions or by using an electronic thermohygrometer.
Waxy maize	A genetic variety of maize, the starch of which is predominately composed of amylopectin (i.e. 99%). When a waxy maize kernel is cut to half, the cut surface has a slight waxy appearance, and when stained with iodine due to the absence of the linear amylose fraction, the waxy maize stains red as opposed to blue. Such waxy starches are more freeze stable than starches which contain amylose. This stability can be further improved by chemical modification. *See also* **Starch** and **Modified starch**.
Wet milling	The process which separates out the four parts of the maize kernel – fibre, germ (oil), gluten (protein) and starch.

Wort sugar	Sugar which is added to the wort, whilst it is still in the copper. *See* **Copper sugar**.
XOS	Abbreviation for xylo-oligosaccharide.
Xylan	A yellow, water-soluble, gummy polysaccharide found in plant cell walls, and yields xylose upon hydrolysis.
Xylitol	A sugar alcohol, made by the hydrogenation of xylose.
Xylose	A sugar derived from hemi-cellulose. Whilst it is absorbed from the jejunum of the human digestive tract, it is not metabolised by the body, and is excreted unchanged in the urine.
Yucca	An alternative name for cassava, from which tapioca starch is derived. *See also* **Cassava** and **Manioca**.
Zea mays	The botanical name for maize (corn in the United States). It is a member of the grass family, the Gramineae.
Zein	A corn protein of the prolamine class. It contains no lysine and no tryptophan. It is soluble in alcohol and is sometimes used as a glaze or as a barrier coating for sweets, pills and tablets.
Zymase	An old collective name for the complex enzymes present in yeast, which converts sugars to alcohol.

Appendix A
Simple analytical information

A.1 Introduction

A lot of analytical information can be obtained for free, by just reading both the ingredients declaration and nutritional information panels on the product's packaging – it can save a lot of time providing that the information is correctly interpreted. Some will say that this approach is more of a 'guestimation', rather than an accurate chemical analysis. Whilst it might be a 'guestimation', it is a good starting point for any future work whether it is a more detailed analysis or for product development, and if the product declaration does not mention a bulk sweetener, then there should be no object in analysing for one, unless there is a reason for concern.

A.2 The ingredient declaration panel

Thanks to legislation, a lot can be learnt about the ingredients in a product due to the labelling regulations. These regulations require the ingredients in the product to be listed in order of magnitude, so that the largest ingredient will be the first named, with the rest of the ingredients placed in descending order. Where a product is composed of two or more other products, some manufacturers will obligingly list most of the ingredients of each part separately, as well as listing all the ingredients individually. For example, the manufacturers of a simple jam sponge might list the ingredients of the sponge, and then the ingredients of the jam. This sort of labelling can be very helpful. Another useful piece of labelling is when the manufacturer states the actual percentage of an ingredient in the declaration. This percentage will then act as a 'marker' for other ingredients. Other manufacturers might prefer to treat the two parts as one product and to list all the ingredients in order of magnitude in one single declaration, which is not so helpful for the would be analyst as it could be difficult to determine which ingredient came from which part of the product.

The ingredient declaration is based on the ingredients which go into the mix, and not what is actually in the end product, so analysis of the end product might not agree with the declaration. For example, sucrose in an acid medium like in a cola drink, would be inverted into dextrose and fructose, yet the ingredient declaration would have correctly stated that sucrose was used.

The EU labelling regulations permit starch derived fructose to be legally called a glucose syrup. High pressure liquid chromatography (HPLC) analysis would be required to confirm whether or not the syrup is a fructose syrup or a conventional glucose syrup and its exact sugar spectrum.

Attention should also be given to the exact wording of the ingredients to make sure that the interpretation is correct. For example, if the product is declared as sugar or glucose, it would suggest that the solid has been used and similarly if sugar syrup or glucose syrup has been declared then the liquid form will have been used. Always check if the ingredients are expressed on the basis of grams per 100 grams of product or grams per 100 ml for liquids. Solid products are generally expressed in grams per 100 grams, whilst soft drinks are usually in grams per 100 ml.

The nutritional label can also be a useful source of analytical information, but care again must be taken in how this information is interpreted. A good example is the term 'sugar' – this can mean sucrose as well as the fructose, dextrose and maltose, which are present in glucose syrups. Large molecular weight sugars present in glucose syrups such as maltotriose and higher cannot, under EU legislation, be classified as 'sugar'. They have to be called starch.

Other sources of information can be the product information sheet.

Consider the following label declarations.

Example 1. Raspberry Sponge Sandwich (Sponge cake sandwich (61%) with Raspberry flavoured filling (15%), and Vanilla flavoured Buttercream (21%)).

Ingredients: Sugar, Wheat Flour, Raspberry Flavoured Apple Jam, (Glucose Syrup; Apples; Sugar; Colours – Anthocyanins, Curcumin; Gelling Agent – Pectin; Acidity Regulators – Citric Acid, Sodium Citrate; Flavouring), Butter, Water, Egg, Vegetable Oil and Hydrogenated Vegetable Oil, Soya Protein, Humectant – Glycerine, Soya Flour, Emulsifiers – Mono and –Di-Glycerides of Fatty Acids and Polyglycerol Esters of Fatty Acids, Maltodextrin, Salt, Modified Maize Starch, Preservative – Potassium Sorbate, Colour – β-Carotene, Dried Egg White, Maize Starch.

In this example, the manufacturer has not only complied with the legislation by listing the individual ingredients but has also optionally stated that the product is made up of three parts – sponge, raspberry filling and buttercream. Interestingly, the way in which the ingredients of the raspberry filling have been listed provides the bases for a formulation to make the filling. By analysing the raspberry filling using HPLC, the type of syrup which has been used could be determined.

Example 2. Cereal and Hazelnut Bar, covered with Milk Chocolate.

Ingredients: Milk Chocolate (25%) (Milk, Sugar, Cocoa Mass, Cocoa Butter, Emulsifiers – E442 and E476, Flavourings), Oats (17%), Invert Sugar Syrup, Bran Flakes (9%) (Wheat Malt), Glucose Syrup, Hazelnuts (9%), Hydrogenated Vegetable Oil, Crispy Cereal (8%) (Rice, Wheat Flour, Sugar, Malt, Salt), Honey (2%), Glycerine, Salt, Molasses, Emulsifiers – E471 and Soya Lecithin.

In this example because the invert comes after the 17% of oats and above the 9% of bran flakes, it would suggest that the invert content is between 9% and 17%. Similarly, because the glucose syrup declaration is in between the bran flakes and hazelnuts, both of which are declared at 9%, it must be assumed that the glucose syrup inclusion is also 9%.

Whilst sugar analysis of the cereal part of the bar would give an amount of 'sugars' present, it would be difficult to decide which sugar came from which sweetener, that is did the dextrose come from the invert, the honey or the glucose syrup.

Example 3. Marmalade.

Ingredients: Glucose Fructose Syrup, sugar Solution, Glucose Syrup, Oranges, Gelling Agent (Pectin), Citric Acid, Cane Molasses, Acidity Regulator (Sodium Citrate), Orange Oils.

Nutritional Information (Typical values per 100 grams): Carbohydrates – 62.3 grams; Protein – 0.2 grams; Fat – trace; Energy – 1076 kJ/253 kcal.

This is an interesting label. Glucose fructose syrup could mean that the fructose was either from an HFGS syrup, or that the fructose came from inulin. One thing it does establish beyond any shadow of doubt is that a glucose syrup blend was the major ingredient. By declaring the sucrose content as a solution, this would give the impression that sucrose was the second largest ingredient. In fact, it only means that the solution was the second largest ingredient, unfortunately, it does not indicate how much sucrose was in the solution.

Example 4. Yoghurt containing fruit.

Ingredients: Yoghurt, Red Cherry (14%), Fructose, Sugar, Cherry Puree (2%), Rice Starch, Natural Flavouring, Concentrated Lemon Juice, Concentrated Blackcurrant Juice.

Nutritional Information (Typical values per 100 grams): Carbohydrate 13.8 grams, of which sugars are 12.3 grams; protein 4.7 grams; fat 0.1 grams; fibre 0.5 grams; sodium 0.1 grams (salt equivalent 0.2 grams). Energy value 320 kJ/75 kcal.

The nutritional information suggests that the total amount of both fructose and sugar is 12.3 grams. The ingredients label would suggest that as fructose is before sugar, the fructose is present in the larger amount of the two. A possible alternative could be that equal amounts of both fructose and sucrose have been used, and the reason why fructose comes before sucrose is purely for alphabetical reasons. Some of the sugars could also have originated from the fruit, fruit puree and the juice.

A.3 Does it contain glucose syrup?

So what analyses should be carried out, if there is not an ingredient declaration? The way one carries out the analysis of a product will depend upon whether the product is solid or liquid, what information is required and why? Is the analysis being carried out so as to copy an existing product? Is the analysis to find out more about a particular ingredient, for example, which DE glucose syrup has been used and how much, or how much sucrose has been used? Or is the analysis required following a customer complaint?

When analysing products which contain glucose syrups or sucrose, the following basic approach will give an indication of approximate inclusion rates. Where a more accurate analysis is required, the methods of the Association of Official Analytical Chemists (AOAC) should be used. (The AOAC changed its name in 1991 to AOAC International).

1. Determining the type of syrup in a liquid, for example a soft drink
This can be achieved using HPLC to determine the sugar spectrum of the syrup and therefore the type – fructose, maltose or a conventional glucose syrup.

The refractive index solids will give a good indication of the solids of the drink.

2. Determining the type of syrup in a solid

Fortunately, glucose syrup and sucrose are very soluble in water. Therefore, a known weight of the solid is stirred in hot water. The solids are filtered out and the weight of liquid obtained is noted, together with the solid content, which can be quickly determined by an RI solids measurement. From these two figures, the amount of syrup or sugar used can be calculated. HPLC analysis of the liquid will give an indication of the type and proportion of sugars present. An alternative to dispersing the solid in water is to use a Soxhlet water extraction. To obtain maximum extraction when using a Soxhlet extraction procedure, the sample should be in the form of small pieces.

If any fruit is present, it should be separated from the rest of the product because fruit normally contains sugars which these sugars will also be extracted with the other sugars which will result in a higher sugar figure.

3. Fats

Fats can be determined in a solid by a Soxhlet solvent extraction. If a Soxhlet apparatus is not available, then a possible alternative would be to cut the solid into small pieces and shake the product with a solvent (e.g. petroleum ether) in a separating funnel.

For a toffee, this can be dissolved in hot water, and the 'emulsion' can be transferred to a separating funnel, and shaken with petroleum ether. To determine the sugar content of the toffee, both the volume and solids of the water fraction are determined to calculate the sugars present, whilst HPLC sugar analysis of the water fraction will give details of the sugar composition. The solvent fraction is then evaporated on a water bath to give the amount of fat.

For an emulsion such as an ice cream, the fats can be determined using the Rose–Gottlieb method, with either a modified Dreschel tube or a Mojonnier tube.

The fat or oil can be identified from the melting point if it is a fat, or by its refractive index if it is a liquid.

4. Solids

In the case of a health bar or a similar product which contains solids, after both the water and fat soluble ingredients have been extracted from the same sample, the insoluble fraction will remain. If cereals, fruit pieces and vine fruits are present, then the relative proportions can be determined by hand sorting the fraction and weighing each fraction. The following analytical schematic (figure A.1) can be adapted for other products.

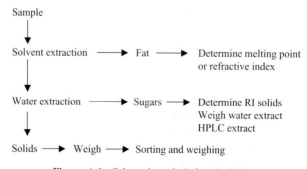

Figure A.1 Schematic analysis for a health bar.

5. The amount of some ingredients allowed in a product are controlled by legislation, so look at the relevant legislation, to get a 'ball park figure'. This is particularly relevant when looking for food additives such as preservatives, anti-oxidants, etc.

6. The pH and moisture content are also easy and useful determinations to carry out.

7. Does the analysis add up to 100%? *N.B.* There are some products such as bakery products which legally allow for a discrepancy due to losses during processing.

8. Remember, even when the analysis is complete, and the quantity of each ingredient is known, the analysis will not show how the ingredients were put together to make that particular product.

A.4   What HPLC sugar analysis can tell

One of the most powerful analytical tools available to the glucose syrup industry is HPLC. In its simplest form, this analytical technique will determine the different types of sugars present in a glucose syrup, or other sugars present in a mixture. It can also determine the amount of different alcohols or polyhydric alcohols in a sample. Typical alcohol could be ethyl alcohol, and a typical polyhydric alcohol could be sorbitol. Results are usually expressed on a 'dry basis'.

One HPLC method separates the different sugars into monosaccharides, disaccharides, trisaccharides, and so on. This means that maltose, sucrose and lactose will all be expressed as one sugar because they are all disaccharides. In order to tell which sugar you are looking at, it is necessary to treat part of the sample with an enzyme. The enzyme invertase will break down sucrose into dextrose and fructose, whilst the enzyme lactase will break down lactose into dextrose and galactose. By comparing the HPLC results before and after treating the sample with enzyme, it is possible to determine the ratio of the various sugars.

Additionally, in a sample which contains both sucrose and a glucose syrup, it is possible to determine both the sugar spectrum and DE of the glucose syrup, also the amount of sucrose which has been inverted during processing (process inversion), and the total reducing sugars present in the sample.

The advantage of this approach is that in a glucose syrup, there is only dextrose, fructose, maltose, maltotriose, etc and so the analytical system will be set up to look for these types of sugar, and by using water as the carrier, and a refractive index detector, the higher sugars will come out first, with dextrose and fructose coming out last. Therefore, when dextrose and fructose have come out on the trace, you can be sure that **all** the sugars have been eluted.

There are many other HPLC sugar detecting systems which can be used equally well. Because there are several different HPLC systems, using different solvents and different methods of detection, different end results can occur. Additionally, whilst the HPLC equipment will record results to two places of decimals, this is a reflection upon the performance of the equipment, rather than on the ability of the analyst. Ask the question – were the solutions made up to an accuracy of two decimal places? Reporting to one place of decimals is probably acceptable.

Example

A hard-boiled sweet is made from sucrose and glucose syrup. What can HPLC sugar analysis tell? (Despite the above comments on accuracy of reporting HPLC results, the following calculations will be reported to two places of decimals, purely to illustrate the mathematics involved.)

The hard boiled sweet is made into an approximate 15% solution depending upon the analytical technique, using deionised water and the minimal amount of heat (so as to prevent any extra process inversion due to the presence of any acid, e.g. citric acid in the sweet). The solution is divided into two parts. One part is called the 'initial sample'. The other part is treated with the enzyme invertase, and is called the 'inverted sample'.

To make the inverted sample, adjust the pH of the solution to 4.0–5.0. Add some invertase and place in a water bath at 60°C for 1 hour. Gently swirl the solution every fifteen minutes. Cool to room temperature. The inverted solution is now ready for HPLC analysis. Inject the initial sample into the HPLC, followed by the inverted sample.

Initial sample		Inverted sample	
Sugars		Sugars	
Fructose	0.79%	Fructose	43.81%
Dextrose	2.76%	Dextrose	46.81%
Maltose and sucrose	88.95%	Maltose	3.19%
Maltotriose	1.54%	Maltotriose	1.82%
Higher sugars	5.83%	Higher sugars	4.37%

(a) Sucrose

$$= \text{Fructose after inversion} \times 2$$
$$= 43.81 \times 2$$
$$= 87.62\% \text{ DB}$$

(b) Glucose syrup

$$= 100 - \text{Sucrose}$$
$$= 100 - 87.62$$
$$= 12.38\% \text{ DB}$$

(c) Glucose syrup sugar spectrum.

Only use the results of the inverted sample and express the sugars as a percentage, but exclude all the fructose and an equal amount of dextrose from the calculations. The reason for excluding equal amounts of dextrose and fructose is because these two amounts would have originated from the sucrose, and not the glucose syrup.

The (Dextrose − Fructose) + Maltose + Maltotriose + Higher sugars are now expressed as a percentage of the total sugars, viz,

$$= 3.0 + 3.19 + 1.82 + 4.37 = 12.38 \; (N.B. \text{ This total of } 12.38 \text{ corresponds with the amount of glucose syrup found in (b) above.})$$

$$= \frac{3.0 \times 100}{12.38} + \frac{3.19 \times 100}{12.38} + \frac{1.82 \times 100}{12.38} + \frac{4.37 \times 100}{12.38}$$

Therefore, the sugar spectrum of the glucose syrup used was as follows:

Dextrose   = 24.23%
Maltose   = 25.77%
Maltotriose   = 14.70%
Higher sugars = 35.30%

(d) The DE of the glucose syrup.

This can be calculated from the sugar spectrum, by multiplying each sugar % by a factor. The factor for dextrose and fructose is 1.0; maltose is 0.58; maltotriose is 0.395, and higher sugars is 0.20.

$$\begin{aligned}
24.23 \times 1 &= 24.23 \\
25.77 \times 0.58 &= 14.95 \\
14.70 \times 0.395 &= 5.80 \\
35.30 \times 0.20 &= \underline{7.06} \\
& 52.00
\end{aligned}$$

Therefore, DE of the glucose syrup is 52.00.

Based on the DE and the sugar spectrum, a quick calculation would suggest that a 50/50 blend of a 42 DE and a 63 DE would produce a close match.

(e) Percentage of sucrose inverted.

When sucrose is heated with an acid, some or possibly all of the sucrose could be inverted. This is known as 'process inversion', that is when sucrose is broken down into dextrose and fructose during processing. Therefore, if fructose is detected by HPLC, then this would suggest that process inversion has occurred. In this example, the process inversion can be calculated as follows:

$$\% \text{ Sucrose process inversion} = \frac{\text{Initial fructose} \times 2 \times 100}{\text{Total sucrose}}$$

$$= \frac{0.79 \times 2 \times 100}{87.62}$$

$$= 1.80\%$$

N.B. There is always the possibility that if the fructose content is high, that the manufacturer has added 're-work' back to the process. 'Re-work' is material which has been rejected by quality control, and usually consist of deformed sweets which would otherwise be wasted. As the material has already been through the process once, further process inversion will occur during subsequent processing.

(f) Total reducing sugars.

The total reducing sugars in this sweet will come from the reducing sugars in the glucose syrup, and the additional dextrose and fructose resulting from the process inversion of the sucrose.

(a) % Reducing sugars from the glucose syrup.

Since DE is a measure of reducing sugars, and as this sweet contained 12.38% glucose syrup with a DE of 52, it follows that the reducing sugars from the glucose syrup will be

$$\frac{12.38 \times 52}{100} = 6.44\%.$$

(b) % Reducing sugars from the process inversion of the sucrose.

The sweet contained 87.62% of sucrose, of which 1.8% was inverted. Therefore the reducing sugars due to the process inversion will be

$$\frac{1.8 \times 87.62}{100} = 1.58\%.$$

(c) Therefore total reducing sugars will be $6.44 + 1.58 = 8.02\%$.

(g) Molecular weight.

Knowing both the composition and DE of the blend, it is now possible to calculate the molecular weight of the blend, and hence the effect which the blend will have on either the boiling point, the freezing point depression or the osmotic pressure.

$$\text{Molecular weight} = \frac{18{,}000}{\text{DE}}$$

$$= \frac{18{,}000}{60.6}$$

$$= 297$$

Appendix B
Simple calculations

B.1 Introduction

When working with syrups, it is often necessary to adjust the solids or change the sugar spectrum of a syrup. The following examples will explain how this can be carried out.

Please note that in some of the calculations, the numbers have been either rounded up or down, which has resulted in the calculations not being mathematically exact, but the resulting errors are so small that the results are accurate enough for most practical purposes.

B.2 Adjusting syrup solids

The solids of a syrup can be either increased by adding solids or reduced by the addition of water or a liquid containing lesser solids, and one of the easiest ways to calculate the amount to add is by using the Pearson Square technique. Pearson's Square is a very simple but versatile method, and can also be used with concentrated fruit juices, milk, chemical solutions, or any other type of liquid.

Basically, the Pearson's Square is a rectangle, with two diagonals. Each of the four corners of the rectangle, and where the diagonals cross are labelled with a letter, viz:

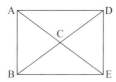

A = Percentage of solids in the original syrup.
B = Percentage of solids in the diluting liquid.
C = Required percentage solids in the final syrup.
D = The difference between B and C.
E = The difference between A and C.

Example 1. Diluting a concentrated syrup with a syrup of lesser solids.

How much syrup, with a solids of 75%, must be added to 900 grams of syrup with a solids of 82% to make a syrup with a solids of 78%?

$$A = 82 \quad D = 78 - 75 = 3$$
$$B = 75 \quad E = 82 - 78 = 4$$
$$C = 78$$

$$\text{Therefore, weight of syrup to add} = \frac{\text{Original weight} \times E}{D}$$
$$= \frac{900 \times 4}{3}$$
$$= 1200 \text{ grams of syrup at 75\% solids}$$

The total amount of syrup at 78% solids would now be $900 + 1200 = 2100$ grams.

Example 2. Increasing the solids of a dilute syrup by the addition of a more concentrated syrup.

How much syrup at 82% solids is required to increase the solids of 800 grams of syrup at 70% solids to 75%?

$$A = 70 \quad D = 7$$
$$B = 82 \quad E = 5$$
$$C = 75$$

$$\text{Therefore, weight of syrup to add} = \frac{\text{Original weight} \times E}{D}$$
$$= \frac{800 \times 5}{7}$$
$$= 571 \text{ grams at 82\% solids}$$

The total amount of syrup at 75% solids would now be $800 + 571 = 1371$ grams.

Example 3. Diluting a concentrated syrup with water to a specific solids content.

How much water would be required to reduce the solids of 750 grams of syrup at 81% solids to 75%?

$$A = 81 \quad D = 75$$
$$B = 0 \quad E = 6$$
$$C = 75$$

$$\text{Therefore, amount of water to add} = \frac{\text{Original weight} \times E}{D}$$
$$= \frac{750 \times 6}{75}$$
$$= 60 \text{ grams of water}$$

The total amount of syrup at 75% solids would now be $750 + 60 = 810$ grams.

Example 4. Increasing the solids of a dextrose syrup by the addition of crystalline dextrose monohydrate.

How much dextrose monohydrate must be added to increase the solids of 500 grams of a dextrose syrup at 60% solids to 75%?

Whilst dextrose monohydrate is a white free-flowing crystalline powder, typically, it will contain about 8.5% moisture, which must be taken into account. The solubility of dextrose at 20°C is only 48%, therefore the solution will have to be heated to increase the dextrose solubility. A dextrose solution containing 75% dextrose will be a supersaturated solution, and upon cooling, it will form a solid lump.

$$A = 60 \qquad D = 16.5$$
$$B = 91.5 \text{ (i.e. } 100 - 8.5 = 91.5) \qquad E = 15$$
$$C = 75$$

Therefore, $\dfrac{500 \times E}{D} = \dfrac{500 \times 15}{16.5}$

$$= 454.5 \text{ grams of anhydrous dextrose monohydrate is required.}$$

The above calculation is based on using anhydrous dextrose, but commercially dextrose contains 8.5% moisture, which must be taken into account, therefore the 454.5 grams must be increased by a further 8.5%, that is by 81.8 grams, to give a total dextrose monohydrate addition of 536.3 grams. The total amount of syrup will now be 1036.3 grams (i.e. 500 + 536.3 grams).

Example 5. Determining the final solids of a syrup blend made from syrups with different solids.

500 grams of syrup at a solids of 71% is mixed with 250 grams of syrup at a solids of 82%. What will be the final solids?

Calculations:

$$\dfrac{\text{Weight of syrup} \times \text{Syrup solids}}{100} = \text{Syrup solids}$$

Therefore, $\dfrac{500 \times 71}{100} = 355$ grams of syrup solids

and $\dfrac{250 \times 82}{100} = 205$ grams of syrup solids.

Therefore, total weight of syrup solids $= 355 + 205 = 560$ grams.
Total weight of syrup mix $= 500 + 250 = 750$ grams. Expressing the total syrup solids as a percentage of the total syrup mix, viz:

$$\dfrac{560 \times 100}{750} = 74.7\% \text{ solids}$$

B.3 Altering the sugar spectra of a glucose syrup blend

Example 1. How to calculate the sugar spectrum and solids of a blend made from different syrups.

A syrup blend is made up of three different syrups. What will be the sugar spectrum and solids of the blend?

A blend is made up of three syrups:

(a) 200 grams of a maltose syrup (70% maltose) at 80% solids.
(b) 300 grams of high-fructose glucose syrup, containing 42% fructose at 71% solids. Usually referred to as HFGS 42.
(c) 350 grams of a 42 DE acid converted glucose syrup (conventional Confectioners Glucose Syrup) at 82% solids.

Calculations

1. (a) Convert the different amounts of 'as is' syrup into dry syrup solids, and then (b) express each amount of syrup solids as a percentage of the total blend.
   (a) Dry syrup solids.

$$\text{Dry syrup solids} = \frac{\text{Syrup solids} \times \text{weight of syrup}}{100}$$

$$\text{\% Maltose solids} = \frac{80 \times 200}{100} = 160 \text{ grams}$$

$$\text{\% HFGS 42} = \frac{71 \times 300}{100} = 213 \text{ grams}$$

$$\text{\% 42 DE} = \frac{82 \times 350}{100} = 287 \text{ grams}$$

Therefore, total amount of syrup solids will be $160 + 213 + 287 = 660$ grams.
   (b) % Dry syrup solids.

$$\text{\% Maltose syrup} = \frac{160 \times 100}{660} = 24.2\%$$

$$\text{\% HFGS 42} = \frac{213 \times 100}{660} = 32.3\%$$

$$\text{\% 42 DE} = \frac{287 \times 100}{660} = 43.5\%$$

2. Look at the sugar spectrum for each syrup, which can be found in the syrup specification provided by the supplier. The percentage of each sugar given in the syrup specification will be on a dry basis, and will add up to 100% for each syrup.

A chart is now constructed so that each sugar in each syrup can be multiplied by the percentage present in the blend, viz:

	Maltose (24.2%)	HFGS 42 (32.3%)	42 DE glucose (43.5%)	Sugar spectrum of blend
Fructose	No fructose in maltose syrup	42% of 32.3 = 13.6 grams	No fructose in 42 DE	13.6%
Dextrose	2% of 24.3 = 0.5 grams	52% of 32.3 = 16.8 grams	19% of 43.5 = 8.3 grams	25.6%
Maltose	70% of 24.3 = 17.0 grams	4% of 32.3 = 1.3 grams	14% of 43.5 = 6.1 grams	24.4%
Maltotriose	20.3% of 24.0 = 4.9 grams	1% of 32.3 = 0.3 grams	11% of 43.5 = 4.8 grams	10.0%
Higher sugars	8% of 24.3 = 1.9 grams	1% of 32.3 = 0.3 grams	56% of 43.5 = 24.3 grams	26.5%

As the total amount of syrup used was $200 + 300 + 350 = 850$ grams, and the total solids are 660 grams, then solids of the blend would be as follows:

$$\frac{660 \times 100}{850} = 77.6\%$$

And the sugar spectrum of the blend would be as follows:

$$
\begin{aligned}
\text{Fructose} &= 13.6\% \\
\text{Dextrose} &= 25.6\% \\
\text{Maltose} &= 24.4\% \\
\text{Maltotriose} &= 10.0\% \\
\text{Higher sugars} &= 26.5\%
\end{aligned}
$$

Example 2. How to make a syrup blend with a specific sugar spectrum.

When making a syrup blend with a specific sugar spectrum, there are several points to bear in mind.

(a) Is the blend to match an existing sugar spectrum or fermentability profile?
(b) Has the blend got to impart specific properties, such as sweetness, or viscosity to the end product?

(c) What syrups and equipment are available for making the blend?
(d) Keep the syrup blend as simple as possible. Ideally, try to match two or three of the major sugars, using two syrups to provide the majority of the required sugars, with the possibility of using a third syrup for 'final adjustments' to meet the required sugar spectrum.
(e) It can be difficult to obtain an exact match for a sugar spectrum when making a blend. The reason for this is the variability of the sugar spectrum of the original syrups – commercially, the sugar spectra of most syrups will have a spread of about four percent. Additionally, when making a blend on a large scale, there is not always the same degree of accuracy available in the measuring of ingredients, as there is in an analytical laboratory. It is for these reasons that the final sugar spectrum of a blend will often be a very close match, with a possible spread of about four percent or less for each sugar.
- Sweetness will come from fructose containing syrups.
- Viscosity will come from low DE syrups containing lots of higher sugars.
- Maltose will come mainly from maltose-rich syrups.
- Whilst dextrose will come from high DE syrups, a 63 DE syrup could also be used in some blends where both maltose and dextrose are required.

For example, a customer requires a syrup with the following sugar spectrum, viz:

Fructose	= 18%
Dextrose	= 29%
Maltose	= 37%
Maltotriose	= 10%
Higher sugars	= 6%

What syrups will be required and what syrups are available?

This sort of blend will require the use of syrups containing fructose, maltose and dextrose, such as HFGS 42 for the fructose and Maltose 70 for the maltose. Some of the dextrose would come from the HFGS 42, with additional dextrose coming from 95 DE syrup. Whilst both Maltose 50 and 40 DE syrup contain both dextrose and maltose, they are unsuitable for this blend because their higher sugars content are too high.

Typical sugar spectra for HFGS 42, Maltose 70 and 95 DE syrups are as follows:

	HFGS 42	Maltose 70	95 DE
Fructose	42%	No fructose	No fructose
Dextrose	52%	2%	93%
Maltose	4%	70%	5%
Maltotriose	1%	20%	1%
Higher sugars	1%	8%	1%

1. Required 18% fructose.

   HFGS 42 contains 42% fructose, therefore 18% of fructose will come from 43 parts of HFGS 42. With this fructose there will also be 22.4 parts of dextrose.

2. Required 37% maltose.

   Maltose 70 contains 70% maltose, therefore 37% of maltose will come from 53 parts of Maltose 70, but maltose will also come from both the 42 HFGS and the 95 DE syrup. These additional sources of maltose will therefore have to be taken into account.

3. Required dextrose 29%.

   As already mentioned, 22.4 parts of the required 29 parts of dextrose will come from the HFGS 42, but 1 part will come from the Maltose 70 leaving about 5.6 parts to come from the 95 DE syrup. Now 6 parts of 95 DE will provide the 5.6 parts of extra dextrose. This means that the amount of Maltose 70 syrup to use would be 100 − 43 (HFGS 42) − 6 (95 DE syrup) = 51.

	43% of HFGS 42	51% of Maltose 70	6% of 95 DE Syrup	**Sugar spectrum of blend**
Fructose	43% of 42 = 18.1	No fructose	No fructose	**18.1%**
Dextrose	43% of 52 = 22.4	51% of 2 = 1.0	6% of 93 = 5.6	**29.0%**
Maltose	43% of 4 = 1.7	51% of 70 = 35.7	6% of 5 = 0.3	**37.7%**
Maltotriose	43% of 1 = 0.4	51% of 20 = 10.0	6% of 1 = 0.05	**10.5%**
Higher sugars	43% of 1 = 0.4	51% of 8 = 4.1	6% of 1 = 0.05	**4.5%**

B.4 How to calculate equivalent sweetness values?

Equivalent sweetness values are used when one sweetener is being replaced by another, but keeping the overall sweetness unchanged.

Example

How much fructose will give the same sweetness as 100 grams of sucrose?

The sweetness value of sucrose is 100, and for fructose, it is 150.

$$\text{Equivalent sweetness} = \frac{\text{Sucrose sweetness}}{\text{Fructose sweetness}} = \frac{100}{150} = 0.67$$

Therefore, only 67 grams of fructose will give the same sweetness as 100 grams of sucrose. Because less carbohydrate is used to give the same sweetness, it follows that less calories will be present and therefore it can be legally claimed that there has been a calorie reduction.

The sweetness values of different glucose syrups and artificial sweeteners can be found under Sugar Data.

B.5 Relationship between density, volume and weight of glucose syrups

Some glucose syrup users prefer to measure the amount of syrup which they use by volume, whilst others prefer to weigh out the syrup. So how do you convert volume to weight, and weight to volume?

Most glucose syrup specifications will state the density of the syrup at a specific temperature. Typically, the density of a commercially supplied syrup will vary from 1.35 to 1.46, at 20°C, depending upon both the type and solids of the syrup.

Example

A 42 DE glucose syrup has a density of 1.421.

1. What weight of syrup should be taken to give 100 litres?

$$\text{Density} = \frac{\text{Mass}}{\text{Volume}}$$

If the density of the syrup is 1.421, and the volume is 100 litres, therefore by substituting these numbers, we have

$$1.421 = \frac{\text{Mass}}{100}$$

Therefore, the weight to use would be as follows:

$$= 1.421 \times 100 = 142.1 \text{ Kg}$$

2. How many litres will be required to give 100 Kg?

$$1.421 = \frac{100}{\text{Volume}}$$

Therefore, volume $= 100 \times 1.421$

$= 70.3$ litres

B.6 How much syrup is required to obtain a given weight of syrup solids?

$$\text{Weight of syrup to take} = \frac{100 \times \text{Required weight of syrup solids}}{\text{Syrup solid}}$$

B.7 Brix, RI and RI Solids, % Solids and Baumé

Since these five terms cause a lot of confusion with syrup users, here is a brief clarification:

(a) **Brix**. This term is unfortunately used very loosely within the food industry, which has lead to considerable confusion when discussing solids. Brix is only applicable to sucrose solutions and is used by the sucrose industry to measure the amount of sucrose solids in a sucrose solution using a Brix hydrometer, which is calibrated to read % w/w of sucrose solids at 20°C. Because of the viscosity of sucrose syrups, the Brix measurements are usually made at an elevated temperature and then corrected back to 20°C.

Despite the Brix hydrometer measuring % w/w, the correct term for reporting results is 'degrees Brix' and not '% Brix'.

(b) **RI**. The RI is a measure of the refractive index of a liquid, in this case a syrup, as measured using a refractometer. Knowing the refractive index of a syrup, the solids of the syrup can then be determined from tables.

(c) **RI Solids**. The RI solids refers to the solids in a solution which have been determined using a refractometer. Most refractometers are calibrated to read sucrose, and since different sugars have different refractive indices, it is necessary to use tables to convert the RI reading into glucose syrup solids.

(d) **% Solids (or % Dry Substance)**. This is a measure of the solids in a syrup as determined by either drying the syrup in an oven or by a Karl Fischer titration and is the true dry substance in a syrup.

(e) **Baumé**. Baumé is used by the glucose industry to measure the specific gravity of a glucose syrup, and by reference to tables the solids can be obtained.

$$\text{Specific gravity} = \frac{145}{145 - \text{degrees Baumé}}$$

The glucose industry normally measures the Baumé of a glucose syrup using a hydrometer with a 145 modulus and at a temperature of 60°F (15.6°C). It is important when reporting Baumé readings to quote both the modulus of the hydrometer and the temperature at which the readings were taken. (The modulus is a constant used to convert specific gravity into degrees Baumé).

Because of the viscosity of glucose syrups, the Baumé measurements are often made at a higher temperature. When measurements are made at a higher temperature, the Baumé reading is normally reported with the temperature details like 'Baumé at 140°F/60°F'. This means that the temperature of the glucose syrup was 140°F when the reading was made, and related back to water at 60°F.

The confusion arises when different industries talk about solids which have been determined by different methods. The sucrose industry use degrees Brix, whilst the glucose industry use RI solids or % solids. When replacing sucrose with a glucose syrup or vice versa, to avoid any discrepancies, it is important to always use % dry solids.

B.8 Recipe costings

Whenever a change is made to a recipe, it is advisable to check the effect which that change will have on the cost of the product. Some changes might be a necessity – an ingredient is no longer available; the change makes production easier or there is a cost saving to be achieved. Whatever the reason, always check the effect on the final cost, and always compare like with like. In the case of glucose syrup, because it contains water, it is necessary when comparing the cost of a recipe change to always compare the price per kilogram of ingredient on a dry for dry basis – dry sucrose with dry syrup solids. The following example is for a typical ice cream, where part of the sucrose has been replaced with a blend of HFGS containing 42% fructose, and a 42 DE glucose syrup.

Ingredients	Control		Syrup trial	
	% As is	% Dry basis	% As is	% Dry basis
SMP*	2.20	2.20	2.20	2.20
Whey powder	2.20	2.20	2.20	2.20
Vegetable fat	6.70	6.70	6.70	6.70
Sucrose	13.30	13.30	4.40	4.40
HFGS 42 – 71.0% solids	–	–	6.19	4.40
42 DE Glucose Syrup – 81.0% solids	–	–	5.45	4.40
Stabiliser	0.50	0.50	0.50	0.50
Emulsifier	6.70	6.70	6.70	6.70
Water	68.40	68.40	65.66	68.4
Flavour and colour	As required	As required	As required	As required

*Skimmed milk powder.

Since both HFGS and 42 DE glucose syrup contain water, it is necessary to take this water into consideration in the calculations:

(a) The dry solids addition from both the HFGS (42 % fructose) and the 42 DE glucose syrups.
(b) The amount of water which the syrups will contribute to the recipe. This extra water must therefore be subtracted from the amount of water in the original recipe, so as to keep the total solids of both the control and trial recipes the same and are therefore comparing like with like.

The cost of each recipe can then be worked out and compared as follows:

Ingredients	Price per DS kilogram	% Inclusion DB	Cost of control recipe	Cost of trial recipe
SMP		× 2.20 =		
Whey powder		× 2.20 =		
Vegetable fat		× 6.70 =		
Sucrose		× 13.30 =		_____
Sucrose		× 4.40 =	_____	
HFGS (42 FX)		× 4.40 =	_____	
42 DE Glucose Syrup		× 4.40 =	_____	
Stabiliser		× 0.50 =		
Water		× 66.80 =		
Flavour		× 0.06 =		
		Total cost of mix		

B.9 Colligative properties

Colligative properties are physical properties of a substance, which depend upon the concentration of molecules in that substance, rather than the weight. These physical properties influence boiling point, freezing point depression, and osmotic pressure, also vapour pressure. In summary, when a solute is added to a solvent,

- the boiling point is raised,
- the osmotic pressure is increased,
- the freezing point is lowered,
- the vapour pressure is lowered.

B.9.1 How to calculate boiling point elevation

Water, at sea level, that is at atmospheric pressure, boils at 100°C, but if the water contains a substance such as a sugar dissolved in the water, then the solution will boil at a higher temperature.

$$\text{Boiling point elevation} = \frac{\text{Boiling point constant} \times \text{Weight of solute in grams} \times 1}{\text{Average molecular weight} \times \text{Weight of solvent in kilogram}}$$

The boiling point constant (or Ebulliscopic constant) for water is 0.52 K·Kg/mol (this constant is obtainable from reference tables).

Example

What will be the boiling point of a solution containing 20% weight/weight of 42 DE glucose syrup solids, compared to a similar weight of sucrose or a 63 DE syrup solids?

Boiling point constant	= 0.52.
Weight of solute (i.e. syrup solids)	= 200 grams
Average molecular weight of 42 DE Glucose Syrup	= 429
Average molecular weight of 63 DE Glucose Syrup	= 285
Average molecular weight of sucrose	= 342
Weight of solvent in kilogram	= 0.800 Kg

1. 42 DE Glucose Syrup boiling point elevation $= \dfrac{0.52 \times 200}{429 \times 0.800} = \dfrac{104}{342} = 0.30°C$

2. 63 DE Glucose Syrup boiling point elevation $= \dfrac{0.52 \times 200}{285 \times 0.800} = \dfrac{104}{228} = 0.46°C$

3. Sucrose boiling point elevation $= \dfrac{0.52 \times 200}{342 \times 0.800} = \dfrac{104}{274} = 0.38°C$

Therefore, the boiling point of the 42 DE syrup solution will be 100.30°C. The boiling point of the 63 DE syrup solution will be 100.46°C, whilst the boiling point of the sucrose solution will be 100.38°C.

This change in boiling point, due to using different sweeteners, is important in the production of high boiled sweets, and should be allowed if the type of syrup is changed or the ratio of sucrose to glucose syrup is changed.

Whilst theoretically, the elevation of boiling point can be calculated for low solids using the above formula, it is not suitable for very high solids. When syrups are heated, the solids increase, so does the viscosity of the syrup, and this increase in viscosity slows down the rate at which the moisture leaves the syrup. It is for this reason that calculating the change of boiling point at very high solids is not particularly accurate, therefore boiling to a specific temperature is one way of reaching a final solids end point, but it is nevertheless important to check the final solids using a refractometer, and in the case of a high boiled sweet by a Karl Fischer titration. Determining the boiling point experimentally has the advantage of sorting out any problems with the equipment.

In the above calculations, it has been assumed that an open pan at sea level is used for the cooking, but it is important to remember that boiling point is also affected by pressure. This change in pressure can be due to the difference in altitude above sea level, or simply by deliberately boiling under vacuum. In both cases, the effect is the same, namely that the boiling action occurs at a lower temperature.

For every 500 feet (152 metres) above sea level, the boiling point will be lowered by 0.6°C (1.0°F). Again, these changes in boiling point should be taken into consideration.

Boiling under vacuum is the normal practice in most modern production units and is used to good effect when evaporating syrups, thereby allowing the syrup solids to be increased, but at the same time preventing the syrup from getting hot, which would result in the syrup becoming coloured. Additionally, less heat is required, thereby offering a cost saving.

B.9.2 How to calculate freezing point depression

$$\text{Freezing point depression} = \frac{\text{Cryoscopic constant} \times \text{Weight of solute} \times 100}{\text{Average molecular weight} \times \text{Weight of solvent}}$$

The cryoscopic constant for 10 grams of water is 186 K·Kg/mol for 100 grams of water, it will be 18.6 K·Kg/mol, and for 1000 grams (1 Kg), it will be 1.86 K·Kg/mol. (This constant is obtainable from reference tables.)

Example

What is the freezing point depression of a solution containing 10% w/w of 42 DE glucose syrup solids compared to a similar weight of sucrose or dextrose?

Cryoscopic constant	= 18.6
Weight of solute (i.e. syrup solids)	= 10 grams
Average molecular weight of 42 DE Glucose Syrup Solids	= 429
Average molecular weight of sucrose	= 342
Average molecular weight of dextrose	= 180
Weight of solvent (i.e. water minus weight of syrup solids)	= 90 grams

1. 42 DE syrup freezing point depression $= \dfrac{18.6 \times 10 \times 100}{429 \times 90} = \dfrac{18{,}600}{38{,}610} = 0.48°C$

2. Sucrose Freezing Point Depression $= \dfrac{18.6 \times 10 \times 100}{342 \times 90} = \dfrac{18{,}600}{30{,}780} = 0.60°C$

3. Dextrose freezing point depression $= \dfrac{18.6 \times 10 \times 100}{180 \times 90} = \dfrac{18{,}600}{16{,}200} = 1.15°C$

Since water freezes at 0°C, the 42 DE syrup solution will freeze at −0.48°C. The sucrose will depress the freezing point to −0.60°C, whilst the dextrose has the greatest effect on depressing the freezing point to −1.15°C. It is for this reason that dextrose is used in soft-serve ice creams. By using different quantities and different types of sweetener, it is possible to alter both the texture and sweetness of a frozen product.

B.9.3 How to calculate osmotic pressure

The units for expressing osmotic pressure can be either atmospheres or milliOsmols. Atmospheres are usually used for industrial applications, whilst milliOsmols are usually used for medical related applications.

Whilst 22.4 atmospheres equals 1000 milliOsmols (mOsm), differences will occur when one set of figures is compared with the another, if one solution is a molar solution and the other a molal solution. Always compare molar with molar and molal with molal.

A molar solution contains grams of solute per litre of solution, but a molal solution will contain grams of solute per kilogram of solution. Generally, a one molal solution will be more dilute than a molar solution, and will have a lower osmotic pressure.

(a) Osmotic pressure calculated as atmospheres of pressure (based on a molar solution)

$$\text{Osmotic pressure} = \frac{\text{Weight of solute in grams} \times \text{Gas constant} \times \text{Absolute temperature}}{\text{Molecular weight of solute} \times \text{Volume}}$$

Example

What is the osmotic pressure of a 10% sucrose solution, compared with the osmotic pressure of a 10% 28 DE syrup and a 10% dextrose solution at 27°C.

$$\begin{aligned}
\text{Molecular weights – sucrose} &= 342 \\
\text{Molecular weights – 28 DE syrup} &= 643 \\
\text{Molecular weights – Dextrose} &= 180
\end{aligned}$$

$$\begin{aligned}
\text{Absolute temperature} &= °C + 273 \\
&= 27 + 273 \\
&= 300 \text{ degrees absolute}
\end{aligned}$$

Gas constant = 0.082 J/(K·mol) (joules per Kelvin per mol)

Volume = 1.0 litres

1. $10\% \text{ Sucrose} = \dfrac{100 \times 0.082 \times 300}{342 \times 1} = \dfrac{2460}{342}$
   = 7.19 atmospheres or (32.0 mOsm)

2. $10\% \text{ 28 DE syrup} = \dfrac{100 \times 0.082 \times 300}{643 \times 1} = \dfrac{2460}{643}$
   = 3.83 atmospheres or (17.0 mOsm)

3. 10% Dextrose $= \dfrac{100 \times 0.082 \times 300}{180 \times 1} = \dfrac{2460}{180}$

$= 13.67$ atmospheres or
(60.7 mOsm)

This means that a 10% dextrose solution would penetrate a fruit, quicker than a 10% sucrose solution, because it has a higher osmotic pressure.

N.B. If the sucrose solution is acidic, then the sucrose will be inverted to give dextrose and fructose. Both of these sugars have a lower molecular weight than sucrose, and therefore the osmotic pressure of the 'inverted sucrose' solution will be greatly increased. This increase in osmotic pressure will affect both the process and end product.

(b) Osmotic pressure calculated in milliOsmols (based on a molal solution)

$$\text{Osmolality} = KnM$$

where $K =$ osmotic coefficient (see note 1)
$N =$ Number of particles into which the molecule dissociates (see note 2)
$M =$ Molality (see note 3)

Notes

1. The osmotic coefficient is taken as being one when there is total dissociation with no water binding.
2. Dissociation is the process whereby a molecule is split into simpler fragments which may be smaller molecules, atoms, free molecules or ions. The dissociation number therefore is the number of smaller molecules into which the molecule split. Because sugars do not dissociate, the number of particles for dextrose, sucrose, fructose, etc, would be one. For substances which dissociate such as sodium chloride (NaCl), it would be two, namely sodium and chloride. For potassium sulphate (K_2SO_4), it would be three, namely two molecules of potassium and one of sulphate.
3. A molal solution contains the gram molecular weight of a substance in 1000 grams of water to give a weight/weight solution. Therefore, a molal solution of dextrose would contain 180 grams of dextrose dissolved in 1000 grams of water.

The above osmolality formula can be simplified as follows:

$$\text{Osmolality (in mOsm)} = \dfrac{\text{Weight of solids in 1000 grams} \times 1000}{\text{Molecular weight} \times \text{Dissociation number}}$$

Using the above formula, the osmotic pressures in milliOsmols for a 10% solution of sucrose, 28 DE glucose syrup and dextrose solutions would be as follows:

(a) 10% Sucrose $= \dfrac{10 \times 1000}{342 \times 1} = 29.2$ mOsm

(b) 10% 28 DE syrup = $\dfrac{10 \times 1000}{643 \times 1}$ = 15.6 mOsm

(c) 10% Dextrose = $\dfrac{10 \times 1000}{180 \times 1}$ = 55.6 mOsm

These two examples demonstrate how the osmotic pressure will be different, depending on whether a molar or molal solution is used for its determination.

Acknowledgement is made to Advance Instruments, Inc., USA and their UK supplier Vitech Scientific Ltd for their advice in preparing these notes on the determination of osmotic pressure.

Appendix C
Sugars data

C.1 Approximate % sugar spectra of different glucose syrups

Product	DE	Fructose	Dextrose	Maltose	Maltotriose	Higher sugars
Maltodextrin	5	–	0.9	0.9	1.0	97.2
	10	–	0.6	2.8	4.4	92.2
	15	–	1.3	4.1	6.0	88.6
	18	–	1.5	6.0	8.3	84.2
Glucose syrup	28	–	2.0	10.0	16.0	72.0
	35	–	11.0	12.0	14.0	63.0
	42	–	19.0	14.0	11.0	56.0
	50	–	26.0	17.0	18.0	39.0
	55	–	30.0	24.0	10.0	36.0
	63	–	34.0	33.0	10.0	23.0
	95	–	93.0	5.0	1.0	1.0
Dextrose	100	–	100.0	–	–	–
Maltose 70%	–	–	2.0	70.0	19.0	9.0
Maltose 50%	–	–	5.0	50.0	20.0	25.0
Maltose 41%	–	–	8.0	41.0	15.0	36.0
Maltose 25%	–	–	5.0	28.0	25.0	42.0
HFGS 80%	–	80.0	18.0	1.0	0.5	0.5
HFGS 55%	–	55.0	41.0	2.0	1.0	1.0
HFGS 42%	–	42.0	52.0	4.0	1.0	1.0

Because of the different ways in which glucose syrups can be made using acid, acid enzyme, enzyme enzyme or by blending, the above sugar spectra should only be considered as indicative.

Where a manufacturer of a syrup quotes a DE, it generally refers to a mid-point within a DE range. For example, the DE range of a 42 DE syrup is usually 40.0–44.0 DE.

C.2 Theoretical molecular weights

Product	Molecular weight
Sucrose	342
Maltodextrin – 5 DE	3600
Maltodextrin – 10 DE	1800
Maltodextrin – 15 DE	1200
Maltodextrin – 18 DE	1000
Glucose syrup – 20 DE	900
Glucose syrup – 25 DE	720
Glucose syrup – 28 DE	643
Glucose syrup – 33 DE	545
Glucose syrup – 37 DE	486
Glucose syrup – 42 DE	429
Glucose syrup – 49 DE	367
Glucose syrup – 55 DE	327
Glucose syrup – 63 DE	286
Dextrose anhydrous – 100 DE	180
Dextrose monohydrate – 100 DE	198
High maltose syrup – 70% maltose	380
High maltose syrup – 50% maltose	448
High maltose syrup – 41% maltose	470
High maltose syrup – 25% maltose	511
Maltose	342
Maltotriose	504
Maltotetraose	667
Fructose	180
High-fructose glucose syrup – 80% FX	182
High-fructose glucose syrup – 55% FX	185
High-fructose glucose syrup – 42% FX	190
Invert sugar solids	180
Lactose	342
Galactose	180
Maltulose	342
Trehalose	342
Erythritol	122
Isomalt	344
Sorbitol	182
Lactitol	344
Lactulose	342
Mannitol	182
Maltitol	344
Xylitol	152
Polydextrose	Over 1000
Glycerol (glycerine)	92
Ethanol	46

Taking the molecular weight of starch to be 18,000, then the approximate molecular weight of a glucose syrup or a maltodextrin can be calculated by dividing 18,000 by the DE of the product:

$$\text{Molecular weight} = \frac{18,000}{\text{DE}}$$

The molecular weight of a sugar can also be determined using a cryoscope.

When one mole of a non-ionic substance is added to one kilogram of water, then the freezing point will be depressed by 186°C.

C.3 Sweetness values

Since there are no chemical tests for sweetness, the following values are subjective values, and do not take into account any synergistic effects with other ingredients.

Sugar	Sweetness
Sucrose	100
Maltodextrin – 5 DE	6
Maltodextrin – 10 DE	11
Maltodextrin – 15 DE	17
Maltodextrin – 19 DE	20
Glucose syrup solids – 20 DE	23
Glucose syrup solids – 25 DE	28
Glucose syrup – 28 DE	30
Glucose syrup – 33 DE	40
Glucose syrup – 37 DE	45
Glucose syrup – 42 DE	50
Glucose syrup – 49 DE	56
Glucose syrup – 55 DE	62
Glucose syrup – 63 DE	70
Dextrose – 100 DE	80
High maltose syrup – 70% maltose	50
High maltose syrup – 50% maltose	50
High maltose syrup – 41% maltose	50
High maltose syrup – 25% maltose	50
Maltose	50
Fructose	150 (120–170)
High-fructose glucose syrup – 90% FX	135 (110–160)
High-fructose glucose syrup – 55% FX	105 (100–110)
High-fructose glucose syrup – 42% FX	95 (90–100)
Invert sugar solids	100
Lactose	40
Galactose	30

Trehalose	45
Erythritol	70
Isomalt	55
Isomaltulose	48
Lactitol	40
Maltitol	90
Mannitol	70
Sorbitol	50
Xylitol	100
Glycerine	80
Polydextrose	3
Alitame	200,000
Acesulfame K	20,000
Aspartame	18,000
Cyclamate	3,000
Glycyrrhizin	7,500
Hernandulicin	100,000
Monellin	175,000
Neohesperidin DC	180,000
Neotame	800,000
Saccharin	30,000
Stevioside	30,000
Sucralose	60,000
Tagatose	90
Thaumatin	250,000
Trealose	45

C.4 Approximate sugar spectra of domestic sweeteners

Because there are many commercial variations of these products, the following sugar spectra must only be considered as indicative, and not absolute sugar spectra. These products are usually used in a recipe because of the flavour which they impart to the end product, rather than for their sweetness contribution. They are also used in foods because of their humectant properties, which is due to the dextrose and fructose content. Because these products contain dextrose and fructose, they will have the same properties as any other reducing sugar, namely reacting with proteins to give a brown colour – the well known Maillard reaction.

Golden Syrup

Fructose	27%
Dextrose	31%
Sucrose	38%
Other sugars	4%

Honey

N.B. As honey is a naturally occurring sweetener, there will be slight variations in its sugar composition. Generally, there is more fructose than dextrose in honey. This predominance of fructose increases the sweetness of honey.

Fructose	49%
Dextrose	40%
Sucrose	1%
Other sugars	10%

Invert

N.B. There are several different grades of invert, depending upon how far the sucrose has been inverted. The following sugar analyses are for a typical fully inverted and partially inverted syrup. Because of the synergy between the different proportions of sugars in an invert, the perceived sweetness of an invert can vary.

Fully inverted

Fructose	47%
Dextrose	50%
Sucrose	–
Other sugars	3%

Partially inverted

Fructose	24%
Dextrose	25%
Sucrose	48%
Other sugars	3%

Molasses

Fructose	11%
Dextrose	9%
Sucrose	63%
Other sugars	17%

N.B. The sugar composition of molasses will vary, depending upon the origin.

Treacle

Fructose	22%
Dextrose	23%
Sucrose	43%
Other sugars	12%

Maple syrup

Produced from the sap of the maple tree (*Acer saccharum*). The sap is 99% sucrose, with the remaining 1% being dextrose and fructose. The sap has a solids content of about 2%, and has to be evaporated to about 66–67%. The characteristic flavour and colour of maple syrup is due to this evaporation process.

Date syrup

The sugar spectrum of a date syrup will vary depending upon the variety of date palm and country of origin. The following is only an approximation.

Fructose	47%
Dextrose	47%
Sucrose	6%

Date palm syrup

Like maple syrup, date palm syrup is produced from the sap of the palm tree (*Phoenix dactyifera*) and is composed mainly of sucrose with about 0.5% of dextrose and fructose. The average solids contents of the sap is about 10%. It is evaporated and sold as either a syrup or a crystalline solid, called 'jaggery'. The production of date palm syrup is frequently not mainstream, but more a cottage industry.

C.5 Typical particle size for different grades of sucrose

Coarse	800–2200 microns
Granulated	450–600 microns
Extra fine	200–600 microns
Caster sugar	150–450 microns
Icing sugar	10–35 microns

The particle size of brown sucrose is generally less than that of granulated sucrose.

The particle size or fineness of powder sucrose is sometimes graded according to the mesh size of the sieve (or bolting cloth). The larger the mesh number, the finer the powder, for example '10X' is finer than '4X'. Bolting cloth is also used for grading flour.

In some particle size specifications, the initials MA and CV might be used. MA refers to the mean aperture size of the sieve, and CV refers to the coefficient of variation. The smaller the CV means that there will be fewer particles present, large or small, which will be outside the specified particle size range.

C.6 Melting points

Product	Melting point (°C)
Sucrose	190
Dextrose monohydrate	83
Dextrose anhydrous	149
Fructose	102
Maltose	160–165
Trehalose	203
Lactose	223
Erythritol	121
Isomalt	145–150
Maltitol	150
Mannitol	165
Sorbitol	97
Xylitol	94

C.7 Glass transition temperatures – T_g values

Product	T_g value (°C)
Sucrose	52
Dextrose	31
Fructose	13
Maltose	43
Lactose	101
Trehalose	79
Erythritol	−42
Isomalt	34
Maltitol	47
Mannitol	−39
Sorbitol	−5
Xylitol	−22
Polydextrose	110

C.8 Solubility – grams per 100 mls.

C.8.1 In water

Sugar	20°C	55°C
Sucrose	67%	73%
Dextrose	48%	73%
Fructose	79%	88%
Maltose	44%	60%

Where there is a mixture of sugars in a solution, the above solubilities do not apply. As a general rule, the solubility of a sugar in a mixture is greater than that of the individual sugar on its own.

C.8.2 In 80% alcohol at 20°C

Sucrose	10 grams per 100 ml
Dextrose	45 grams per 100 ml
Fructose	27 grams per 100 ml

As a general rule, a low DE glucose syrup will be less soluble in alcohol, than a high DE syrup. To increase the alcohol solubility of a low DE syrup, it will be necessary to add water to the alcohol, that is to dilute the alcohol.

Appendix D
Tables

The following tables have been based on the more frequent requests from application chemists, engineers and customers for information which is not readily available. It has been included in the appendix so as to avoid breaking the continuity of the text. In some cases for practical applications, the figures have been rounded up or down to give an approximation.

D.1 Temperature conversion

To convert from °F to °C – find the temperature to be converted in the middle column and read off the conversion in the left-hand column.

To convert from °C to °F – find the temperature to be converted in the middle column and read off the conversion in the right-hand column.

In both cases, temperatures containing a fraction of a degree may be obtained by extrapolation using the nearest temperature and the appropriate fraction (see the end of this table).

General formulae:

$$°F \text{ to } °C \quad C = 5/9\,(F - 32)$$
$$°C \text{ to } °F \quad F = 9/5\,C + 32$$

Table 1 Temperature conversion: Fahrenheit to Celsius and vice versa.

°C	°F	°C		°F
−23.3	−10	14.0	−17.8	0	32.0
−22.8	−9	15.8	−17.2	1	33.8
−22.2	−8	17.6	−16.7	2	35.6
−21.7	−7	19.4	−16.1	3	37.4
−21.1	−6	21.2	−15.6	4	39.2
−20.6	−5	23.0	−15.0	5	41.0
−20.0	−4	24.8	−14.4	6	42.8
−19.4	−3	26.6	−13.9	7	44.6
−18.9	−2	28.4	−13.3	8	46.4
−18.3	−1	30.2	−12.8	9	48.2

(*Continued*)

Table 1 (*Continued*)

°C		°F	°C		°F
−12.2	10	50.0	10.6	51	123.8
−11.7	11	51.8	11.1	52	125.6
−11.1	12	53.6	11.7	53	127.4
−10.6	13	55.4	12.2	54	129.2
−10.0	14	57.2	12.8	55	131.0
−9.4	15	59.0	13.3	56	132.8
−8.9	16	60.8	13.9	57	134.6
−8.3	17	62.6	14.4	58	136.4
−7.8	18	64.4	15.0	59	138.2
−7.2	19	66.2	15.6	60	140.0
−6.7	20	68.0	16.1	61	141.8
−6.1	21	69.8	16.7	62	143.6
−5.6	22	71.6	17.2	63	145.4
−5.0	23	73.4	17.8	64	147.2
−4.4	24	75.2	18.3	65	149.0
−3.9	25	77.0	18.9	66	150.8
−3.3	26	78.8	19.4	67	152.6
−2.8	27	80.6	20.0	68	154.4
−2.2	28	82.4	20.6	69	156.2
−1.7	29	84.2	21.1	70	158.0
−1.1	30	86.0	21.7	71	159.8
−0.6	31	87.8	22.2	72	161.6
0	32	89.6	22.8	73	163.4
0.6	33	91.4	23.3	74	165.2
1.1	34	93.2	23.9	75	167.0
1.7	35	95.0	24.4	76	168.8
2.2	36	96.8	25.0	77	170.6
2.8	37	98.6	25.6	78	172.4
3.3	38	100.4	26.1	79	174.2
3.9	39	102.2	26.7	80	176.0
4.4	40	104.0	27.2	81	177.8
5.0	41	105.8	27.8	82	179.6
5.6	42	107.6	28.3	83	181.4
6.1	43	109.4	28.9	84	183.2
6.7	44	111.2	29.4	85	185.0
7.2	45	113.0	30.0	86	186.8
7.8	46	114.8	30.6	87	188.6
8.3	47	116.6	31.1	88	190.4
8.9	48	118.4	31.7	89	192.2
9.4	49	120.2	32.2	90	194.0
10.0	50	122.0	32.8	91	195.8

(*Continued*)

Table 1 (*Continued*)

°C		°F	°C		°F
33.3	92	197.6	56.7	134	273.2
33.9	93	199.4	57.2	135	275.0
34.4	94	201.2	57.8	136	276.8
35.0	95	203.0	58.3	137	278.6
35.6	96	204.8	58.9	138	280.4
36.1	97	206.6	59.4	139	282.2
36.7	98	208.4	60.0	140	284.0
37.2	99	210.2	60.6	141	285.8
37.8	100	212.0	61.1	142	287.6
38.3	101	213.8	61.7	143	289.4
38.9	102	215.6	62.2	144	291.2
39.4	103	217.4	62.8	145	293.0
40.0	104	219.2	63.3	146	294.8
40.6	105	221.0	63.9	147	296.6
41.1	106	222.8	64.4	148	298.4
41.7	107	224.6	65.0	149	300.2
42.2	108	226.4	65.6	150	302.0
42.8	109	228.2	66.1	151	303.8
43.3	110	230.0	66.7	152	305.6
43.9	111	231.8	67.2	153	307.4
44.4	112	233.6	67.8	154	309.2
45.0	113	235.4	68.3	155	311.0
45.6	114	237.2	68.9	156	312.8
46.1	115	239.0	69.4	157	314.6
46.7	116	240.8	70.0	158	316.4
47.2	117	242.6	70.6	159	318.2
47.8	118	244.4	71.1	160	320.0
48.3	119	246.2	71.7	161	321.8
48.9	120	248.0	72.2	162	323.6
49.4	121	249.8	72.8	163	325.4
50.0	122	251.6	73.3	164	327.2
50.6	123	253.4	73.9	165	329.0
51.1	124	255.2	74.4	166	330.8
51.7	125	257.0	75.0	167	332.6
52.2	126	258.8	75.6	168	334.4
52.8	127	260.6	76.1	169	336.2
53.3	128	262.4	76.7	170	338.0
53.9	129	264.2	77.2	171	339.8
54.4	130	266.0	77.8	172	341.6
55.0	131	267.8	78.3	173	343.4
55.6	132	269.6	78.9	174	345.2
56.1	133	271.4	79.4	175	347.0

(*Continued*)

Table 1 (*Continued*)

°C		°F	°C		°F
80.0	176	348.8	103.3	218	424.4
80.6	177	350.6	103.9	219	426.2
81.1	178	352.4	104.4	220	428.0
81.7	179	354.2	105.0	221	429.8
82.2	180	356.0	105.6	222	431.6
82.8	181	357.8	106.1	223	433.4
83.3	182	359.6	106.7	224	435.2
83.9	183	361.4	107.2	225	437.0
84.4	184	363.2	107.8	226	438.8
85.0	185	365.0	108.3	227	440.6
85.6	186	366.8	108.9	228	442.4
86.1	187	368.6	109.4	229	444.2
86.7	188	370.4	110.0	230	446.0
87.2	189	372.2	110.6	231	447.8
87.8	190	374.0	111.1	232	449.6
88.3	191	375.8	111.7	233	451.4
88.9	192	377.6	112.2	234	453.2
89.4	193	379.4	112.8	235	455.0
90.0	194	381.2	113.3	236	456.8
90.6	195	383.0	113.9	237	458.6
91.1	196	384.8	114.4	238	460.4
91.7	197	386.6	115.0	239	462.2
92.2	198	388.4	115.6	240	464.0
92.8	199	390.2	116.1	241	465.8
93.3	200	392.0	116.7	242	467.6
93.9	201	393.8	117.2	243	469.4
94.4	202	395.6	117.8	244	471.2
95.0	203	397.4	118.3	245	473.0
95.6	204	399.2	118.9	246	474.8
96.1	205	401.0	119.4	247	476.6
96.7	206	402.8	120.0	248	478.4
97.2	207	404.6	120.6	249	480.2
97.8	208	406.4	121.1	250	482.0
98.3	209	408.2	121.7	251	483.8
98.9	210	410.0	122.2	252	485.6
99.4	211	411.8	122.8	253	487.4
100.0	212	413.6	123.3	254	489.2
100.6	213	415.4	123.9	255	491.0
101.1	214	417.2	124.4	256	492.8
101.7	215	419.0	125.0	257	494.6
102.2	216	420.8	125.6	258	496.4
102.8	217	422.6	126.1	259	498.2

(*Continued*)

Table 1 (*Continued*)

°C	°F		°C		°F
126.7	260	500.0	134.4	274	525.2
127.2	261	501.8	135.0	275	527.0
127.8	262	503.6			
128.3	263	505.4	\multicolumn{3}{c}{Fractions for extrapolation:}		
128.9	264	507.2			
129.4	265	509.0	0.06	0.1	0.18
130.0	266	510.1	0.11	0.2	0.36
130.6	267	512.6	0.17	0.3	0.54
131.1	268	514.4	0.22	0.4	0.72
131.7	269	516.2	0.28	0.5	0.90
132.2	270	518.0	0.33	0.6	1.08
132.8	271	519.8	0.39	0.7	1.26
133.3	272	521.6	0.44	0.8	1.44
133.9	273	523.4	0.50	0.9	1.62

°C are expressed to nearest tenth of a degree – except extrapolation fractions in °C which are to nearest hundredth.

D.2 Viscosity of Glucose syrups at different Dextrose equivalents and temperatures.
Reproduced by courtesy of The Corn Refiners Association.

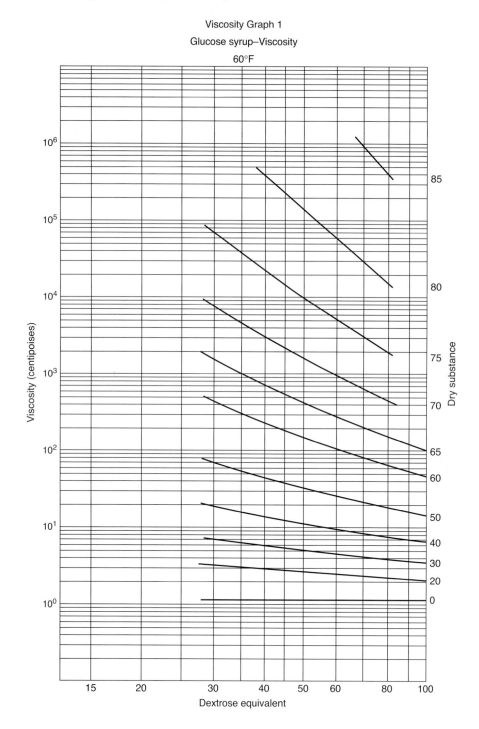

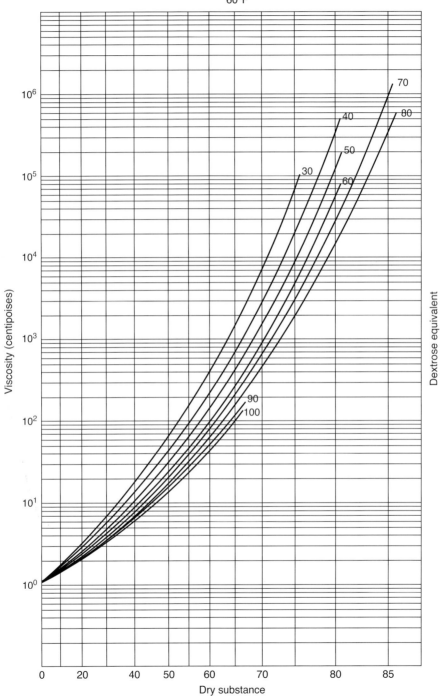

Viscosity Graph 2
Glucose syrup–Viscosity
60°F

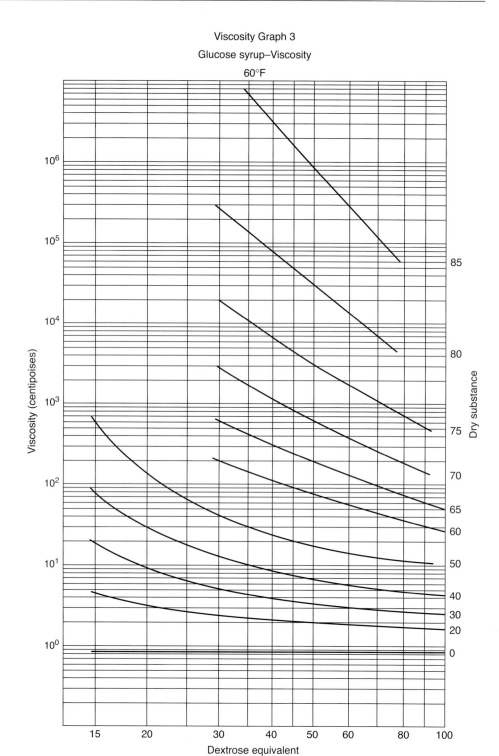

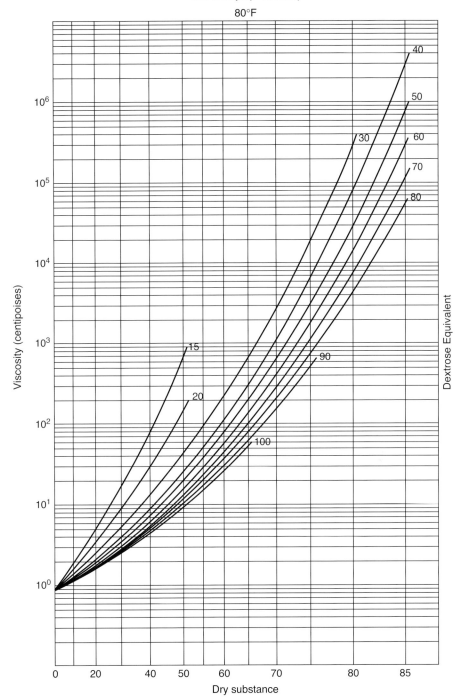

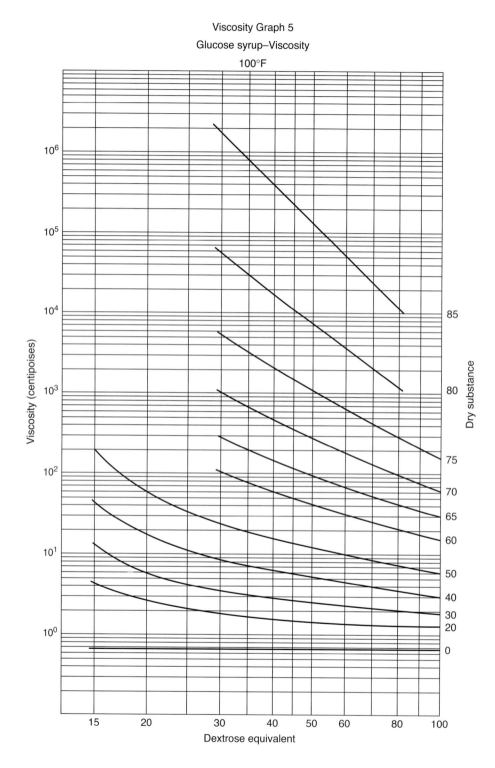

Appendix D: Tables 339

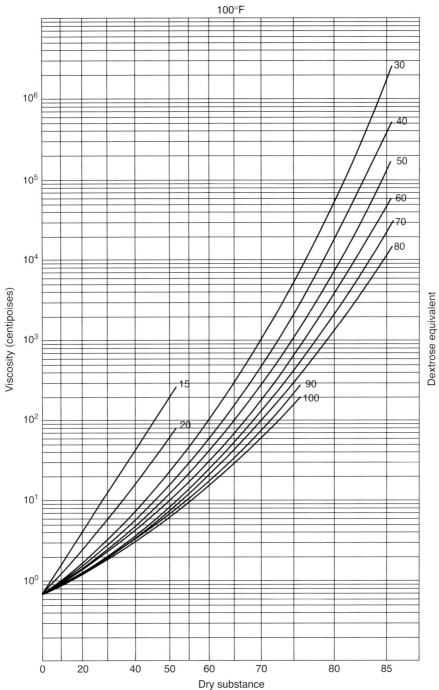

Viscosity Graph 6
Glucose syrup–Viscosity
100°F

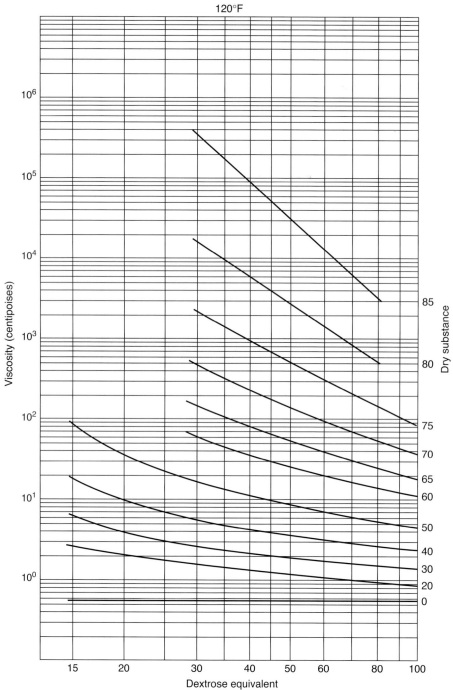

Viscosity Graph 7
Glucose syrup–Viscosity
120°F

Appendix D: Tables 341

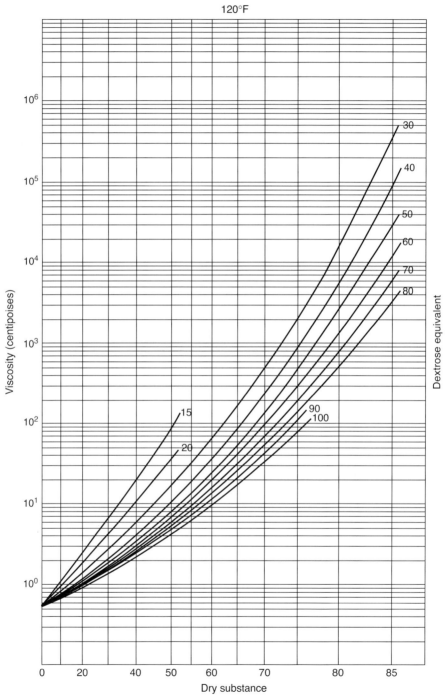

Viscosity Graph 8
Glucose syrup–Viscosity
120°F

Viscosity Graph 9
Glucose syrup–Viscosity
140°F

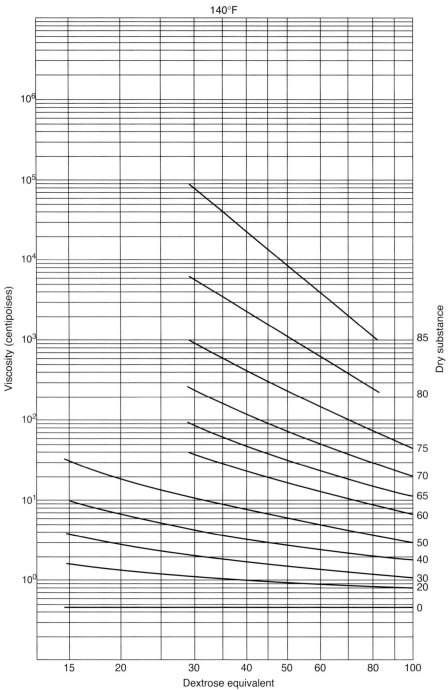

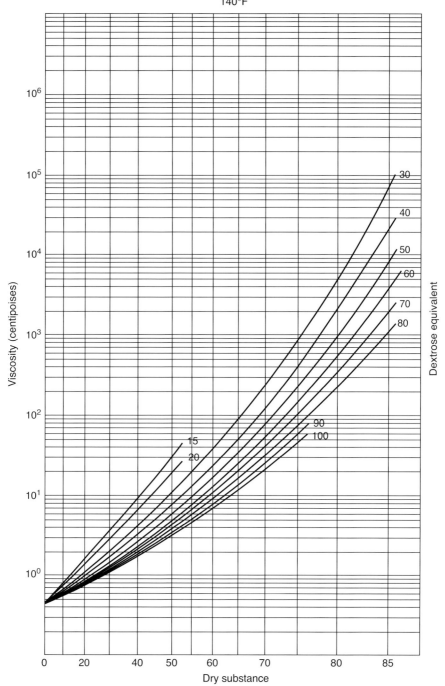

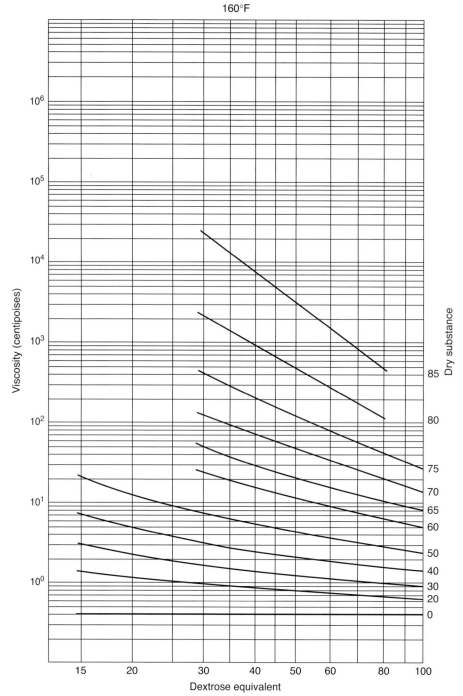

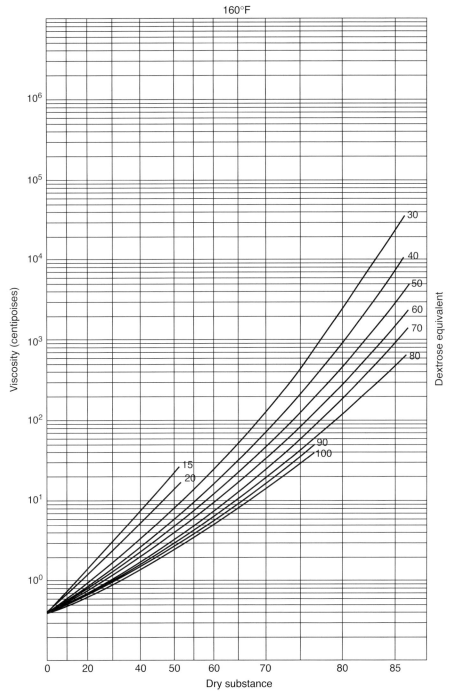

Viscosity Graph 12
Glucose syrup–Viscosity
160°F

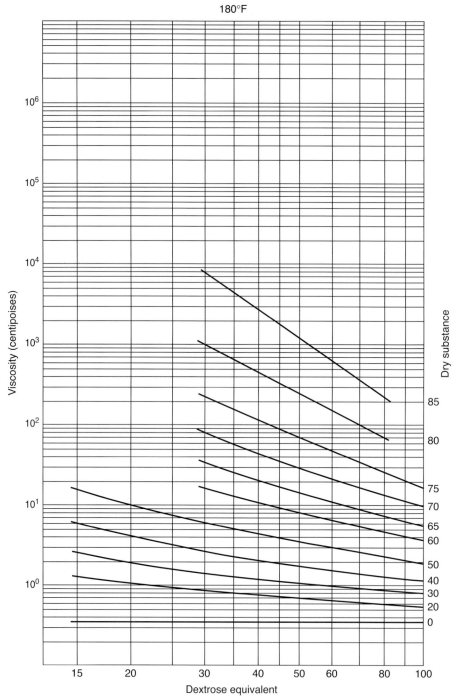

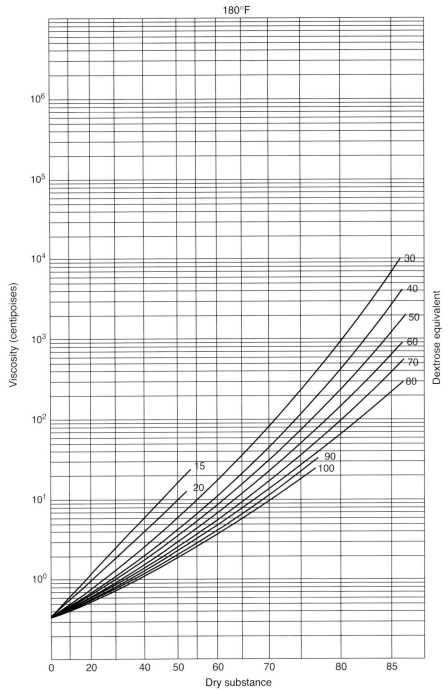

D.3 Maize starch Baumé tables.
Reproduced by courtesy of The Corn Refiners Association.

Baumé @ 15°C (59°F) (145 modulus)	Specific gravity @15.6/15.6°C (60/60°F)	% Starch dry basis	Grams of dry starch per 100 mL
0.1	1.0007	0.178	0.178
0.5	1.0035	0.885	0.887
1.0	1.0069	1.777	1.785
1.5	1.0105	2.666	2.684
2.0	1.0140	3.554	3.595
2.5	1.0176	4.443	4.518
3.0	1.0211	5.331	5.428
3.5	1.0248	6.220	6.363
4.0	1.0285	7.108	7.298
4.5	1.0322	7.997	8.232
5.0	1.0358	8.885	9.179
5.5	1.0396	9.774	10.138
6.0	1.0433	10.662	11.096
6.5	1.0471	11.551	12.067
7.0	1.0508	12.439	13.049
7.5	1.0547	13.328	14.032
8.0	1.0585	14.216	15.015
8.5	1.0624	15.105	16.009
9.0	1.0663	15.993	17.016
9.5	1.0703	16.882	18.034
10.0	1.0742	17.770	19.053
11.0	1.0822	19.547	21.114
12.0	1.0903	21.324	23.199
13.0	1.0986	23.101	25.332
14.0	1.1071	24.878	27.489
15.0	1.1156	26.655	29.682
16.0	1.1242	28.432	31.898
17.0	1.1330	30.209	34.163
18.0	1.1419	31.986	36.452
19.0	1.1510	33.763	38.789
20.0	1.1602	35.540	41.149
21.0	1.1696	37.317	43.558
22.0	1.1791	39.094	46.002
23.0	1.1888	40.871	48.495
24.0	1.1986	42.648	51.011
25.0	1.2086	44.425	53.576

Above 25 Baumé, the water of the starch slurry is totally absorbed by the starch to produce a very damp powder, which cannot be pumped, and is therefore unsuitable for glucose syrup production using today's engineering technology, which is designed to use a starch slurry. It is this slurry/damp powder boundary which is the limiting factor in being able to hydrolyse a starch slurry with solids greater than 42–44%.

D.4 Sucrose Brix table – Brix – % sucrose w/w, specific gravity and Baumé (145 modulus)

(See Sections B.5 to B.7 of Appendix B for relevant calculations).

The solids of sucrose solutions are usually expressed in degrees Brix, which is a measure of the amount of sucrose in a solution, expressed as a percentage on a weight/weight basis.

Degrees Brix or % sucrose w/w	Specific gravity @ 20°C (68°F)	Degrees Baumé modulus 145 @ 20°C (68°F)	Degrees Brix or % sucrose w/w	Specific gravity @ 20°C (68°F)	Degrees Baumé modulus 145 @ 20°C (68°F)
10	1.040	5.57	41	1.184	22.50
11	1.044	6.13	42	1.189	23.04
12	1.048	6.68	43	1.194	23.57
13	1.053	7.24	44	1.199	24.10
14	1.057	7.79	45	1.205	24.63
15	1.061	8.34	46	1.210	25.17
16	1.065	8.89	47	1215	25.70
17	1.070	9.45	48	1.221	26.23
18	1.074	10.00	49	1.226	26.75
19	1.078	10.55	50	1.232	27.28
20	1.083	11.10	51	1.237	27.81
21	1.087	11.65	52	1.243	28.33
22	1.092	12.20	53	1.248	28.86
23	1.096	12.74	54	1.254	29.38
24	1.101	13.29	55	1.260	29.90
25	1.106	13.84	56	1.265	30.42
26	1.110	14.39	57	1.271	30.94
27	1.115	27.93	58	1.277	31.46
28	1.009	15.48	59	1.283	31.97
29	1.124	16.02	60	1.288	32.49
30	1.129	16.57	61	1.295	33.00
31	1.134	17.11	62	1.301	33.51
32	1.139	17.65	63	1.307	34.02

Continued

(*Continued*)

Degrees Brix or % sucrose w/w	Specific gravity @ 20°C (68°F)	Degrees Baumé modulus 145 @ 20°C (68°F)	Degrees Brix or % sucrose w/w	Specific gravity @ 20°C (68°F)	Degrees Baumé modulus 145 @ 20°C (68°F)
33	1.143	18.19	64	1.313	34.53
34	1.148	18.73	65	1.319	35.04
35	1.153	19.28	66	1.325	35.55
36	1.158	19.81	67	1.331	36.05
37	1.163	20.35	68	1.337	36.55
38	1.168	20.89	69	1.343	37.06
39	1.173	21.43	70	1.350	37.56
40	1.179	21.97			

The above table is based on readings at a temperature of 20°C (68°F). Since the specific gravity will change with change of temperature, it is necessary to apply a correction factor to the Brix readings for readings below or above 20°C (68°F).

		Temperature.		Degrees Brix and correction								
		°C	°F	10	20	25	30	35	40	45	50	55	60
Subtract correction		4.4	40	0.5	0.6	0.7	0.8	0.9	0.9	0.9	0.9	0.9	1.0
		10.0	50	0.5	0.5	05	0.5	0.6	0.6	0.6	0.6	0.6	0.6
		15.6	60	0.2	0.2	0.2	0.2	0.2	0.2	0.2	0.2	0.2	0.2
Add corrections		21.1	70	0.1	0.2	0.2	0.2	0.2	0.2	0.2	0.2	0.2	0.2
		26.7	80	0.5	0.6	0.6	0.6	0.6	0.6	0.6	0.6	0.6	0.6
		32.2	90	0.9	1.0	1.0	1.0	1.1	1.1	1.1	1.1	1.1	1.0
		37.8	100	1.3	1.4	1.5	1.5	1.5	1.5	1.5	1.5	1.5	1.5
		48.9	120	2.5	2.6	2.6	2.6	2.6	2.6	2.6	2.6	2.5	2.5
		60.0	140	3.8	3.8	3.8	3.8	3.8	3.8	3.7	3.7	3.6	3.6
		71.1	160	5.1	5.1	5.1	5.1	5.1	5.0	5.0	4.9	4.8	4.8
		83.2	180	6.7	6.5	6.4	6.4	6.3	6.3	6.2	6.1	6.0	5.9
		100.0	212	10.0	9.6	9.4	9.3	9.1	8.9	8.7	8.4	8.2	8.1

For temperatures below 20°C (68°F), subtract the correction.
For temperatures above 20°C (68°F), add the correction.

Example

For a hydrometer reading 50.0 degrees Brix at a temperature of 71.1°C (160°F), the correction is 4.9. Therefore, corrected reading would be $50.0 + 4.9 = 54.9$ degrees Brix.

D.5 Sucrose Brix – refractive indices at 20°C

N.B. The refractometer must be calibrated to read sucrose solids.

Degrees Brix or % sucrose w/w	Refractive indices	Degrees Brix or % sucrose w/w	Refractive indices
10	1.3478	41	1.4018
11	1.3494	42	1.4038
12	1.3509	43	1.4058
13	1.3525	44	1.4078
14	1.3541	45	1.4098
15	1.3557	46	1.4118
16	1.3573	47	1.4139
17	1.3589	48	1.4159
18	1.3605	49	1.4180
19	1.3621	50	1.4201
20	1.3638	51	1.4221
21	1.3655	52	14243
22	1.3672	53	1.4264
23	1.3689	54	1.4286
24	1.3706	55	1.4308
25	1.3723	56	1.4330
26	1.3740	57	1.4352
27	1.3758	58	1.4374
28	1.3776	59	1.4397
29	1.3793	60	1.4419
30	1.3811	61	1.4442
31	1.3830	62	1.4465
32	1.3848	63	1.4488
33	1.3866	64	1.4511
34	1.3885	65	1.4535
35	1.3903	66	1.4558
36	1.3922	67	1.4582
37	1.3940	68	1.4606
38	1.3960	69	1.4630
39	1.3979	70	1.4655
40	1.3998		

The above table is based on readings at a temperature of 20°C (68°F). Since the refractive index will change with temperature, it is necessary to apply a correction factor to the refractive index readings below or above 20°C (68°F).

D.6 Glucose syrup tables – commercial Baumé, DE, % solids – at 60°C (140°F)

Dextrose equivalent	Baumé	% Solids
28.0	41.0	76.0
35.0	43.5	80.7
42.0	43.2	80.7
52.0	43.3	80.7
63.0	43.1	82.0
74.0	41.4	78.9
95.0	38.2	74.4

D.7 Sieve specifications

Most countries have adopted The International Standard for sieves as laid down in ISO 3310 Series. America uses the American Standard ASTM E 11.

In ISO 3310, mesh size is no longer mentioned. The reason is that the number of apertures in a mesh is very dependent upon the diameter of the wire used to make the mesh, and therefore mesh size was considered no longer appropriate.

Table 7.0 International Standard ISO 3310 Series (Sieve Specifications).

International Standard ISO 3310 Series		British Standard BS 410 Series
Wire mesh ISO 3310-1 BS 410-1 Nominal aperture sizes mm	Wire mesh ISO 3310-1 BS 410-1 Nominal aperture sizes μm	Perforated plate ISO 3310-2 BS 410-2 Nominal aperture sizes Round and Square Holes mm
125	900	125
112	850	112
106	800	106
100	710	100
90	630	90
80	600	80
75	560	75
71	500	71
63	450	63
56	425	56
53	400	53

(Continued)

Table 7.0 (*Continued*).

International Standard ISO 3310 Series Wire mesh ISO 3310-1 BS 410-1 Nominal aperture sizes mm	Wire mesh ISO 3310-1 BS 410-1 Nominal aperture sizes μm	British Standard BS 410 Series Perforated plate ISO 3310-2 BS 410-2 Nominal aperture sizes Round and Square Holes mm
50	355	50
45	315	45
40	300	40
37.5	280	37.5
35.5	250	35.5
31.5	224	31.5
28	212	28
26.5	200	26.5
25	180	25
22.4	160	22.4
20	150	20
19	140	19
18	125	18
16	112	16
14	106	14
13.2	100	13.2
12.5	90	12.5
11.2	80	11.2
10	75	10
9.5	71	9.5
9	63	9
8	56	8
7.1	53	7.1
6.7	50	6.7
6.3	45	6.3
5.6	40	5.6
5	38	5
4.75	36	4.75
4.5	32	4.5
4	25	4
3.55	20	Round hole only
3.35		3.55
3.15		3.35

(*Continued*)

Table 7.0 (*Continued*).

International Standard ISO 3310 Series		British Standard BS 410 Series
Wire mesh ISO 3310-1 BS 410-1 Nominal aperture sizes mm	Wire mesh ISO 3310-1 BS 410-1 Nominal aperture sizes μm	Perforated plate ISO 3310-2 BS 410-2 Nominal aperture sizes Round and Square Holes mm
2.8		3.15
2.5		2.8
2.36		2.5
2.24		2.36
2		2.24
1.8		2
1.7		1.8
1.6		1.7
1.4		1.6
1.25		1.4
1.18		1.25
1.12		1.18
1		1.12
1		1

Table 7.1 American Standard ASTM E 11.

Wire mesh series designation	
Standard mm	Alternative
125	5″
106	4.24″
100	4″
90	3 1/2″
75	3″
63	2 1/2″
53	2.12″
50	2″
45	1 3/4″
37.5	1 1/2″

(*Continued*)

Table 7.1 (*Continued*).

Wire mesh series designation	
Standard mm	Alternative
31.5	1 1/4"
26.5	1.06"
25	1"
22.4	7/8"
19	3/4"
16	5/8"
13.2	0.53"
12.5	1/2"
11.2	7/16"
9.5	3/8"
8	5/16"
6.7	0.265"
6.3	1/4"
5.6	No.3 1/2
4.75	No.4
4	No.5
3.35	No.6
2.8	No.7
2.36	No.8
2	No.10
1.7	No.12
1.4	No.14
1.18	No.16
1	No.18
μm	
850	No.20
710	No.25
600	No.30
500	No.35
425	No.40
355	No.45
300	No.50
250	No.60
212	No.70
180	No.80
150	No.100
125	No.120
106	No.140

(*Continued*)

Table 7.1 (*Continued*).

Wire mesh series designation	
Standard mm	Alternative
90	No.170
75	No.200
63	No.230
53	No.270
45	No.325
38	No.400
32	No.450
25	No.500
20	No.635

Table 7.2 Nominal mesh aperture sizes (British Standards)

Mesh	Nominal aperture size (mm)	Mesh	Nominal aperture size (μm)	Mesh	Nominal aperture size (μm)
4	4.00	18	850	85	180
5	3.35	22	710	100	150
6	2.80	25	600	120	125
7	236	30	500	150	106
8	2.00	36	425	170	90
10	170	44	355	200	75
12	140	52	300	240	63
14	1.18	60	250	300	53
16	1.00	72	212	350	45

Mesh is the number of openings per linear inch.
Aperture or screen size is the minimum clear space between the edges of the opening in the screen surface.

Bibliography

The following books have been a useful reference source in the preparation of this book.

Analyses

Advanced Instruments, Inc. (1971). *The Physical Chemistry, Theory and Technique of Freezing Point Determinations*. Advanced Instruments Inc.
Aurand, L.W., Woods, A.E., Wells, M.R. (1987). *Food Composition and Analysis*. AVI Publishing.
Balston, J.N., Talbot, B.E. (1952). *A Guide to Filter Paper and Cellulose Powder Chromatography* (Ed. Jones, T.S.G.). Reeve Angel & Co.
Browne, C.A., Zerban, F.W. (1955). *Physical and Chemical Methods of Sugar Analysis*. John Wiley & Sons, Inc.
Emerton, V, Choi, E. (2008). *Essential Guide to Food Additives*. RSC Publishing.
McCance, R.A., Widdowson, E.M. (1991). *The Composition of Foods*. Royal Society of Chemistry.
Kirk, R.S., Sayer, R. (1991). *Pearson's Composition and Analysis of Foods*. Longman.

Brewing

Bamforth, C. (1998). *Beer – Tap into the Art and Science of Brewing*. Insight Books.
Briggs, D.E., Hough, J.S., Stevens, R. Young, T.W. (1981). *Malting and Brewing Science*. Chapman & Hall.
Daniels, R. (1996). *Designing Great Beers*. Brewers Publications.
Hardwick, W.A. (1994). *Handbook of Brewing*. Marcel Dekker Inc.
Hornsey, I.S. (1999). *Brewing*. Royal Society of Chemistry.
Lewis, M.J., Young, T.W. (1995). *Brewing*. Chapman & Hall.
Wainwright, T. (2000). *Basic Brewing Science*. PTB Publications Ltd.

Cereals

Inglett, G.E. (1970). *CORN: Culture, Processing, Products*. AVI Publishing.
Kent, N.L. (1966). *Technology of Cereals*. Pergamon.
Kent-Jones, D.W., Amos, A.J. (1947). *Modern Cereal Chemistry*. The Northern Publishing Company.
Matz, S.A. (1991). *The Chemistry and Technology of Cereals as Food and Feed*. Van Nostrand Reinhold (AVI Book).
Watson, S.A., Ramstad, P.E. (1987). *CORN: Chemistry and Technology*. American Association of Cereal Chemists, Inc.

Company literature

Citrus Colloids Ltd., Hereford – Pectin Applications.
Grinstead Products, Denmark – Scoopable Ice Cream.
H.P. Bulmer Ltd., Hereford – Pectin Applications.

Company history

Chalmin, P. (1990). *The Making of a Sugar Giant*. Harwood Academic Publishers.
Forrestal, D.J. (1982). *The Kernel and the Bean. The 75-Year Story of the Staley Company*. Simon and Schuster.
Green, D. (1979). *CPC Europe, A Family of Food Companies*. Publications for Companies.
Hugill, A. (1978). *Sugar and All That. A History of Tate & Lyle*. Gentry Books.

Confectionery

Alikonis, J.J. (1979). *Candy Technology*. AVI Publishing.
Edwards, W.P. (2000). *Science of Sugar Confectionery*. Royal Society of Chemistry.
Harris, N., Peterson, M., Crespo, S. (1991). *A Formulary of Candy Products*. Chemical Publishing Co.
Lees, R. (1980). *Faults, Causes and Remedies – In Sweet and Chocolate Manufacture*. Specialised Publications.
Lees, R, Jackson, E.B. (1995). *Sugar Confectionery and Chocolate Manufacture*. Blackie Academic.
Minifie, B.W. (1999). *Chocolate, Cocoa, and Confectionery*. Aspen Publications.

Dictionaries

Abercrombie, M., Hickman, C.J., Johnson, M.L. (1954). *Dictionary of Biology*. Penguin Books.
CRC Handbook of Chemistry and Physics, 83rd edn. (2002–2003).
Sharp, D.W.A. (1984). *The Penguin Dictionary of Chemistry*. Penguin Reference Books.
The Condensed Chemical Dictionary, 10th edn. (1981). Van Nostrand Reinhold, Co., USA.
The Merck Index, 14th edn. (2006). Merck & Co., Inc., Whitehouse Station, NJ, USA.

Fermentation

Dunn, C.G., Prescott, S.C. (1957). *Industrial Mycology*. McGraw-Hill Book Company.
Hockenhull, D.J.D. (Ed.). (1968). *Progress in Industrial Microbiology*. J & A Churchill Ltd.
Yoshiharu, D. (1990). *Microbial Polyesters*. VCH.

Food science

Arbuckle, W.S. (1986). *Ice Cream*. Chapman & Hall.
Ashurst, P.R. (2005). *Chemistry and Technology of Soft Drinks and Fruit Juices*. Blackwell.
Corn Refiners Association Inc. (1971). Symposium Proceedings 'Products of the wet milling industry in food'. Corn Refiners Association, Inc, Washington, DC, USA.

Dijksterhuis, J., Samson, R.A. (2007). *Food Mycology – A Multifacted Approach to Fungi and Food.* CRC Press.
Fennema, O.R. (1987). *Food Composition and Analysis.* AVI Publishing.
Fox, B.A., Cameron, A.G. (1986). *Food Science – A Chemical Approach.* Hodder & Stoughton Educational.
Massel, A. (1968). *Applied Wine Chemistry and Technology.* Heildelberg Publishing.
Mountney, G.J., Gould, W.A. (1988). *Practical Food Microbiology and Technology*, 3rd edn. AVI Publishing.
Pyke, M. (1984). *Food Science and Technology.* John Murray.
Riviere, J. (1975). *Les Application Industrielles de la Microbiologie. Paris.* English translation by Moss, M.O., Smith, J.E. (1977). Surrey University Press.
Shapton, D.A, Shapton, N.F. (1991). *Principles and Practices for the Safe Processing of Foods.* Butterworth Heinmann.
Smith, W.H. (1972). *Biscuits, Crackers and Cookies – Technology, Production, and Management.* Applied Science.
Walford, J. (1980). *Developments in Food Colours.* Applied Science.
Wedzicha, B.L. (1984). *Chemistry of Sulphur Dioxide in Foods.* Elsevier Applied Science Publishers.
Woollen, A. (1969). *Food Industries Manual.* Leonard Hill.

Medical science

Bell, G.H., Emslie-Smith, D, Paterson, C.R. (1980). *Textbook of Physiology.* Churchill Livingstone.
British National Formulary. Number 45. March 2003. British Medical Association and Royal Pharmaceutical Society.
Marshall, W.J. (1990). *Clinical Chemistry.* Gower Medical Publishing.
The British Medical Association Complete Family Health Encyclopedia (1998). Dorling Kindersley.
Yudkin, J. (1985). *The Penguin Encyclopaedia of Nutririon.* Penguin Books.

Starch

Kearsley, M.W., Dziedzic, S.Z. (1995). *Handbook of Starch Hydrolysis Products and Their Derivatives.* Blackie Academic.
Kerr, R.W. (1950). Chemistry and Industry of Starch. Academic Press.
Knight, J.W. (1969). *The Starch Industry.* Pergamon.
Radley, J.A. (1954). *Starch and Its Derivatives.* John Wiley & Sons, Chapman & Hall.
Schenck, F.W., Hebeda, R.E. (1992). *Starch Hydrolysis Products – World Wide Technology, Production and Applications.* VCH.
Whistler, R.L., Pascall, E.F. (1965). *Starch, Chemistry and Technology.* Academic Press.

Sugars

Descotes, G. (1993). *Carbohydrates as Organic Raw Materials 2.* VCH.
Fairrie, G. (1925). *Sugar.* Fairrie.
FAO Agricultural Services Bulletin 101. (1993). *Date Palm Products.*
Grenby, T.H. (1987). *Developments in Sweeteners 3.* Applied Science.

Hough, C.A.M., Parker, K.J., Vlitos, A.J. (1979). *Development in Sweeteners 1*. Applied Science.
Lyle, O. (1941). *Technology for Sugar Refinery Workers*. Chapman & Hall.
Mitchell, H. (2006). *Sweeteners and Sugar Alternatives in Food Technology*. Blackwell.
O'Brien, L. (2001). *Alternative Sweeteners*. Marcell Dekker.
Pancoast, H.M., Junk, W.R. (1980). *Handbook of Sugars*. AVI Publishing.
Alexander, R.J. (1997). Sweeteners Nutritive. Egan Press Handbook Series. Practical Guide for the food industry.

Index

This index should be read in conjunction with the Glossary on page 265.

Abbe, Ernest, 6
Acetone, 183
Acid addition, 28
Acid enzyme hydrolysis, 33
Acid hydrolysis process, 27
Acidulants, 235
Adhesives, 78
Aldehyde group, 20
Amylopectin structure, 22
Amylose structure, 22
Amylum, 12, 14
Antibiotics, 85, 89
Arsenic, 10
Ascorbic acid, 91, 113
Aspartame, 259
Association des Amidonniers et Feculiers, 15
Aureobasidum, sp., 109

Bacillus coagulans, 12
Bakery, *see also* baking
Bakery products, 93, 114, 115
   biscuits fillings, 124
   fondant, 125
   glaze, 78, 129
   hundreds and thousands, 126
   icings, 126
   marzipan, 128
   piping jelly, 129
   sundries, 125
   wafer fillings, 124

Baking, 82, 85, 89
   biscuits, 123
   fermented goods, 120
   marshmallows, 127
   non-fermented goods, 121
   reduced calorie products, 130
von Balling, Karl Joseph, 6
Baumé, Antoine, 6
Baumé hydrometer, 6
Baumé, Explanation, 313
Behr, Dr. Arno, 7
Berzelius, Prof. Jons Jacob, 1
Boiling point elevation, 315
Bonaparte, Napoleon, 2, 13
Breakfast cereals, 94, 130
Brewer's extract, 141
Brewing, 81, 82, 86, 96, 133
   extract values, 143, 144
   fermentable sugars, 81, 135, 137, 138, 144
   high gravity, 140
   low alcohol beer, 96, 138
   process, 134
   sugar, 3
Bibliography, 357
Brix, Adolf F., 6
Brix, Explanation, 313
Brobst, Scobell and Steele, 5
Browning agent, 89
Building industry, 86
Bulking agent, 96
Bulk tank design, 58

Bulk tank installation, 57, 59
Butanol, 183

Calcium ions, 41
Caloric values, 260
Calvin cycle, 17
Cakes, syrup inclusion rates, 122
CAP, *see* Common Agriculture Policy
Caramel colour, 78, 86, 261, 262
    isoelectric point, 262
    type, Ammonia, 261
        Caustic sulphite, 261
        Sulphite ammonia, 261
Caramelisation, 60
Caramels, 77, 159, 163
Carbohydrate absorption, 253, 254
    metabolic problems, 257
    metabolism, 251
Carbon treatment, 30
Carl Zeiss, 6
Cargill, 3, 13, 14
Carrier, 86, 89, 97
Cattle lick blocks, 86
Centrifuge, 30
Cereal coatings, 97
Cerestar, 14
Chewing gum, 78
Chews, 160, 167
Chicago Sugar Refining Company, 7
Chicory, 17, 19
Chiozza, M.M., 7
Chip sugar, 146
Chocolate spread, 210
Cider, 87
Citric acid, 181
Clinton Corn Refining, 11, 12
Coeliac disease, 259
Coliforms, 50
Common Agriculture Policy, 13
Confectioners Glucose, 26
Confectionery, 80, 87, 94, 97, 105, 114, 116, 146
    calorie reduced, 172
    centres, 171
    soft, 84

Conversion, 29
Corn Products, Argo, 7, 9
Corn Products, Belgium, 4
Corn Refiners Association, 61, 64, 334, 348
Cosmetic, 106
Cryoscope, 4, 20
Crystalline dextrose production, 7, 36, 37
Crystalline fructose production, 40, 43

Dairy, 105
    fillings, 98
Dale & Langlois, 7
Date syrup, 326
    palm, 326
Davy, Sir Humphry, 1
Delessert, Benjamin, 3
Dextrose, 1, 20, 35, 88
    absorption, 254
    particle size, 88
    syrup, 20
Diabetes mellitus, 257
Diabetic foods and drinks, 94, 114
Diastatic malt, *see* beta amylase
Digestion, 252
Drinks
    carbonated, 237
    dilutable, 237
    geriatric, 248
    health, 239
        classification, 240
    powdered, 238
    reduced calorie, 238
    soft, 12, 84, 227
        drink recipes, 236
    sports, 100, 239
        formulation, 244
Dry mixes, 98

E/E, *see* enzyme enzyme hydrolysis
EEC, *see* European Commission
    Council Regulation No. 1293/79, 15
    Council Regulation No. 318/2006, 15
Encapsulation, 98

Enzymes, 23, 179
  alpha amylase, 24
  AMG, see glucoamylase
  amyloglucosidase, see glucoamylase
  bacterial amylase, 180
  beta amylase, 24, 34
  beta gluconase, 25, 34
  cellulase, 25
  diastic malt, 24, 33, 34
  fungal amylase, 180
  glucoamylase, 7, 24, 180
  pentanase, 25
  pullulanse, 25, 180
  isomerase, 25, 41, 180
Enzyme enzyme hydrolysis, 26, 39
Erythritol, 89, 107, 117, 183
  absorption, 256
  production schematic, 110
European Commission, 14, 17
European Courts of Justice, 14
Evaporation, 32
Eynon, Lewis, 4
Excise Duty, 3, 4

Fat reduction, 97
Feedstock, see fermentation, industrial
von Fehling, Herman, 4
Fehling's solution, 20
Fermentation, 175, 177
  feedstock, 87, 89
  industrial, 87, 90
  substrate, 176
Fermented drinks, 83
Fibres, absorption, 255
Film forming, 98
Fitton, M.C., 4
Flat sour spores, 53
Flouride, 10
Fondants, 70, 78, 159, 162
Foods
  baby, 98
  geriatric, 98
  invalid, 98
  liquid, 248

  slimming, 249
  snacks, 99
Foundry work, 90
Freezing point depression, 5, 67, 317
Freezing point osmometer, 4
French dressings, 224
Frozen foods, 98
Fructose, 9, 11, 20
  absorption, 254
  crystalline, 13
  diabetes, 257
  intolerance, 258
Fructose glucose syrup, 19
Fructose syrup, 13, 19, 61
  production, 40
Fruit curds, 216
  pie fillings, 214
  preparations, 87, 94
Fudge, 78, 159
Fumaric acid, 181

Galactosaemia, 258
Galactose absorption, 255
Gaucher's disease, 259
Glaze, 79
Glossary, 265
Glucose fructose syrup, 19
Glucose syrup, Applications, 61
Glucose syrup, Definition, 19
Glucose syrup properties, Table of, 63
  bodying agent, 64
  browning reaction, 64
  cohesiveness, 65
  fermentability, 65
  flavour enhancement, 65
  flavour transfer medium, 66
  foam stabilisers, 66
  freezing point depression, 66, 188
  humectancy, 67
  hygroscopicity, 68
  molecular weight and freezing point
    depression, 67
  molecular weight and osmotic pressure,
    69, 205

Glucose syrup properties, Table of (*Cont.*)
   nutritive solids, 68
   osmotic pressure, 68, 205, 241
   prevention of coarse ice crystal formation, 70
   prevention of sucrose crystallisation, 70
   sheen producer, 71
   summary of properties, 73
   sweetness, 71, 92, 188, 189, 207, 232
   sweetness response profile, 72, 231
   viscosity, 72, 73, 334
Glucose syrups
   55 DE, 245
   63 DE, 82, 157, 166, 168, 169, 205
   95 DE, 84
   acid converted, 28 DE, 79
   acid converted 35 DE, 79, 156
   acid converted, 42 DE, 61, 77, 156, 165, 167, 168
   acid enzyme converted 42 DE, 61, 156
   de-ionized, 139
   spray dried, 84, 85, 158
   sugar spectra, 44, 61, 156, 189, 222, 231, 240, 321
   viscosities, 44
Glucose syrup solids, 80, *see also* glucose syrups, spray dried
Glucose syrup specifications, explanation of
   acetaldehyde, 49
   acidity, 47
   ageing colour, 46
   appearance, 45
   Brewers Extract, 49
   chlorides, 48
   colour measurement, 45
   commercial Baume, 50
   conductivity, 49
   corrosion properties, 52
   dextrose equivalent, 46
   disposal, 53
   EBC colour, 49
   ergoline alkaloids, 51
   explosion risk, 52
   fermentable sugars, 49
   fire risk, 52
   first aid requirements, 53
   foam index, 48
   handling procedures, 52
   labelling, 51
   legislation, 52
   metals, 48
   microbiological, 50
   Mycotoxins, 50
   packaging, Availability &, 52
   pH, 46
   protective clothing, 52
   optical density, 46
   refractive index, 47
   R.I. sugar solids, 47
   specific gravity, 49
   starch test, 46
   storage temperatures, recommended, 51
   sugar spectrum, 49
   sulphated ash, 48
   sulphur dioxide, 47
   toxicity, 52
   transmission, *see* optical density
   true solids, 47
   viscosity, 51
Golden Syrup, 324
Gums, 78, 159, 165

Hatch–Slack Cycle, 17
HFGS, 11–13, 17, 91, 157
   production, 40
High boilings, 78, 159, 161
HMF, *see* hydroxyl methyl firfural, 26
Honey, 2, 325
   honey type spread, 209
Hoover, W.J., 61, 63
Horticulture, 90
Hough, Jones & Wadman, 5
HPLC, 5, 20
   calculations, 301
Hydrogenation, 107

Hydrolysed cereal solids, 20
Hydroxy methyl fufural, 26
Hypertonic drinks, *see* Drinks, Health, classification
Hypotonic drinks, *see* Drinks, Health, classification

Ice cream, 67, 79, 87, 94, 99, 115, 116, 185, 186
    fats, 185, 194
    hard, 191
    other types, 195
    overrun, 194
    recipe sheet, 193
    reduced calorie, 201, 202
    soft, 190, 194
    solids, 187, 194
    viscosity, 194
Immobilised enzyme, 11, 12
Inulin, 17, 19
    absorption, 256
Intervention Board, 14
Intravenous drips, 90
Inversion of sucrose, 74, 150, 204, 228, 229
Invert, 12, 325
Inverted sucrose, *see* inversion of sucrose
Ion exchange, 31, 62
Isoglucose, 14
    Quotas, 15, 16
Isomalt, 107
Isomerisation, 40
Isotonic drinks, *see* Drinks, Health, classification

Jams, 83, 94, 106, 115, 116, 203
    bake stable, 212
    biscuit, 212
    diabetic, 217
    domestic, 208
    fillings, 212
    industrial, 211

    jelly, 209
    flan, 212
    marmalades, 209
    recipe calculations, 215
    reduced calorie, 217
    spreadable, 212
Japanese Agency of Industrial Science and Technology, 10
Japanese Fermentation Institute, 10
Jellies, 78, 159, 165
    tablet, 214

Ketchups, 83, 95
Kingsford, Thomas, 6
Kirchoff, Konstantin Gottlieb Sigismund, 1
Koninglije Sholten Honig (KSH), 13

Lactic acid, 181, 195
Lactitol, 107
Lactobacillus bulgaris, 195
Lactose, 107, 195
    absorption, 255
    intolerance, 258
Lane & Eynon Fehling's titration, 4, 20
Lane, Joseph Henry, 4
Langlois & Turner, 7
Liquers, 83
Lollies fruit, 197
    ice, 197
Lysine, 182

Magnesium ions, 41
Maillard reaction, 28, 34, 62, 74
Malt Extract, *see* Brewers Extract
Maltitol, 107, 115
    absorption, 256
Maltodextrins, 95, 158
    absorption, 254
    analyses, 95
Maltose molecule, 21, 104

Maltose syrup, 20, 35, 36, 61, 80, 157
Mambre, Alexandre, 3
Manbre, see Mambre, Alexandre, 3
Manbre and Garton, 3
Mannitol, 107, 116
   absorption, 256
   maple syrup, 326
Margarine, 99
Marshall, Richard, & Kooi, Earl, 7, 9, 10
Marshmallow, 160, 168
Martin, Archer, 5
Mayonnaise, 83, 224
Meat analogues, 99
   preparation, 90
Medical applications, 106, 114
Mesophilic bacteria, 50, 233
Mincemeat, 215
Molasses, 325
   beet, 176
   cane, 176
Molecular weight, average, 69, 322
Monilieaella pollinis, 109
Monosodium glutamate, 181
Moulds, 50, 233
Mousse, 196
Muesli bars, 78, 160, 170
Mycoprotein, 181

Neutralisation, 29
Newkirk, W.B., 7
Novo Industries S/A, 12, 13

Oral hygiene, 115
Oral rehydration, 90, 247
Osmolality, 241, 244
Osmolarity, 241
Osmophilic yeasts, 50
Osmotic pressure, 241, 318

Partridge, Stanley, 5
Paste enzyme enzyme hydrolysis process, 34, 35

Pasteurisation, 188
Peanut spread, 211
Pearson's Square calculations, 305
Pectin, 166
   buffered, 213
PEE, see paste enzyme enzyme hydrolysis
Personal hygiene, 99, 106
Pharmaceuticals, 99, 106, 117, 178
Phenylketonuria, 259
Plastics, 90, 183, 184
Plato, Fritz, 6
Pickles, 95
Polydextrose absorption, 256
Polyols, 107
   absorption, 256
   humectant, 114
   properties, 111
Preservative, 88
Process inversion, *see* inversion of sucrose
Production quota, 14–16
Proust, Louis-Joseph, 2
Pseudomonas hydrophila, 9, 10

Recipe costings, 314
Reducing sugars, 20
R.I. solids, explanation, 313
Runge, Friedlieb F., 5

Salad cream, 224
Sauces, 82, 95, 225
de Saussure, Nicolas Theodore, 1, 4
Sieve specifications, 352
Solubility of sugars, 75, 93, 204
Sorbet, 196
Sorbitol, 113
   absorption, 257
   forms, 109
   hydrogenation, 108
   production, 91
Spray dried, 39
Spray dried glucose syrup, definition, 20

Spreads, 100
Staley, A.E., 5, 7, 12, 13
Starch, 21
   absorption, 255
   gel temperatures, 23
   jellies, 166
   linkages, 23–25
   resistant, absorption, 255
   slurry, 28
   thin boiling, 169
   viscosity, 23
Streptococus thermophilus, 195
Streptomyces genus, 10
Sucrose, 74, 107
   absorption, 254
   Brix tables, 349
   molecular structure, 104
   particle size, 326
   sports drinks, 243
   sugar alcohols, 107
Sugar Regime, 14
Sugars, 187
   glass transition temperatures, 327
   melting points, 327
   solubility, 328
Sulphide spoilage spores, 53
Sulphur dioxide, 28, 41, 62
Surimi, 115
Sweetness values, 323
   how to calculate, 311
Synge, Richard, 5
Syral, 14
Syrup problems, 55
   alcohol, 57
   amine smell, 57
   bagged products, smell, 57
   black specs, 55
   bluey green colour, 57
   brown coloured, 56
   chlorine smell, 57
   fishy smell, 57
   hazy syrup, 55
   iodine smell, 57
   pink colour, 57
   petroleum smell, 57
   solid syrup, 56
   white crystals, 56
Syrups
   dessert, 198, 199
   Ripple, 197
   topping, 199, 200, 201

Tabletting, 91
Takasaki, Professor, 10, 11
Takasaki–Tanabe enzyme process, 11
Tank Room, 3
Tanning, 91
Tartaric acid, 181
Tate & Lyle, 3, 14
Tate & Lyle Americas Inc., 5
Tereos France, 14
Thermophilic anaerobe spores, 53
Toffees, 77, 159, 163
Tomato products, 221-3
Topping syrups, 94
Total sugar production, 38
Total thermophilic count, 53
Toys, 79
Treacle, 326
Trealose, 89
   absorption, 255
   molecular structure, 103
   production schematic, 102
   properties, 105
Treaty of Rome, 14
Tswett, Mikhail, 5, 12
Tunnel Refineries Ltd., 12–14
Turkish Delight, 168

Ultraviolet (UV), 60

Vinegar, 182
Vitamin B 12, 182
Vitamin C, 113, 182

Well drilling, 88

Xanthan gum, 184
Xlitol, 107, 111
D-Xylose, 9

D-Xylulose, 9
Xylose isomerase, 10

Yeast, 50, 181, 233
Yogurts, 195